D1594272

Luck, Logic, and White Lies

Luck, Logic, and White Lies
The Mathematics of Games

Jörg Bewersdorff

Translated by David Kramer

A K Peters
Wellesley, Massachusetts

Editorial, Sales, and Customer Service Office

A K Peters, Ltd.
888 Worcester Street, Suite 230
Wellesley, MA 02482
www.akpeters.com

Copyright © 2005 by A K Peters, Ltd.

All rights reserved. No part of the material protected by this copyright notice may
be reproduced or utilized in any form, electronic or mechanical, including photo-
copying, recording, or by any information storage and retrieval system, without
written permission from the copyright owner.

Library of Congress Cataloging-in-Publication Data

Bewersdorff, Jörg.
    [Luck, logik und Bluff. English]
    Luck, logic, and white lies : the mathematics of games / Jörg Bewersdorff ; trans-
lated by David Kramer.
        p. cm.
    Includes bibliographical references and index.
    ISBN 1-56881-210-8
        1. Game theory. I. Title

QA269.B39413 2004
519.3—dc22
2004053374

Originally published in the German language by Friedr. Vieweg & Sohn Verlag,
D-65189 Wiesbaden, Germany,
as "Jörg Bewersdorff: Glück, Logik und Bluff, 2. Auflage."
© Friedr. Vieweg & Sohn Verlag/GWV Fachverlage GmbH, Wiesbaden 2001

Printed in Canada
09 08 07 06 05   10 9 8 7 6 5 4 3 2 1

Contents

Preface

A feeling of adventure is an element of games. We compete against the uncertainty of fate, and experience how we grab hold of it through our own efforts. —Alex Randolph, game author

The Uncertainty of Games

Why do we play games? What causes people to play games for hours on end? Why are we not bored playing the same game over and over again? And is it really the same game? When we play a game again and again, only the rules remain the same. The course of the game and its outcome change each time we play. The future remains in darkness, just as in real life, or in a novel, a movie, or a sporting event. That is what keeps things entertaining and generates excitement.

The excitement is heightened by the possibility of winning. Every player wants to win, whether to make a profit, experience a brief moment of joy, or have a feeling of accomplishment. Whatever the reason, every player can hope for victory. Even a loser can rekindle hope that the next round will bring success. In this, the hope of winning can often blind a player to what is in reality a small probability of success. The popularity of casino games and lotteries proves this point again and again.

Amusement and hope of winning have the same basis: the variety that exists in a game. It keeps the players guessing for a long time as to how the game will develop and what the final outcome will be. What causes this uncertainty? What are the mechanisms at work? In comparing games like roulette, chess, and poker, we see that there are three main types of mechanism:

1. chance;

2. the large number of combinations of different moves;

3. different states of information among the individual players.

Random influences occur in games involving dice and the mixing of cards. The course of a game, in accordance with its rules, is determined not only by decisions made by the players, but by the results of random processes. If the influence of chance dominates the decisions of the players, then one speaks of games of chance. In games of pure chance, the decision of a player to take part and the size of a player's bet are perhaps his[1] most important decisions. Games of chance that are played for money are generally governed by legal statute.

During the course of most games, there are certain situations in which the players have the opportunity to make decisions. The available choices are limited by the rules of the game. A segment of a game that encompasses just one such decision of a single player is called a *move*. After only a small number of moves, the number of possibilities can already represent an enormous number of combinations, a number so large that it is difficult to recognize the consequences of an individual move. Games whose uncertainty rests on the multiplicity of possible moves are called *combinatorial games*. Well-known representatives of this class are chess, go, nine men's morris, checkers, halma, and reversi. Games that include both combinatorial and random elements are backgammon and pachisi, where the combinatorial character of backgammon is stronger than that of pachisi.

A third source of uncertainty for the players of a game arises when the players do not all have the same information about the current state of the game, so that one player may not possess all the information that is available to the totality of players. Thus, for example, a poker player must make decisions without knowing his opponents' cards. One could also argue that in backgammon a player has to move without knowing the future rolls of the dice. Yet there is a great difference between poker and backgammon: no player knows what the future rolls of the dice will be, while a portion of the cards dealt to the players are known by each player. Games in which the players' uncertainty arises primarily from such imperfect information are called *strategic games*. These games seldom exist in a form that one might call purely strategic. Imperfect information is an important component of most card games, like poker, skat, and bridge. In the board games ghosts and Stratego, the imperfect information is based on the fact that one knows the location, but not the type, of the opponent's pieces.[2] In

---

[1]Translator's note: the German word for player, *Spieler*, is masculine, and so the author of this book could easily write the equivalent of "a player...his move" without too many qualms. Faced with this problem in English, I have decided to stick primarily with the unmarked masculine pronoun, with an occasional "his or her" lest the reader forget that both men and women, boys and girls, can play games.

[2]Ghosts and Stratego are board games for two players in which each player sees only the blank reverse side of his opponent's pieces. At the start, a player knows only his own pieces and the positions of the opposing pieces. In ghosts, which is played on a

Figure P.1. The three causes of uncertainty in games: a player wins through some combination of chance, logic, and bluff.

Diplomacy,[3] and rock–paper–scissors, [4] the players move simultaneously, so that each player is lacking the information about his opponent's current move. How this imperfect information plays out in a game can be shown by considering what happens to the game if the rules are changed so that the game becomes one of perfect information. In card games, the players would have to show their hands. Poker would become a farce, while skat would remain a combinatorially interesting game similar to the half-open two-person variant. In addition to the game rock–paper–scissors, which is a purely strategic game, poker is also recognized as a primarily strategic game. The degrees of influence of the three causes of uncertainty on various games are shown in Figure P.1.

There remains the question whether the uncertainty about the further course of the game can be based on other, as yet unknown, factors. If one investigates a number of games in search of such causes, one generally finds the following:

chessboard with four good and four bad ghosts on the two sides, only the captured figures are revealed. In Stratego, the capturing power of a piece depends on its military rank. Therefore, a piece must be revealed to the opponent at the time of an exchange.

The simple rules of ghosts and a game with commentary can be found in *Spielbox* **3** 1984, pp. 37–39. Tactical advice on Stratego can be found in *Spielbox* **2** 1983, pp. 37 f.

[3]Diplomacy is a classic among board games. It was invented in 1945 by Alan Calhamer. Under the influence of agreements that the players may make among themselves, players attempt to control regions of the board, which represents Europe before the First World War. The special nature of Diplomacy is that the making and abrogating of agreements can be done secretly against a third party. An overview of Diplomacy appears in *Spielbox* **2** 1983, pp. 8–10, as well as a chapter by its inventor in David Pritchard (ed.), *Modern Board Games*, London 1975, pp. 26–44.

[4]Two players decide independently and simultaneously among the three alternatives "rock," "paper," and "scissors." If both players made the same choice, then the game is a draw. Otherwise, "rock" beats (breaks) "scissors," "paper" beats (wraps) "rock," and "scissors" beat (cut) "paper."

- the result of a game can depend on physical skill and performance ability. In addition to sports and computer games, which do not belong to the class of parlor games that we are considering here, Mikado is a game that requires manual dexterity.

- the rules of a game can be partially ambiguous. One arrives at such situations particularly in the learning phase of a complex game. In other cases, doubts arise in the natural course of the game. Thus in the crossword game Scrabble it can be unclear whether a word should be permitted. And even in skat, there are frequently questions raised, if only about minor details.

- an imperfect memory does not make only the game "memory" more difficult. However, this type of uncertainty is not an objective property of the game itself.

In comparison to chance, combinatorial richness, and differing informational states, these last phenomena can safely be ignored. None of them can be considered a typical and objective cause of uncertainty in a parlor game.

Games and Mathematics

If a player wishes to improve his prospects of winning, he must first attempt to overcome his degree of uncertainty as much as possible and then weigh the consequences of his possible courses of action. How that is to be managed depends, of course, on the actual causes of the uncertainty: if a player wishes to decide, for example, whether he should take part in a game of chance, then he must first weigh the odds to see whether they are attractive in comparison to the amount to be wagered. A chess player, on the other hand, should check all possible countermoves to the move he has in mind and come up with at least one good reply to each of them. A poker player must attempt to determine whether the high bid of his opponent is based on a good hand or whether it is simply a bluff. All three problems can be solved during a real game only on a case-by-case basis, but they can also be investigated theoretically at a general level. In this book, we shall introduce the mathematical methods that have been developed for this and provide a number of simple examples:

- games of chance can be analyzed with the help of probability theory. This mathematical discipline, which today is used in a variety of settings in the natural sciences, economics, and the social sciences, grew out of a 17th-century desire to calculate the odds in a game of chance.

- there is no unified theory for the combinatorial elements in games. Nonetheless, a variety of mathematical methods can be used for answering general questions as well as solving particular problems.

- out of the strategic components of games there arose a separate mathematical discipline, called *game theory*, in which games serve as a model for the investigation of decison-making in interactive economic processes.

For all three game types and their mathematical methods, the computer has made possible applications that formerly would have been unthinkable. But even outside of the development of ever faster computers, the mathematical theory itself has made great strides in the last hundred years. That may surprise those unversed in modern mathematics, for mathematics, despite a reputation to the contrary, is by no means a field of human endeavor whose glory days are behind it.

Probability theory asks questions such as, which player in a game of chance has the best odds of winning? The central notion is that of *probability*, which can be interpreted as a measure of the certainty with which a random event occurs. For games of chance, of course, the event of interest is that a particular player wins. However, frequently the question is not who wins, but the amount of the winner's winnings, or score. We must then calculate the average score and the risk of loss associated with it. It is not always necessary to analyze a game completely, for example, if we wish only to weigh certain choices of move against each other and we can do so by a direct comparison. In racing games governed by dice, one can ask questions like, how long does it take on average for a playing piece to cover a certain distance? Such questions can become complicated in games like snakes and ladders, in which a piece can have the misfortune to slip backward. Even such a question as which squares in the game Monopoly are better than others requires related calculational techniques. It is also difficult to analyze games of chance that contain strong combinatorial elements. Such difficulties were first overcome in the analysis of blackjack.

Combinatorial games, such as the tradition-rich chess and go, are considered games with a high intellectual content. It was quite early in the history of computational machines that the desire was expressed to develop machines that could serve as worthy opponents in such games. But how could that be accomplished? Indeed, we need computational procedures that make it possible to find good moves. Can the value of a move be somehow uniquely determined, or does it always depend on the opponent's reply? In any case, the current state of technology for search procedures and computational techniques is impressive. An average chess player no longer has a ghost of a chance against a good chess program. And it is not only chess that has been the object of mathematical interest. Winning strategies have been found for many games, some of them surprisingly simple. For other games it has been determined only which player theoretically should win, without a winning strategy actually being found. Some of these games possess properties that make it doubtful whether such a strategy will ever be found.

It is a task of *game theory* to determine how strategic games differ fundamentally from combinatorial games and games of chance. First, one needs a mathematical definition of a game. A game is characterized by its rules, which include the following specifications:

- the number of players.

- for each game state, the following information:

  - whose move it is;

  - the possible moves available to that player;

  - the information available to that player in deciding on his move.

- for games that are over, who has won.

- for random moves, the probabilities of the possible results.

Game theory arose as an independent discipline in 1944, when out of the void there appeared a monumental monograph on the theory of games. Although it mentions many popular games such as chess, bridge, and poker, such games serve game theory only as models of economic processes. It should not be surprising that parlor games can serve as models for real-life interactions. Many games have borrowed elements of real-life struggles for money, power, or even life itself. And so the study of interactions among individuals, be it in cooperation or competition, can be investigated by looking at the games that model those interactions. And it should come as no surprise that the conflicts that arise in the games that serve as models are idealized. That is just as inevitable as it is with other models, such as in physics, for example, where an object's mass is frequently considered to be concentrated at a single point.

About This Book

We have divided the book into three parts to reflect our division of games into three types, and so we investigate mathematically in turn the chance, combinatorial, and strategic elements of games. Each of the three parts encompasses several chapters, each of which considers a specific problem—generally a game or game fragment.

In order to reach as broad an audience as possible, we have not sought the generality, formalism, and completeness that are usual in textbooks. We are more concerned with ideas, concepts, and techniques, which we discuss to the extent that they can be transferred to the study of other games.

Due to the problem-oriented selection of topics, the mathematical level differs widely among the different chapters. Although there are frequent references to earlier chapters, one can generally read each chapter independently of the others. Each chapter begins with a question, mostly of a rhetorical nature, that attempts to reveal the nature and difficulty of the problem to be dealt with. This structure will allow the more mathematically sophisticated readers, for whom the mathematical treatment will frequently be too superficial and incomplete, to select those parts of greater mathematical interest. There are many references to the specialist literature for those who wish to pursue an issue in greater depth. We have also given some quotations and indications of the mathematical background of a topic as well as related problems that go beyond the scope of the book.

We have placed considerable emphasis on the historical development of the subject, in part because recent developments in mathematics are less well known than their counterparts in the natural sciences, and also because it is interesting to see how human error and the triumph of discovery fit into a picture that might otherwise seem an uninterrupted sequence of great leaps forward. The significance of the progress of mathematics, especially in recent decades, in the not necessarily representative area of game theory, can be seen by a comparison with thematically similar, though often differing in detail of focus, compilations that appeared before the discovery of many of the results presented in this book:

- René de Possel, *Sur la théorie mathématique des jeux de hasard et de réflexion*, Paris 1936. Reprinted in Hevre Moulin, *Fondation de la théorie des jeux*, Paris 1979.

- R. Vogelsang, *Die mathematische Theorie der Spiele*, Bonn 1963.

- N. N. Vorob'ev, *The Development of Game Theory* (in Russian), 1973. The principal topic is game theory as a mathematical discipline, but

this book also contains a section on the historical development of the
theories of games of chance, and combinatorial and strategic games.[5]

- Richard A. Epstein, *The Theory of Gambling and Statistical Logic*,
  New York 1967 (expanded revised edition, 1977).

- Edward Packel, *The Mathematics of Games and Gambling*, Washington 1981.

- John D. Beasley, *The Mathematics of Games*, Oxford 1989.

- *La mathématique des jeux, Bibliothèque pour la Science*, Paris 1997.
  Contributions on the subject of games from the French edition of *Scientific American*, some of which have been published in the editions
  of other countries.

Acknowledgments

I would like to express my thanks to all those who helped in the development of this book: Elwyn Berlekamp, Richard Bishop, Olof Hanner, Julian
Henny, Daphne Koller, Martin Müller, Bernhard von Stengel, and Baris
Tan kindly explained to me the results of their research. I would like to
thank Bernhard von Stengel additionally for some remarks and suggestions
for improvements and also for his encouragement to have this book published. I would also like to thank the staffs of various libraries who assisted
me in working with the large number of publications that were consulted in
the preparation of this book. As representative I will mention the libraries
of the Mathematical Institute of Bonn, the Institute for Discrete Mathematics in Bonn, as well as the university libraries in Bonn and Bielefeld.
Frauke Schindler, of the editorial staff of Vieweg-Verlag, and Karin Buckler
have greatly helped in minimizing the number of errors committed by me.
I wish to thank the program director of Vieweg-Verlag, Ulrike Schmickler-Hirzebruch, for admitting into their publishing progam a book that falls
outside of the usual scope of their publications. Last but not least, I thank
my wife, Claudia, whose understanding and patience have at times in the
last several years been sorely taxed.

---

[5] The author wishes furthermore to express his gratitude for observations of Vorob'ev
for important insights that have found their way into this preface.

Preface to the Second Edition

The happy state of affairs that the first edition has sold out after two years has given me the opportunity to correct a number of errors. Moreover, I have been able to augment the references to the literature and to newer research. I wish to thank Hans Riedwyl, Jürg Nievergelt, and Avierzi S. Fraenkel for their comments.

Finally, I wish to direct readers to my web site, http://www.bewersdorff-online.de/, for corrections and comments.

Preface to the Third Edition

Again, I have alert readers to thank for calling a number of errors to my attention: Pierre Basieux, Ingo Briese, Dagmar Hortmeyer, Jörg Klute, Norbert Marrek, Ralph Rothemund, Robert Schnitter, and Alexander Steinhansens. In this regard I would like to offer special thanks to David Kramer, who found errors both logical and typographical in the process of translating this book into English.

The necessity to make changes to the content of the book arose from some recently published work, especially that of Dean Allemang on the misère version of nim games, and of Elwyn Berlekamp on the idea of environmental go. I have also gladly complied with the request of readers to include newer approaches to game tree search procedures.

—Jörg Bewersdorff[6]

---

[6]I would be glad to receive comments on errors and problems with the text at joerg.bewersdorff@t-online.de. Questions are also welcome, and I will answer them gladly within the constraints of time and other commitments.

Part I

Games of Chance

I

Dice and Probability

With a pair of dice, one can throw the sum 10 *either as the combination*
5 + 5 *or* 6 + 4. *The sum* 5 *can also be obtained in two ways, namely, by*
1 + 4 *or* 2 + 3. *However, in repeated throws, the sum* 5 *will appear more*
often than 10. *Why?*

Although we are exposed in our daily lives to a variety of situations involving chance and probability, it was games of chance that provided the
primary impetus for the first mathematical investigations into this domain.
Aside from the fact that there is a great attraction in discovering ways of
winning at the gaming table, games of chance have the advantage over the
rough and tumble of real-life events that chance operates in fixed and precise ways. Thus the odds, dictated by the laws of probability, of throwing a
six, say, are much simpler to calculate than the odds that a bolt of lightning
will strike the Eiffel Tower on July 12 of next year. The reason for this
is primarily that the situation in a game of chance is reproducible under
identical conditions, with the consequence that theoretical results can be
checked by experiments that can be repeated as often as one likes, if the
results are not already well known as facts of common experience.

The first systematic investigation of games of chance began in the middle of the 17th century. To be sure, there was some sporadic research earlier;
indeed, in the 13th century, the probabilities of the various sums that can
be obtained with a pair of dice were correctly determined.[1] This fact is
particularly noteworthy in that over the following centuries, many incorrect

---

[1]R. Ineichen, Das Problem der drei Würfel in der Vorgeschichte der Stochastik,
Elemente der Mathematik **42**, 1987, pp. 69–75; Ivo Schneider, *Die Entwicklung der*

analyses of the same problem appeared. The first to create a systematic, universal approach to the description of problems of chance and probability was Jacob Bernoulli (1654–1705), with his *Ars coniectandi* (*The Art of Conjecture*). According to Bernoulli, its object was "to measure the probability of events as precisely as possible, indeed, for the purpose of making it possible for us in our judgments and actions always to choose and follow the path that seems to us better, more worthy, more secure, or more advisable."[2] Bernoulli had in mind not only games of chance, but also the problems of everyday life. His belief in the necessity of a mathematical theory of probability is alive and well even today. This was formulated with admirable concision by the renowned physicist Richard Feynman (1918–1988): "The theory of probability is a system for making better guesses."

Of central importance in Bernoulli's theory is the notion of *probability*, which Bernoulli called a "degree of certainty." This degree of certainty is expressed by a number. As inches and centimeters measure length, so a probability measures something. But what exactly does it measure? That is, what sort of objects are being measured, and what qualities of those objects are the subject of measurement?

Let us first take a single die. We can describe the result of throwing the die by saying, "the result of throwing the die is equal to 5," or "the result is at most 3." Depending on what was actually thrown, such a statement may be either true or false. To put it another way, the *event* described by the statement made may occur, or may not occur, as the result of a single experiment. The extreme case of an impossible event, such as that represented by the statement, "The result of throwing the die is 7," never occurs. In contrast, a certain event, for example that described by the statement, "The result is between 1 and 6," occurs in every trial of the experiment.

The events may be considered objects that can be measured with probabilities. What is measured with respect to an event is the degree of certainty, or the degree of likelihood, with which the event will occur in a single trial.

But how is this degree of certainty to be measured? To measure means to compare. Thus we measure length by comparing the item to be measured

Wahrscheinlichkeitstheorie von den Anfängen bis 1933, Darmstadt 1988, p. 1 and 5–8 (annotated references). A historical overview of the development of the calculation of probabilities can also be found in the appendix to the textbook by Boris Vladimirovich Gnedenko, *Theory of Probability*, Berlin 1998.

[2]See the comprehensive reprint, Jacob Bernoulli, *Wahrscheinlichkeitsrechnung*, Ostwalds Klassiker der exakten Wissenschaften, volume 107, Frankfurt am Main 1999, p. 233.

with a ruler. In the case of probabilities, measuring is not so simple. On the one hand, the objects to be measured are intangible, and on the other hand, in contrast to measures such as speed, temperature, or brightness, what we wish now to measure cannot be directly observed. Nonetheless, it is intuitively clear how one might estimate the degree of certainty for an event: simply perform the deed! That is, one throws the die, and the more often, the better. The greater the frequency with which the event occurs, the more likely it is that the event will occur in a single throw of the die. To state this numerically, the measurement is encapsulated in the *relative frequency*, which is the quotient of the number of times the event occurred divided by the total number of trials. For example, if in 6000 throws, the event of throwing at least a 5 occurred 2029 times, then the relative frequency would be $2029/6000 \approx 0.338$. We have thus measured the degree of certainty of throwing at least a 5 to be 0.338. A second measurement with the same or different number of throws would be unlikely to yield the same result, though we may suppose that the result would be similar. A conclusive value is therefore not to be obtained in this way, and even specifying what it would mean to make a precise measurement is problematic. The only events that are exactly measurable are the certain events, which always have relative frequency 1, and the impossible events, whose relative frequency is always 0.

If one wishes to compare the degree of certainty of different events, one does not necessarily have to resort to experiment. It is possible to use the idea of symmetry. For example, since the six faces of the die are geometrically identical, we may assume that the corresponding events have equal likelihood. That is, the six possible events relating to the outcome of throwing a die have the same probability. On a probability scale, ranging from 0 for impossible events through 1 for certain events, the probability for each of the six results of throwing a die, precisely one of which will occur in a particular trial, is $\frac{1}{6}$. Bernoulli stated it thus: "Probability is namely the degree of uncertainty, and it differs from it as a part from a whole."

The event of throwing at least a 5 comprises the events of throwing a 5 and throwing a 6. Therefore, the probability of this event should be assigned the value $\frac{2}{6} = \frac{1}{3}$. Similarly, the probability of throwing an even number is equal to $\frac{3}{6} = \frac{1}{2}$.

Probabilities can thus be computed just as they are for dice whenever a system of equal probabilities (an equiprobable system) is under consideration. In 1812, Pierre Simon Laplace (1749–1824) declared, in his *Essai philosophique sur les probabilités*, that outcomes are equiprobable whenever "we have the same uncertainty about their occurrence" and "have no reason to believe that one of these events is more likely to occur than another." If the possible results of a random experiment are equiprobable in

		2nd die					
		1	2	3	4	5	6
1st die	1	1-1	1-2	1-3	1-4	1-5	1-6
	2	2-1	2-2	2-3	2-4	2-5	2-6
	3	3-1	3-2	3-3	3-4	3-5	3-6
	4	4-1	4-2	4-3	4-4	4-5	4-6
	5	5-1	5-2	5-3	5-4	5-5	5-6
	6	6-1	6-2	6-3	6-4	6-5	6-6

Table 1.1. The 36 combinations of two dice.

this sense, then the probability of an event, according to Laplace, can be defined as follows: the number of outcomes in which the event occurs, that is, those that are "favorable" for the event, divided by the total number of possible outcomes. If $\mathcal{A}$ is an event, then Laplace's definition can be expressed with the formula

$$\text{probability of event } \mathcal{A} = \frac{\text{number of favorable outcomes for } \mathcal{A}}{\text{number of possible outcomes}}.$$

We have already mentioned the close connection between the relative frequency in the context of a series of trials and the notion of probability: both use the measuring scale from 0 to 1, and in the case of impossible and certain events, their measurements are the same. If a series of trials takes place "ideally," in the sense that equiprobable events occur with the same frequency, then the notions of relative frequency and probability coincide. Bernoulli discovered another relation, which is of considerable interest: the *law of large numbers*. It states that in long series of trials, the relative frequencies are approximately equal to the associated probabilities. This is also justification for stating that probabilities of events truly measure the degree of certainty as intuitively understood. For example, if the probability of winning a game is greater than that of losing, then if one plays frequently enough, one is highly likely to win more often than lose. Bernoulli's law of large numbers even makes assertions about how closely probabilities and relative frequencies agree. We shall return to this topic later on.

In the case of a single die, symmetry is the reason for considering the six possible values as equiprobable and thus for assigning the same probability to each of the six events. Thus there is no reason—in the sense of Laplace— why one value should appear more frequently than another. In the case of two dice, there are 36 combinations of the two values (Table 1.1). What is important—and this was omitted from our earlier discussion—is that combinations like 2-3 and 3-2 are different. In practice, of course, the difference is not often considered, especially when two identical dice are thrown. However, if the dice are of different colors, then there will be no trouble in distinguishing 2-3 from 3-2.

Are these 36 combinations equiprobable in the sense of Laplace? We note first that it is insufficient simply to rely on symmetry. It is possible that there are some dependencies between the values on the two dice, as is the case when two cards are drawn from a pack of cards. If a card is drawn from a standard deck of 52 cards, then the probability of drawing any of the 13 card values is equal to $4/52 = 1/13$. However, if a second card is drawn without replacing the first, then the result for this second card has a different set of probabilities. Thus it is less probable that the same value as that of the first card will be drawn, since only 3 of the remaining 51 cards have that value. On the other hand, each of the remaining 12 values has probability $4/51$ of being drawn.

The reason for the change in probabilities is that as a result of drawing the first card, the state of the pack of cards has changed. Something similar in the case of a pair of dice is less plausible, since their state, in contrast to that of the cards, does not depend on previous events: dice have no "memory." In the sense of Laplace, then, no matter how the first die lands, there is no reason to think that the results of the second die do not all have the same probability. Thus all 36 combinations of two dice may be viewed as equiprobable.

The question posed can now be answered: using the Laplacian approach, we see that four of the 36 combinations yield the sum 5, namely, 1-4, 4-1, 2-3, and 3-2. The sum 10, however, is achieved with only three combinations: 4-6, 6-4, and 5-5. Therefore, the event of throwing the sum of 5 with two dice is more likely than that of throwing 10.

2

Waiting for a Double 6

If one wagers that one will throw at least one 6 in four throws of a single die, then experience shows that a win is more likely than a loss. However, what if instead, one wagers on the variant of throwing at least one double 6 in a certain number of throws? How many attempts should one require in order to make such a wager favorable? The following reasoning seems to make sense: Since a double 6 is one of 36 equiprobable combinations, and therefore only one-sixth as likely as obtaining a 6 with a single die, one would require six times as many attempts. Thus it would appear that the wager of throwing at least one double 6 in 24 throws of a pair of dice is a favorable one. Should one actually make such a bet?

Something like the above thought process likely went through the mind of Chevalier de Méré (1607–1684), who, in the judgment of Blaise Pascal (1623–1662), was "rather a clever fellow," though to be sure, no mathematician ("a great shortcoming"). Among de Méré's principal occupations was wagering in games of chance, and the following observation baffled him: whereas with a single die, four throws suffice to make it worthwhile to bet on obtaining at least one 6, with two dice, six times as many throws do not suffice. Thus the obvious conclusion that one should multiply the number of necessary trials by the factor representing the degree of the smaller probability is not to be relied upon.

De Méré, who could not explain his bad luck, turned to Pascal in 1654 for help. Pascal, who at the time was carrying on a correspondence with his colleague Pierre de Fermat (1601–1665) on the subject of odds in games of chance, considered de Méré's problem. And so this episode, together with a portion of the correspondence, has come down to us.[1] This correspondence is generally considered the beginning of the subject of mathematical probability, even though a unified theory centered on the notion of probability was conceived only later, by Jakob Bernoulli. One might add that de Méré's problem caused Pascal and Fermat no difficulty whatsoever. An explanation for de Méré's observation is simply that one should compare the number of possible outcomes with the number of outcomes that lead to a win.

Thus there are $6 \times 6 \times 6 \times 6 = 1296$ ways of throwing a single die four times. In the sense of Laplace, all 1296 outcomes are equiprobable, and therefore have the same probability. One loses the wager if no 6 is thrown. For that to happen, there are five possibilities for each throw, leading to $5 \times 5 \times 5 \times 5$ losing combinations, against which are to be counted $1296 - 625 = 671$ winning combinations, so that the probability of a win is $671/1296 \approx 0.518$, somewhat greater than the probability of losing, which is $625/1296 \approx 0.482$.

With 24 throws of two dice there are astronomically many possibilities, namely, 36^{24}, which is a 38-digit number. The probability of losing is $35^{24}/36^{24}$, which is more easily calculated in the form $(35/36)^{24} \approx 0.5086$. This time, the probability of winning is smaller than that of losing, which is 0.4914, just as de Méré apparently experienced.

The formula for computing probability, going back to Laplace, where for an event the number of favorable outcomes is divided by the total number of all possible outcomes, is simple in principle, but is often unmanageable in practice, such as in our example, in which there are astronomically many combinations. In such cases it is more practicable to use the formulas provided by the multiplication law and the addition law. Both laws make statements about the probability of events that occur in a logical relation to one another. Thus we have the *multiplication law* for independent events:

> If the occurrence or nonoccurrence of an event does not influence the probability of another event—one says that these events are *independent*—then the probability of both events occurring is equal to the product of the individual probabilities.

For example, the probability of obtaining two even numbers with the throw of two dice is $1/2 \times 1/2 = 1/4$. Of course, one also obtains this result if one calculates the number of favorable combinations: with a single die, an even number is obtained with probability $1/2$, that is, in three out of the six

---

[1]Ivo Scheider, *Die Entwicklung der Wahrscheinlichkeitstheorie von den Anfängen bis 1933*, Darmstadt 1988, p. 3, 25–40.

possible outcomes. Therefore, in $3 \times 3 = 9$ equiprobable combinations, both values are even, which yields the probability $9/36 = 1/4$. Of importance here is that the favorable outcomes of both events can be combined as equiprobable events only because the two dice do not influence each other's outcomes.

If one throws a pair of dice once, then the probability of not obtaining a double 6 is $35/36$. Therefore, the probability of not obtaining a double 6 in 24 throws of the dice is $(35/36)^{24}$. How does one parlay this probability of losing into the probability of winning? For this, we summon the *addition law*:

> If two events are mutually exclusive, that is, both events cannot occur in a single trial, then the probability that one of the events will occur in a single trial is equal to the sum of the individual probabilities.

For example, the probability of obtaining either an even number or a 5 in one throw of a single die is $3/6 + 1/6 = 4/6 = 2/3$. One can throw either a 2, 4, 6, or 5. Since the number of favorable outcomes is added, the probabilities are also added. A special case of the addition law is that in which two events are complementary; that is, they are mutually exclusive, but together constitute a certain event. The sum of their probabilities is always 1. Thus for example, the probability of obtaining at least one double 6 in 24 throws of two dice is equal to $1 - (35/36)^{24}$.

With the addition and multiplication laws at our disposal, we can obtain an even more interesting view of de Méré's problem: if the probability of an event is p, then in a series of m trials, the probability that the event will occur at least once is given by the formula $1 - (1-p)^m$. To have at least an even chance of winning, this value must be equal to at least $1/2$. This happens when the number m of trials is at least[2]

$$\frac{\ln 2}{- \ln(1 - p)}.$$

This fraction is equal approximately to $\ln 2/p$, with the natural logarithm $\ln 2 = 0.6931\ldots$, where the exact value is obtained by dividing by $1 + p/2 + p^2/3 + p^3/4 + \cdots$.[3]

---

[2]The condition $1 - (1-p)^m \geq 1/2$ is converted to the form $(1-p)^m \leq 1/2$, and then logarithms are taken. Note that both logarithms are negative.

[3]The power series of the natural logarithm is given by

$$\ln(1 - p) = -p - \frac{p^2}{2} - \frac{p^3}{3} - \frac{p^4}{4} - \cdots.$$

This correction is especially important when the probability p is not very small. For example, in the case $p = 1/6$, one should divide by 1.094. On the other hand, with smaller probabilities, the approximation $\ln 2/p$ can be used with no problem, so that the required number of trials grows approximately in inverse proportion to the probability, just as de Méré apparently hypothesized as a general law.

Thus de Méré's intuition was not wholly in error. Moreover, his false conclusion was often in later years outdone by others. For example, it has been frequently conjectured that with three throws of a single die or with 18 throws of a pair of dice, the probability of winning, that is, obtaining at least one 6 in the first case and at least one double 6 in the second, is already 50%. It was easily overlooked that some of the results can occur more than once in the course of the 18 throws, so that fewer than half of the possible results occur. One case, in which de Méré's error was outdone in a spectacular way, is mentioned by the American gambling expert John Scarne in his book *Complete Guide to Gambling*.[4] In 1952, a gambler, imagining that he had the advantage, is said to have lost \$49 000 betting on obtaining at least one double 6 in 21 throws of the dice. In fact, the probability of winning, $1 - (35/36)^{21} \approx 0.4466$, is considerably less than the gambler must have believed.

---

[4] John Scarne, *Complete Guide to Gambling*, New York 1974, p. 16.

3

Tips on Playing the Lottery:
More Equal Than Equal?

A statistical analysis of 1433 rounds of the German lottery, from October 1955 to the beginning of 1983, revealed—without taking into account the "bonus" numbers—that in 76.4% of the rounds, at least one of the numbers between 1 and 10 was among the winning six numbers drawn. Therefore, players whose bets contained none of the numbers between 1 and 10 had, on the basis of this fact, no chance of winning with "six numbers correct" in 76.4% of the rounds. Should one therefore always select at least one of the numbers between 1 and 10 in one's bet?

This lottery, which in Germany and a number of other countries is played in the form "6 out of 49," is one of the most popular games of chance today. And not only for the public, but for governments as well, whose profit of roughly half of ticket sales is assured even before the numbers are drawn. Lotteries began in the 16th century, in the city of Genoa, where in that era five senators were chosen yearly by lot. Moreover, one could place bets on the outcome from among the 110 candidates. In time, the game became independent of the election, and was played for its own sake. Instead of names, numbers were used. Even the regimes of the former Iron Curtain countries could not resist the onslaught of the lottery.[1] Even there, the game originally branded as capitalistic took hold.

---

[1] The development of the lottery in the German Democratic Republic is described in Wolfgang Paul, *Erspieltes Glück: 500 Jahre Geschichte der Lotterien und des Lotto*, Berlin 1978, pp. 190–192.

Because of its popularity, the lottery has been the subject of a number of publications. In a book on lotteries,[2] the question cited at the beginning of this section is discussed:

> From such a point of view, one must admit that the lottery is illogical. If one thinks about it, it is quite simple. All the numbers, or all "starting numbers,"[3] those from 1 to 44, do not have the same odds.
>
> Because that is so, all series of six numbers do not have the same odds.
>
> ... Those who play the lottery and therefore create lists of numbers that begin with a number greater than 10, are throwing away three-quarters of their chances of choosing six correct numbers. Even if fortune were smiling on them. They can hope to choose six winning numbers in just under one-fourth of the drawings, since their selection of six numbers is simply incomplete. Players who choose high numbers are like those who buy one-fourth of a raffle ticket and expect to be able to win the entire jackpot. It is simply impossible.

One is almost ready to believe that all one's future lottery tickets should contain at least one of the numbers between 1 and 10. On the other hand, every number, and hence every set of lottery numbers, is theoretically equiprobable, as formulated by Laplace. And why should the numbers from 1 to 10, and not other groups of ten numbers, such as

- 34 to 43 or

- $4, 9, 14, 19, 24, 29, 34, 39, 44, 49$, or

- $11, 16, 17, 22, 23, 25, 29, 32, 36, 48$,

play a special role? All well and good, but perhaps these speculations are nothing more than dull theory. After all, the results of a statistical analysis cannot simply be ignored, can they? But is it really as extraordinary as it appears? And can the statistical result really be an argument for such a recommendation?

Let us forget for a moment that the statistical analysis has been done. What sort of result was to be expected? That is, how great is the probability, and thus the relative frequency, that a drawing of lottery numbers would contain at least one of the numbers from 1 to 10? To obtain an answer, one could program a computer to count all of the possibilities and to

---

[2] Rolf B. Alexander, *Das Taschenbuch vom Lotto*, Munich 1983, pp. 26, 68 f.
[3] By "starting number" is meant the smallest of the six chosen numbers.

keep track of how many are "favorable." A simpler method exists, thanks
to some formulas from the field of combinatorics, a subdiscipline of mathematics that deals with the variety of ways that objects can be combined or
arranged. The simplest situation that one could imagine is the completely
independent combining of characteristics, such as the result of throwing
two dice: each result of one die can be combined with the result of the
second, so that there are $6 \times 6 = 36$ combinations, where we distinguish
such combinations as 2-6 and 6-2.

The situation becomes somewhat more complex when we start shuffling
the cards. How many ways are there to arrange a specified number of
different cards? If we are dealing with three cards, which we may as well
label 1, 2, 3, then the following arrangements are possible:

$$1\ 2\ 3 \quad 1\ 3\ 2 \quad 2\ 3\ 1 \quad 2\ 1\ 3 \quad 3\ 1\ 2 \quad 3\ 2\ 1.$$

We see, then, that three cards can be arranged in six ways. We say that
there are six *permutations*. With four cards, there are 24 permutations, and
with five cards, the number climbs to 120. To discover what these numbers
are, we do not, fortunately, have to write down all the permutations. For
example, with five cards, there are five possibilities for choosing the first
card. Once the first card has been selected, the second card can be chosen
from the remaining four cards. For the third card, there are three choices,
and for the fourth, there are two. At the end, we must choose the one
remaining card with a single "choice." The number of permutations of five
cards or five of any set of distinct objects is therefore equal to $5 \times 4 \times 3 \times 2 \times 1 = 120$.

The notion of permutation is so important that it is interpreted as a
mathematical operation in its own right. This operation is called *factorial*
and is indicated by an exclamation point. Thus $n!$ (read "n factorial")
stands for the number of permutations that can be formed with n distinct
objects. In analogy to our example of $n = 5$, we can calculate n factorial
with the formula

$$n! = n \times (n-1) \times (n-2) \times \cdots \times 4 \times 3 \times 2 \times 1,$$

which for the numbers $n = 1, 2, 3, 4, 5, 6$ yields the values

$$1! = 1, \quad 2! = 2, \quad 3! = 6, \quad 4! = 24, \quad 5! = 120, \quad 6! = 720.$$

It turns out to be useful to assign the value 1 to zero factorial. That is,

$$0! = 1.$$

The 32 cards of a skat deck can be shuffled in $32! = 32 \times 31 \times \cdots \times 4 \times 3 \times 2 \times 1$ different ways, which is a 35-digit number, and thus, according to current cosmological theory, much greater than the number of seconds that have elapsed since the Big Bang that got the universe going. That number possesses a mere 18 digits:

$$32! = 263\,130\,836\,933\,693\,530\,167\,218\,012\,160\,000\,000.$$

However astronomically large this number may be, it is vanishingly small in comparison to the 52! permutations of a rummy or poker deck: 52! is a 67-digit number, about the number, according to some estimates, of atoms in the entire universe.[4]

In the German lottery, six out of 49 numbers are drawn, if we ignore the bonus numbers for now. Analogous drawings are made in other games as well. In poker, each player receives five of the 52 cards, while in skat, one receives ten out of the 32 cards in the deck. What is common to all these situations is that from a set of distinct objects, a fixed number of those objects is selected randomly. The order in which the items are selected is irrelevant. In such cases, one speaks of *combinations*.

The number of possible combinations can be determined in a manner similar to that by which permutations are calculated: for the first ball, there are 49 possibilities. When the second number is chosen, there are 48 ways in which that can occur. Thus there are 49×48 ways of drawing the first two numbers. Each time a ball is chosen, the number of ways that the next number can be selected is diminished by 1. Therefore, there are altogether $49 \times 48 \times 47 \times 46 \times 45 \times 44$ sequences of six lottery numbers, where we note that some of the sequences are identical in the collection of numbers drawn, though not in the order in which they were chosen. We can state this more precisely: each set of six lottery numbers can appear in precisely $6! = 720$ different drawing sequences. That is, the number of *combinations* is

$$\frac{49 \times 48 \times 47 \times 46 \times 45 \times 44}{6!} = 13\,983\,816.$$

Thus there are just under 14 million possible sets of numbers that one can bet on. It follows that the probability of choosing "six correct" is about 1 in 14 million. That in spite of this minuscule probability, one or more winners

---

[4]One can see how rapidly the factorial function grows with the aid of Stirling's formula, which provides an approximation to the factorial function. Stirling's formula reads

$$n! \approx \left(\frac{n}{e}\right)^n \sqrt{2\pi n},$$

where the relative error is quite small for large values of n. The quality of the approximation can be described by saying that the quotient $n!$ divided by the Stirling approximation lies between $e^{1/(12n+1)}$ and $e^{1/(12n)}$. For example, for $n = 32$, one obtains the approximation 2.6245×10^{35}, which is a mere 0.26% too small.

are announced almost every week is due solely to the enormous number of lottery tickets sold, which greatly exceeds the number of players, many of whom place multiple bets.

The general formula for the number of combinations can be written down by analogy with the lottery formula: if k objects are chosen from a set of n distinct objects, then there are

$$\frac{n(n-1)(n-2)\cdots(n-k+1)}{k!}$$

different possible selections. This fraction, which always reduces to an integer, is called the *binomial coefficient*. It is written

$$\binom{n}{k},$$

and is read as "n choose k" to reflect the fact that we are choosing k objects from a set of n objects. The number of possible lottery combinations is

$$\binom{49}{6},$$

which yields the number calculated above.

Pascal's Triangle

The collection of all binomial coefficients can be arranged in a diagram that is known as Pascal's triangle

$$
\begin{array}{ccccccccccc}
 & & & & & 1 & & & & & \\
 & & & & 1 & & 1 & & & & \\
 & & & 1 & & 2 & & 1 & & & \\
 & & 1 & & 3 & & 3 & & 1 & & \\
 & 1 & & 4 & & 6 & & 4 & & 1 & \\
1 & & 5 & & 10 & & 10 & & 5 & & 1 \\
\cdots & & & & & \cdots & & & & & \cdots
\end{array}
$$

In the diagram, the binomial coefficient $\binom{n}{k}$ is located in the $(n+1)$st row, in the $(k+1)$st position. For example, $\binom{4}{2} = 6$ is to be found as the third value in the fifth row. What

makes Pascal's triangle interesting is that all of the values can
be computed without any multiplication, since every number in
the triangle is the sum of the two numbers lying immediately
above it. We can easily explain why this is so: to select k cards
from a deck of n cards, choose either the first card and then
the remaining $k-1$ from the remaining $n-1$, or do not choose
the first card, and choose all k cards from the remaining $n-1$
cards. This way of proceeding corresponds to the equality

$$\binom{n}{k} = \binom{n-1}{k-1} + \binom{n-1}{k}.$$

With binomial coefficients in our bag of tricks, it is now a simple matter
to calculate lottery probabilities. From among the almost 14 million possible combinations of six lottery numbers, there are only $\binom{39}{6} = 3\,262\,623$
that are formed from the 39 numbers from 11 to 49. That is, the probability that all six numbers played are greater than or equal to 11 is 0.2333.
Based on the law of large numbers, one expects over time that the number
of drawings in which at least one number is in the range 1 to 10 will be
about 76.67%. The result of the statistical analysis, 76.4%, is therefore
anything but unusual.

If the statistical result is nothing unusual, what does that say about
the proposed recommendation always to select at least one number in the
range from 1 to 10? Forget it! It is based on an erroneous conclusion.
The statement that a ticket without one of the numbers from 1 to 10 has
an almost 77% chance of losing because with that probability one of the
numbers from 1 to 10 will be chosen is simply of no interest. It says no
more than that the probability of choosing the six correct numbers without
following the recommendation is less than 0.2333. But that is clear, for we
know already that the probability of six correct is much less than that,
namely, 0.0000000715.

Still not convinced? Let us imagine that we had chosen the numbers
22, 25, 29, 31, 32, 38. Since we can watch the numbers being drawn on
television, we ask a friend to write down the numbers as they are drawn.
Still unconvinced whether it was a good idea not to choose at least one
number between 1 and 10, we then ask our friend, "Is one of the numbers
between 1 and 10 among the chosen?" In just under 77% of the cases, our
dreams of riches are shattered with the answer yes. So far, so good. Yet in
another 23% of the cases—and here the author cited above is in error—our

hope of winning is not the same, but in fact, it increases. Now we need only to have chosen six correct numbers out of 39, instead of out of 49.

Now that we have convinced ourselves that the suggestion from the book is unfounded, the question remains what a lottery player who does follow the suggestion should expect. We should first mention, of course, that such combinations of numbers are not "better," as the author supposes, but neither are they "worse," that is, any less probable, than the others. In that respect, following the recommendation does no harm. However, if one considers that the amount that one wins in the lottery depends on how many players have purchased a winning ticket, the picture changes considerably.[5] Every combination of numbers that contains frequently chosen numbers and combinations of numbers is therefore less favorable, because if that combination wins, it is more likely than otherwise to be won by a number of players. For example, it is known that many lottery players derive their numbers from dates, in which the numbers from 1 to 12 and from 1 to 31 appear. Other preferences are based on numerological considerations, such as the lucky number 7. Even the way the numbers are distributed on the lottery form has an influence on the numbers chosen.

The Winning Categories in the Lottery

We can use the binomial coefficients to determine with ease the chances of winning in one of the other lottery winning categories. For example, one chooses exactly 4 correct numbers when

- 4 of the 6 numbers wagered and
- 2 of the 43 numbers not wagered

[5]The amounts won can vary by a great deal. In the group of winners who chose "six correct," it has twice happened that the payouts were particularly small. In the drawing of 18 June 1977, there were 205 supposed "lucky duckies" who had all bet on the six winning numbers $9, 17, 18, 20, 29, 40$. However, the payout to each was not millions, but a pitiful 30 737.80 German marks. (Translator's note: about $13 100 at 1977 exchange rates). What happened? It turns out that many players, particularly in the north by northwestern areas of Germany, had gotten into the habit of betting on the numbers that had won the week before in the Netherlands lottery. This strategy showed itself to be a grave error, not because the chances of an exact match of the Dutch numbers was any less likely than any other combination, but because too many other players had the same idea. In another drawing, from 23 January 1988, there were 222 winners. The reason here was apparently the regularity of the numbers: $24, 25, 26, 30, 31, 32$.

are drawn. Thus if we combine the number of combinations for the winning numbers and those for the nonwinning numbers, we obtain

$$\binom{6}{4}\binom{43}{2} = 15 \times 903 = 13\,545$$

possibilities. This yields a probability of 0.00097, or about 1/1032. So not even one in every thousand players chooses "four correct." The following table shows the probabilities of winning in one of the various winning categories:

Win Class	Combinations	Probability
6 Correct	1	1/14 million
6 Correct + Bonus	6	1/2.3 million
5 Correct	252	1/55 491
4 Correct	13 545	1/1032
3 Correct + Bonus	17 220	1/812
3 Correct	229 600	1/61
loss (all the rest)	13 723 192	0.981

Because of the additional "bonus number" (a number chosen between 0 and 9), the highest winning category is subdivided into two classes, with the odds ratio 9:1. Thus the probability of landing in the top category (6 numbers out of 49 plus the bonus number) comes to only 1 in 140 million. In spite of increasing sales, not least due to the reunification of Germany, a winner in the highest category often does not appear for several weeks. The unclaimed cash is added to the jackpot for the next drawing. In 1994, the jackpot reached its highest value of about $24 million.

It is possible to obtain indirectly some information about the most frequently chosen numbers by studying the amounts distributed in the weekly drawings to see which numbers yield higher payouts, and which lower. However, there are many influences on these numbers, such as more- and less-favored partial combinations and variations from drawing to drawing, so that only limited information can be obtained.[6] Much more informative

---

[6]Heinz Klaus Strick, Zur Beliebtheit von Lottozahlen, *Praxis der Mathematik* **33:1**, 1991, pp. 15–22; Klaus Lange, *Zahlenlotto*, Ravensburg 1980, pp. 61–110.

was a study done in 1993 in the state of Baden-Württemberg, in which all bets were examined.[7] Among these players' bets, 80.7% had at least one number between 1 and 10. One may conclude that such numbers are so frequently chosen that combinations with numbers less than or equal to 10 are to be avoided. One would be doing the book of this study too much honor to believe that the popularity of those numbers is due to its recommendation. The cause is undoubtedly the preference for dates, as we mentioned above.

A Game of Poker

In the game of poker, each player receives 5 cards from a pack of 52. Thus there are

$$\binom{52}{5} = \frac{52 \times 51 \times 50 \times 49 \times 48}{1 \times 2 \times 3 \times 4 \times 5} = 2\,598\,260$$

different poker hands. If out of these almost 2.6 million combinations one wished, say, to determine how many contain "two pairs," it would be perhaps best to proceed as follows: a hand of two pairs consists of five cards, with three different values, of which two appear twice. An example is

- 4 of hearts, 4 of clubs, jack of hearts, jack of spades, queen of spades.

A hand of two pairs is uniquely determined by the following attributes:

---

[7]Karl Bosch, *Lotto und andere Zufälle*, Braunschweig 1994, pp. 201 ff.; Karl Bosch, *Glücksspiele: Chancen und Risiken*, Munich 2000, pp. 57–70. This study looked at almost 7 million sets of lottery numbers, so that each of the almost 14 million combinations was to be expected about 0.5 times, on average. However, there were some combinations that were wildly more popular, 24 of which appeared over one thousand times. The most frequent selection consisted of the numbers $7, 13, 19, 25, 31, 37$, which was chosen 4004 times! It was not the numbers themselves, all of which, except for 25, are primes, that appears to be cause of their popularity. More telling, these numbers form an almost perfect diagonal on the betting form, starting at the upper right-hand corner. We can project that throughout Germany, this sequence of numbers has been chosen over 30 000 times. Few of the players have likely suspected how small their prospects were.

Similar results in other countries were obtained by H. Riedwyl. See Hans Riedwyl, *Zahlenlotto: Wie man mehr gewinnt*, Bern 1990; Norbert Henze, Hans Riedwyl, *How to Win More*, Natick 1998; Hans Riedwyl, Gewinnen im Zahlenlotto, *Spektrum der Wissenschaft* **3** 2002, pp. 114–119.

- the two values of the pairs (in our example, 4 and jack),

- the value of the singleton (queen),

- the two suits of the lower-valued pair (hearts, clubs),

- the two suits of the higher-valued pair (hearts, spades),

- the suit of the singleton (spades).

The number of possible combinations can now be obtained by multiplying together the numbers of cards that satisfy these criteria. One must keep in mind, though, that not all of the characteristics are independent of one another. In particular,

- there are $\binom{13}{2}$ possibilities for the values of the two pairs.

- there are then 11 choices for the singleton.

- there are $\binom{4}{2} = 6$ choices for the suits of the first pair.

- there are also 6 choices for the suits of the second pair.

- finally, there are 4 choices for the suit of the singleton.

Altogether, there are then $78 \times 11 \times 6 \times 6 \times 4 = 123\,552$ hands consisting of two pairs. If we choose five cards at random from a well-shuffled deck, then the probability of obtaining a hand with two pairs is equal to

$$\frac{123\,552}{2\,598\,960} = 0.04754.$$

Thus about one in every 25 poker hands contains two pairs.

The following table contains the numbers of each type of poker hand. There also appear in the table the values for the variant "poker dice," which uses five dice with the symbols 9 of spades, 10 of diamonds, jack, queen, king, and ace of clubs.

Poker Hand	Combinations with	
	5 cards	5 dice
Five of a Kind		6
Royal Straight Flush	4	
Straight Flush	36	
Four of a Kind	624	150
Full House	3744	300
Flush	5108	
Straight	10 200	240
Three of a Kind	54 912	1200
Two Pairs	123 552	1800
One Pair	1 098 240	3600
Remainder	1 302 540	480

Further Literature on Lotteries

[1] Norbert Henze, 2000mal Lotto am Samstag: gibt es Kuriositäten? In: *Jahrbuch Überblicke der Mathematik*, Braunschweig, 1995, pp. 7–25.

[2] Glück im Spiel, *Bild am Sonntag Buch*, Hamburg circa 1987, pp. 6–29.

[3] Ralf Lisch, *Spielend gewinnen? Chancen im Vergleich*, Berlin 1983, pp. 38–54.

[4] Günter G. Bauer (ed.), Lotto und Lotterie, Homo Ludens: der spielende Mensch, *Internationale Beiträge des Institutes für Spielforschung und Spielpädagogik an der Hochschule "Mozarteum" Salzburg* **7**, Munich 1997.

4

A Fair Division: But How?

Two players, Jill and Jack, engage in a game of chance that stretches over a number of rounds in each of which the chances of winning are 50:50. The entire stake is won by the first player to win four rounds. The score stands at 3 games for Jill to 2 for Jack, and then suddenly, due to some event, the game has to be called off. The players agree to divide the stake fairly based on the state of the game. But what share is fair?

The *division problem* belongs among the classical problems of probability theory. It was discussed in detail in the previously mentioned correspondence between Fermat and Pascal.[1] But there were even earlier attempts at finding a just solution,[2] where the usual suggestion was to divide the spoils according to the number of rounds won, that is, in shares of 3 to 2 in our example. This corresponds to usual business practice according to which, for example, profits earned in common are divided in proportion to investment. Other authors, however, took the opinion that the ratio of division should take into account the wins that have not yet taken place. Thus in our example, for the game to be decided, Jill must win only one game, while Jack must win two. This could lead to a division in the ratio 2 to 1.

Both Fermat and Pascal solved the division problem, using two different generally applicable procedures whose results always agree. Pascal describes his solution in a letter of 29 July 1654, which focuses on the odds

---

[1] Ivo Schneider, *Die Entwicklung der Wahrscheinlichkeitstheorie von den Anfängen bis 1933*, Darmstadt 1988, pp. 3 f., 25–40.

[2] Ivo Schneider, pp. 2 f., 9–24.

of each player winning if the game were to be continued. Thus a single additional round leads to a score of 4 to 2 or a tie of 3 to 3. In the first case, Jill wins everything, while in the second case it is clearly a just solution to divide the stake evenly. Therefore, Jill should certainly receive at least half the stake, and for the remaining half, the chances of each player to win it are equal. Therefore, if the score is 3 to 2, a just division should be in the ratio 3 : 1; that is, Jill receives 75%, while Jack gets 25%.

Using this result, we may analyze other situations. For a score of 3 to 1, in which after one more round the score will stand at 4 to 1 or 3 to 2, Jill should receive 87.5% of the stake, which is the average of 100% (for the 4 to 1 possibility) and 75% (for 3 to 2).

The principle behind Pascal's argumentation is to associate a division ratio with each possible score. These division ratios can be calculated one after the other, where, as we have just demonstrated, the order in which they are calculated is the reverse of the chronology of the game. The two division shares, such as 0.75 and 0.25, are nothing but the probabilities of winning associated with the two players. That is, the stake is divided according to each player's probability of winning.

An elegant idea for calculating the two probabilities of winning directly is due to Fermat. It was mentioned briefly in the above-mentioned letter of Pascal. In this case as well, the basic idea is that additional games will be played. However, this time, the number played is that necessary to produce a winner regardless of the games' outcomes. In the example that we are considering, two additional rounds are played—fictional rounds, that is— even if the first round has already determined the winner. There are four possible results that these two rounds can produce, and all of them have the same probability, which is therefore 1/4. Only in the last of the cases shown in Table 4.1 does Jack win the match. Therefore, Jack's probability of winning is only 1/4, while Jill wins with probability 3/4.

Of course, one can carry out an analogous set of calculations for other scores. If Jill must win n rounds, while Jack must win m, then an additional $n + m - 1$ additional fictive rounds need to be played. After that many rounds—and possibly earlier—one of the players will have won, while the

Next Round	Round After Next
Jill wins	Jill wins
Jill wins	Jack wins
Jack wins	Jill wins
Jack wins	Jack wins

Table 4.1. Possible courses of the game in two additional rounds.

other will not have been able to win a sufficient number of games. Since for each round there are exactly two possible outcomes, the $n + m - 1$ fictive rounds lead to 2^{n+m-1} different equiprobable courses that the game might take. How many of these bring victory to Jill? That is, how many of these game sequences are there in which Jill wins at least n of the rounds?

This is a purely combinatorial question, and it can be answered with the assistance of the binomial coefficients introduced in the preceding section. If k is the number of rounds won by Jill, then there are $\binom{n+m-1}{k}$ ways of distributing the k winning rounds among the $n + m - 1$ rounds. Since Jill must win at least $k = n$ rounds to achieve victory, there are altogether

$$\binom{n+m-1}{n} + \binom{n+m-1}{n+1} + \binom{n+m-1}{n+2} + \cdots + \binom{n+m-1}{n+m-1}$$

possibilities. If this number of favorable outcomes is divided by 2^{n+m-1}, the number of all possible outcomes, then one obtains the desired probability. Jill, who wins the match with that probability, should receive exactly that portion of the stake when the match is broken off.

If Jill has to win four rounds, and Jack only three, then Jill's share, on the basis of six fictive rounds, will be equal to

$$\frac{\binom{6}{4} + \binom{6}{5} + \binom{6}{6}}{64} = \frac{22}{64} = 0.34375.$$

The principle behind the repeated rounds in a game of chance is that of a series of trials, in which an experiment, namely, a single game of chance, is repeated over and over, with repetition independent of the others. A victory by one player is then simply an event that may or may not occur. The number of rounds won is measured as a frequency with which a single event can be observed within a series of trials. Of course, the probability of a particular event need not be exactly $1/2$, as was the case in the example presented above. To get a look at the general situation, let us assume that the probability of a particular event occurring in an individual trial is p.

How great, then, for example, is the probability that in six attempts the event is observed precisely two times? In considering all possible series of six trials in which the event occurs exactly twice, their number is precisely the number of ways of selecting two favorable trials from among the six trials, which is equal to $\binom{6}{2} = 15$. Since the individual trials are independent of one another, the probability of each of these 15 series of trials can be calculated with the help of the multiplication law. For example, the probability that the event occurs at the first and third trials, but not otherwise, is

$$p(1 - p)p(1 - p)(1 - p)(1 - p) = p^2(1 - p)^4.$$

By considering all the sequences that lead to the same final result, we obtain the probability that the event occurs exactly twice:

$$\binom{6}{2}p^2(1-p)^4 = 15p^2(1-p)^4.$$

For example, the probability of obtaining exactly two 6s in six throws of a die is

$$15 \times \left(\frac{1}{6}\right)^2 \times \left(\frac{5}{6}\right)^4 = 0.2009.$$

In general, the probability of obtaining an event k times in a series of n trials is equal to $\binom{n}{k}p^k(1-p)^{n-k}$. On account of this formula, the frequencies of the event are said to be *binomially distributed*.

The formulas of the *binomial distribution* create a connection between the abstract probability, whose value was obtained from considerations of symmetry, and the frequency of events in a series of trials, which can actually be measured. Because of the indeterminate nature of random processes, this connection is not free of uncertainty. That is, the statements that we can make are themselves expressed in terms of probabilities. However, the degree of uncertainty can be largely overcome by making statements about long series of trials and a large number of possible frequencies. For example, one can calculate the probability of obtaining at least 900 and at most 1100 6s with 6000 throws of a die by adding up the associated 201 binomial terms, obtaining 0.9995.[3] Thus it is almost a certainty that in 6000 throws, the number of 6s will be between 900 and 1100.

Of much greater significance than the quantitative result is the underlying principle, namely, the *law of large numbers*: in the course of a series of trials, the relative frequencies of an event approach ever more closely and with ever greater certainty the probability of the event. We shall have much more to say about the law of large numbers in the next chapter.

---

[3]Fortunately, there is a simpler way to obtain this result. However it is based on more advanced mathematics and so is not discussed here. We will have more to say on this issue in Chapter 13.

5

The Red and the Black:
The Law of Large Numbers

*After ten red numbers in a row have won at the roulette wheel, the crowd
shifts its bets to the black. The reason is clear: after such a large number of
consecutive red numbers, one expects the "law of averages" to kick in. After
all, the law of large numbers states that over time, the balance between the
red and the black should average out. But wait a moment! Each spin of
the roulette wheel should be independent of the others, since the ball has
no more "memory" then dice have. It follows that even after red has won
ten times in a row, the odds for the next turn of the wheel to produce black
should be the same as those of producing red. How can this paradox be
explained?*

The main attraction in every European casino is the roulette tables. There
hardly seems to be much variety in betting on one of 37 numbers or a group
of them and then watching to see which number the ball lands on when the
wheel is spun. But if one thinks in those terms, one fails to appreciate the
heady atmosphere that surrounds the roulette table. One is immediately
taken by the sumptuous interior of the casino and the elegant clothing
draping the backs of the guests; the large sums of money in play, shoved
nonchalantly across the table in the form of chips; the heavy carpets and
curtains that dampen the hubbub of the crowds; all dominated by the cry
of the croupier and the clattering of the ball as the wheel turns. Roulette
is a pure game of chance, whose odds are in many aspects symmetric: it
makes no difference whatsoever whether one bets on 17, 25, or 32. And the

simple odds of winning with red, black, even, odd, "1 to 18," or "19 to 36" are equiprobable. However, a player can control the risk factor. In betting on a single number, the payoff is $1 + 35$ times the amount wagered, though the odds of winning are correspondingly small, at $1/37$.[1] With the simple wagers the situation is reversed: the probability of winning are relatively high, at $18/37$, but the payoff is only twice the amount wagered.

Just as in dice games, the probabilities in roulette depend on the symmetries of the game, and thus on more or less abstract considerations. These probabilities are connected to what happens in the real world only through the law of large numbers, which is surely among the most important laws for the calculation of probabilities: in a series of trials, the relative frequencies of an event approach the probability of the event arbitrarily closely. For example, the proportion of red numbers over a long series of roulette games moves generally in the direction of $18/37$. The law of large numbers thus creates a link between theory and practice, that is, between the abstract notion of probability on the one hand and the experimentally determinable relative frequencies on the other.

As simple and plausible as the law of large numbers may seem, it is invoked incorrectly again and again. This is especially true in the situation in which an event is over- or underrepresented in a series of trials that is underway. How is it possible for such an imbalance to equalize, as predicted by the law of large numbers? It seems more likely that compensation in the opposing direction should have to take place. But is such compensation truly necessary? That is, after an excess of red numbers in roulette, can equalization take place only if thereafter, fewer red numbers appear than are to be expected based on the laws of probability? Let us take, for example, a number of series of 37 spins, each of which should yield, on average, according to the law of large numbers, 18 red numbers. But if in the first series, 25 red numbers appear, then red has exceeded the theoretical average by 7. If in the second series red appears 23 times, then the situation is even worse, since now the excess is $7 + 5 = 12$. A countervailing compensation has not taken place. Nevertheless, the relative frequency has moved in the direction of $18/37$, namely, from $25/37$ to $25 + 23/74 = 24/37$.

The explanation in actually quite simple: the law of large numbers predicts only that the *relative* frequencies move in the direction of the probabilities. A relative equalization takes place whenever an "outlier" is followed by a series that is less unbalanced. Since outliers remain the exception, the relative equalization is always very probable. Yet despite

---

[1] In European roulette, the wheel contains the numbers 0 through 36. The 0 is special, and colored green. Translator's note: Half of the remaining numbers are red, half black.

the relative equalization, the absolute imbalance can increase, as we saw in our example.

One might well ask what it is that legitimates our speaking of the law of large numbers as a *law*. There are two reasons:

- one may observe empirically that in series of trials in which a single experiment is repeated, the relative frequencies of an event have a definite value toward which they move, even if the series is again started from the beginning. In this sense, one speaks of a stability of frequencies. The law of large numbers attempts to explain this empirical observation, and it does this by stating that the definite values in question are the probabilities of the event.

- from the mathematical point of view, the law of large numbers arises from the basic assumptions of probability theory, which means primarily the addition and multiplication laws. These allow us to determine how rapidly and with what degree of likelihood the relative frequencies will approach the probabilities. This can be stated relatively complexly, though precisely, with the formulas of the binomial distribution, which we met in the previous chapter.[2] In this sense, the law of large numbers was proved mathematically around the year 1690, by Jakob Bernoulli.[3]

Thus theory and practice coincide, which is just what is required of a model in the exact sciences. Here the probability model makes statements that are more precise than the corresponding observations. But these predictions, such as those made by the formulas of the binomial distribution, are confirmed in practice when series of trials are made.

There are various ways of formulating how quickly and with what degree of certainty the relative frequencies will approach the probability. A far-reaching statement, known as the *strong law of large numbers*, was discovered (that is, derived mathematically from the axioms of probability theory) in a special case in 1909 by Émile Borel (1871–1956), and in 1917 in general form by Francesco Cantelli (1875–1966):[4]

---

[2]Quantitative interpretations of the law of large numbers will be more easily derived when we have more fully developed the concepts and methods of probability theory. We will therefore avoid a more precise formulation at this juncture.

[3]Ivo Schneider, *Die Entwicklung der Wahrscheinlichkeitstheorie von den Anfängen bis 1933*, Darmstadt 1988, pp. 118–124; Herbert Meschowski, *Problemgeschichte der Mathematik*, volume II, Zurich 1981, pp. 185–187.

[4]Information on the origins of the strong law of large numbers and a 1928 controversy over who had priority of discovery can be found in E. Senteta, On the History of the Strong Law of Large Numbers and Boole's Inequality, *Historia Mathematica* **19**, 1992, pp. 24–39.

If two numbers are specified, one the maximal deviation for the relative frequency, and the other the largest permissible probability of error, then there exists a minimal number of trials after which in "most" sequences of trials of that length, the relative frequency does not exceed that maximal deviation. Stated more precisely, the event that a series of trials will exceed the maximal deviation at least once at some point beyond that minimal number has probability at most the given probability.

It is important to emphasize that the probability relates to the event that comprises the deviations from *all* following trials. For example, if the (largest permissible) probability is 0.01, and the maximal deviation is 0.001, then with probability 0.99, a roulette *permanence*[5] will behave in such a way that from the minimal number of trials, *all* relative frequencies of "red" will be in the range from $18/37 - 0.001$ to $18/37 + 0.001$. Permanences with outliers beyond that minimal number appear with a probability of at most 0.01.

The question that we have posed has actually already been answered. Independence and the law of large numbers are not contradictory. That would be the case only if the law of large numbers predicted an absolute equalization. But there is no law of absolute equalization. Even after ten "red" numbers in a row, one can continue with confidence to bet on red, if one so desires, even at the risk of being considered an incorrigible ignoramus by the other gamblers. And one *will* be so considered, as a sampling from a book on roulette makes clear, a book that, we should mention, was published by a reputable firm:[6]

> The mathematicians of centuries past made a simple assertion: "In roulette, every spin of the wheel is new. Each future event is in no way influenced by those that came before." If that were true, then the roulette problem could be solved mathematically. But since that is not the case, one cannot solve the problem with the aid of mathematics alone.
>
> Namely, if each spin were new, as the mathematicians assert, if chance were truly to know no law, then how is it possible that approximately 500 black and 500 red numbers appear at the roulette table? Why not 1000 black one day, and 1000 red the next? And why has our CDC 6600 supercomputer (the same model that NASA uses), provided with 37 random numbers,

---

[5] A roulette permanence is jargon for a sequence of numbers obtained from the spins of the wheel.

[6] Thomas Westerburg, *Das Geheimnis des Roulette*, Vienna 1974.

generated in a matter of minutes a permanence of 5 000 000
spins in which after each sequence of a thousand numbers it
prints out the result, which is roughly divided into 500 red
and 500 black? Nonetheless, the mathematicians cling to their
beliefs, following the traditions of the previous century: A series
of 1000 black or red is not impossible! It is now our turn to smile
on them with pity.

We see, then, that there are two main objections raised against mathematical probability theory: the spins of the roulette wheel are not truly
independent of one another, and the predictions made on the basis of probability theory do not correspond with reality, for example, in sequences of
1000 red or black numbers.

It is definitely worth our while to give serious thought to these objections. The starting point, that spins of the roulette wheel do not influence
one another, cannot be proved mathematically at all. That the ball has no
"memory," as we like to say, is nothing more than a conclusion based on
our experience with mechanical processes. A crumpled feather changes its
internal state in addition to its outward form, and so it "knows" that it
"wants" to return to its original state. In the realm of atomic physics the
situation in more complex. Does the nucleus of a radioactive isotope split
when an internal clock—invisible to us—tells it that its time is up? Or is
it a matter of pure chance, a nucleus that throws the dice, so to speak, and
when a certain number comes up, spontaneously splits?

Such suppositions could make the acceptance of such a model plausible,
but nothing more. In contrast, the only way to verify a model consists
in checking the predictions of the model through empirical observation
to determine whether observation corresponds to prediction. And such a
correspondence occurs only when one assumes that the individual results of
the wheel are independent of one another, that is, when the multiplication
law can be applied to the different spins of the wheel. Thus in 5000 series
of trials of 1000 red and black numbers each, the distribution of the colors
will behave as predicted by the formulas of the binomial distribution. The
mathematics does not "cling" to unproved facts. Indeed, the results must
prove themselves over and over, every day, in applied statistics as well as
in the casino. And that is precisely what they do.

To be sure, the author of the book on roulette cited above is not alone.
A very nice example is to be found in the last lines of Edgar Allan Poe's
(1809–1849) "The Mystery of Marie Roget," from the year 1842:

Nothing, for example, is more difficult than to convince the
merely general reader that the fact of sixes having been thrown

twice in succession by a player at dice, is sufficient cause for betting the largest odds that sixes will not be thrown in the third attempt. A suggestion to this effect is usually rejected by the intellect at once. It does not appear that the two throws which have been completed, and which lie now absolutely in the Past, can have influence upon the throw which exists only in the Future. The chance for throwing sixes seems to be precisely as it was at any ordinary time—that is to say, subject only to the influence of the various other throws which may be made by the dice. And this is a reflection which appears so exceedingly obvious that attempts to controvert it are received more frequently with a derisive smile than with any thing like respectful attention. The error here involved—a gross error redolent of mischief—I cannot pretend to expose within the limits assigned me at present; and with the philosophical it needs no exposure. It may be sufficient here to say that it forms one of an infinite series of mistakes which arise in the path of Reason through her propensity for seeking truth in detail.

6

Asymmetric Dice:
Are They Worth Anything?

Can a single die that is irregular in form or material nonetheless be used as a valid replacement for a symmetric die? That is, with an asymmetric die, can one of the six numbers 1 to 6 be chosen randomly in such a way that all six results are virtually equiprobable?

In our previous investigations we have always assumed our dice to be absolutely symmetric. The Laplacian model of probability allows nothing else! In practice, though, it is unrealistic to expect that a normal die will not be at least a bit asymmetric, even leaving aside the possibility of nefarious manipulation.

In casinos, on the other hand, in games like craps, precision dice are used, which are machined to a tolerance of 0.005 millimeters.[1] In order to achieve this degree of precision, the edges and corners of casino dice are not—in contrast to garden-variety dice—rounded off. Even the holes forming the dots on the dice have been taken into account. They are filled with a material of the same density as that of the dice themselves. To make manipulation difficult, they are made of a transparent material. Moreover, they are numbered and inscribed with the casino's monogram. Dice that are no longer usable are so marked.

But what about dice that do not even begin to satisfy this Platonic ideal, perhaps because they have been loaded with a small piece of metal? In such cases, the Laplacian model no doubt fails to be satisfied, but the

---

[1]John Scarne, *Complete Guide to Gambling*, New York 1974, p. 261. Manipulation of dice and other forms of deceit are discussed starting on page 307.

situation is not hopeless, since the laws of probability can be extended to account for asymmetric random processes. What, then, is the probability of throwing a 6 with an asymmetric die?

In practice, there is but one way to determine this probability: throw the die, and the more often, the better.[2] Just as with a symmetric die, one can observe the relative frequencies. That is, as a series of trials progresses, the variation in the relative frequencies of an event becomes less and less. These frequencies approach a particular value. If one repeats the series of trials, then in spite of differences due to chance, a similar picture results. In particular, the relative frequencies again approach the same limiting value; that is, the deviations between the relative frequencies of both series of trials become arbitrarily small as the number of trials increases. These limiting values are invariants of the structure of the die. They are concrete constants associated with the die used in the experiment.

One may observe analogous results in all random experiments, so that frequent repetition of a random experiment can be used as a measuring device. What is measured is probability, and indeed, that is true by definition. That is, just as is done in physics, one defines the notion of probability by establishing a method by which it can be measured. The fact that the probability cannot be exactly determined as it can in the case of a symmetric die is completely acceptable. After all, when physical quantities are measured, there is always a certain amount of "experimental error." In reality, the series of trials need not be actually carried out, in which case the probabilities are treated as unknown quantities that can be theoretically determined.

Thus every die, no matter how misshapen, possesses six fundamental probabilities $p_1, p_2, \ldots, p_6$. Other than the fact that these probabilities need not all be $1/6$, most of the familiar properties from the Laplacian model hold here as well: all probabilities are numbers between 0 and 1, where 0 is the probability of the impossible event, and 1 the probability of the certain event. Furthermore, the addition and multiplication laws remain valid. Thus the probability of obtaining an even number is $p_2 + p_4 + p_6$. Moreover, the sum of the six probabilities is equal to the probability of a certain event, yielding the identity $p_1 + p_2 + \cdots + p_6 = 1$. If one throws the die twice, then those are two independent events. As a result of the multiplication law, the probability of obtaining a three with the first throw and a six in the second is $p_3 \times p_6$.

Now that we know how to handle asymmetric dice, we can solve the problem posed at the beginning of this chapter. To simplify the situation,

---

[2]Approaches to a geometric solution are discussed by Robert Ineichen, Der schlechte Würfel: ein selten behandeltes Problem in der Geschichte der Stochastik; *Historia Mathematica* **18**, 1991, pp. 253–261.

let us first consider a slightly bent coin, whose condition leads us to suspect that the events "heads" and "tails" may not be equiprobable. We denote the probability of heads by p, with the result that the probability of tails is $1 - p$, which we may abbreviate as q. In the ideal case of a balanced coin, we would have the relation $p = q = 1/2$.

Yet even in the case of our bent coin, it is possible to use it to create 1 : 1 odds. One simply tosses the coin a large number of times, counts the frequency with which heads appears, and makes the decision based on whether the number of heads was even or odd. As an example, let us suppose that we have an asymmetric coin for which the probability of tossing heads is 0.6. Already with the second toss, a considerable equalization of the probabilities has taken place:

- the probability that the frequency of heads is even (two heads or two tails) is
$$0.6 \times 0.6 + 0.4 \times 0.4 = 0.52.$$

- the probability that the frequency of heads is odd (heads–tails or tails–heads) is
$$0.6 \times 0.4 + 0.4 \times 0.6 = 0.48.$$

After three tosses, the numbers are 0.504 and 0.496, and after four tosses, 0.5008 and 0.4992. And what if the initial probabilities are something altogether different? After all, in the case of a real coin, the probabilities are unknown. In that case, we proceed from a general random experiment that can end with two possible results, which we can call "yes" and "no." If the associated probabilities are $1 + d/2$ and $1 - d/2$, then the (possibly negative) number d is a measure of the deviation from symmetry. That is, the smaller the absolute value of d, the less the experiment is likely to deviate from the symmetric ideal case. If two independent yes–no random decisions whose deviations from symmetry are given by d and e are made one after the other, this leads to the following probabilities:

- two yes or two no:
$$\frac{(1+d)(1+e)}{4} + \frac{(1-d)(1-e)}{4} = \frac{1+de}{2};$$

- one yes and one no:
$$\frac{(1+d)(1-e)}{4} + \frac{(1-d)(1+e)}{4} = \frac{1-de}{2}.$$

Thus the measure of deviation from symmetry in the total experiment is equal to the product $d \times e$ of the individual measures d and e. In our example of the bent coin with probabilities $0.6 = 1 + 0.2/2$ and 0.4, this measure was 0.2. Therefore, in multiple flips of the coin, the probability of a yes decision is

- $\left(1 + 0.2^2\right)/2 = 0.52$ for two tosses,

- $\left(1 + 0.2^3\right)/2 = 0.504$ for three tosses,

- $\left(1 + 0.2^4\right)/2 = 0.5008$ for four tosses,

- $\left(1 + 0.2^5\right)/2 = 0.50016$ for five tosses,

and so on.

Thus even with such an unfair coin, a fair, equiprobable decision can be made. The only condition is that each of heads and tails can appear. Thus in a $90 : 10$ situation, which means that $d = 0.8$, after 20 tosses, the probabilities are 0.5058 and 0.4942.

We shall not consider the more complicated case of a die in great detail. However, the result is largely analogous. We begin with a die whose basic probabilities are $p_1, p_2, \ldots, p_6$, all in the range from $1-d/6$ to $1+d/6$, where d is a number in the range from 0 to 1. In order to obtain a result among the numbers $1, 2 \ldots, 6$, one could write down the six numbers in a circle, and move a counter around the circle the number of steps equal to the number that appears on the die. One obtains the same result by considering only the remainder upon dividing the sum of the numbers that appear on the die by six. The longer one tosses the die, the more the probabilities of the different results approach one another. It can be shown[3] that after n throws, the probabilities for the six individual fields all lie in the range from $1 - d^n/6$ to $1 + d^n/6$. As in the case of the coin, this leads slowly but surely to an equal distribution (up to a small error) of the probabilities. An asymmetry in the die has thus been overcome. In contrast to the case of the coin, the six basic probabilities cannot be too large. If a die is so loaded that one event has a probability of $1/3$ or greater, then it is no longer certain that the procedure described will lead to the advertised result.

---

[3]One can proceed as in the case of a coin. That is, one considers two random experiments with the results $1, 2, \ldots, 6$. If the probabilities of a random decision lie between $1-d/6$ and $1+d/6$, and that of a second decision between $1-e/6$ and $1+e/6$, then the probabilities of the combined experiment lie between $1 - de/6$ and $1 + de/6$. In the calculations that follow, one represents the probabilities in the form $1+d_1/6, \ldots, 1+d_6/6$, with

$$d_1 + \cdots + d_6 = 0 \quad \text{and} \quad |d_1|, \ldots, |d_6| \leq d$$

(and analogously for the second experiment).

7

Probability and Geometry

Suppose that one tosses a stick into the air in a room whose floor is made of parallel boards of uniform width, and bets are made as to whether the stick will cross one of the lines between two boards, one player betting that it will, the other that it will not. What are the odds for each of these players?

This question goes back to a quotation from the year 1777. It is known as Buffon's needle problem and was first presented in 1773 by Comte de Buffon (1707–1788) before the Paris Académie des sciences. An additional condition is that the length of the stick should not exceed the distance between two of the boards.

Buffon's needle problem is doubtless one of the classics of probability theory. It is considerably different in kind from the questions that we have been examining thus far. Although there are only two results, namely, that the stick crosses a line or does not, the symmetries and equiprobabilities that underlie the Laplacian model involve geometric data for which the number of possible cases is infinite:

- every angle between the stick (or its extension if necessary) and the lines between the boards is equiprobable.

- the midpoint of the stick can land with equal probability at any point on the floor, which translates into the distance from the midpoint to the nearest line. That is, every possible distance between 0 and the achievable maximum, namely, half the distance between neighboring lines, is equiprobable.

The mathematical consequences of these dual and mutually independent possibilities are more complex than those that we have seen. Because there are infinitely many angles and distances, we cannot take the approach that these are all equiprobable events. The infinite number of possibilities results in the probability for any single angle or single distance being equal to 0, which is not to say that the associated event is impossible.

One obtains positive probabilities by associating events with entire regions, that is, intervals. For example, if we divide the circle into six equal segments, then each segment possesses the probability 1/6. That is, in a single experiment, the probability of the stick landing with an angle belonging to one of the six segments is 1/6. In general, one can find a probability for an arbitrary interval that depends only on the geometric "size" of the interval, by which we mean the relative size in relation to the entire interval. The addition law for such "geometric" probabilities then amounts to nothing more than adding lengths.

But how is one to treat such geometric probabilities mathematically? For example, in Buffon's needle problem, how can the assumptions about the equiprobabilities of angles and distances lead to the desired probability? Why, based on empirical experiment, is that probability equal to $2/\pi = 0.6366$ in the case of a stick of length equal to the distance between boards?

Before we delve into the not-so-simple needle problem, we would like to consider a similar question, one that also goes back to Buffon: a coin of radius r is tossed, and it lands on a tiled floor, the tiles of which are squares of side a. What is the probability that the coin does not touch a line between tiles (we consider these lines to have zero width)?

This problem is easier to the extent that only one geometric value is needed to describe the result of tossing the coin, namely, the point on the floor on which the center of the coin lands. Here the situation is the same for each tile (see Figure 7): the event that the coin does not touch a gap occurs when the center of the coin lies within a square centered on the tile whose sides, parallel to those of the tile, are at a distance the radius of

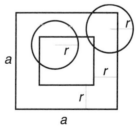

Figure 7.1. The ways in which a coin can land on a tiled floor.

the coin r from those of the tile. Such a square exists only if the length a of the tiles is greater than the diameter $2r$ of the coin. Since every point on the floor is an equiprobable landing site, the probability that no gap is touched by the coin is equal to the ratio of the area of the smaller square to that of the tile, that is, $(a - 2r)^2/a^2$.

By changing the experiment a bit, we can interpret other surface areas as probabilities. For example, if we provide our square tiles with a design in the form of an inscribed circle, then the probability that a random point chosen on a tile will be inside the circle is equal to the ratio of the area of the circle to that of the square, that is,

$$\frac{\pi \left(\frac{a}{2}\right)^2}{a^2} = \frac{\pi}{4} = 0.7854.$$

This has two significant consequences relating to the law of large numbers:

- as a probability, the number $\pi/4$ can be approximated by the relative frequencies occurring in a series of random experiments. Consequently, the number π can also be experimentally approximated using random experiments.

- the experimental nature of calculating areas is quite universal. Instead of circles, we could provide our tiles with other regions. The relative frequencies that occur in a series of trials approach the area to arbitrary precision (with an arbitrary degree of certainty).

In contrast to the situations that we have investigated thus far, in Buffon's needle problem we require two random parameters, namely, the angle between stick and line, and the distance from the midpoint of the stick to the nearest line. Though the mathematical details are more complicated, the principle of calculating the relaltion between two areas can again be used (see "Calculating Buffon's Needle Problem").

Calculating Buffon's Needle Problem

Let L denote the length of the stick, a the distance between two lines (the width of a floorboard), ϕ the angle between stick and line, and x the distance between the midpoint of the stick and the nearest line. Then we see from the left-hand figure that a line is intersected exactly when the inequality

$$\frac{L}{2} \sin \phi \geq x$$

is satisfied. If we now represent all the equiprobable pairs of values for x (from 0 to $a/2$) and ϕ (from 0 to π) in a rectangle, as depicted in the right-hand figure, then the pairs of values that represent a "hit" are those in the region between the sine curve and the horizontal axis. The associated area can be easily computed with a definite integral. The desired probability is again obtained from the relative area of this region to that of the entire rectangle. It is equal to $2L/a\pi$.

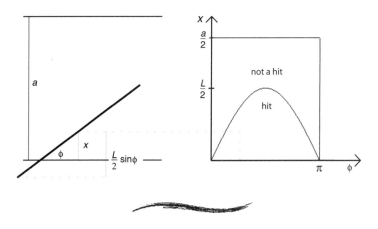

8

Chance and Mathematical Certainty: Are They Reconcilable?

Can the sequence of digits in the decimal expansion of a number such as $\pi = 3.14159265358\ldots$ be just as random as the results of a sequence of throws of the dice?

An important characteristic of randomness is that its results should be unpredictable. Thus the sequence of decimal digits of the number π are not, in fact, random. For even if particular digits of π are unknown, they can be calculated as needed. However, if one did not know the context of the sequence of digits

$$3, \quad 1, \quad 4, \quad 1, \quad 5, \quad 9, \quad 2, \quad 6, \quad 5, \quad 3, \quad 5, \quad 8, \quad \ldots,$$

then the digits would appear to possess a random nature. That is, a formula that would allow one to reproduce the digits of the sequence would be so complicated that one could hardly hope to find such a formula without the background knowledge of the source of the sequence.

On the other hand, how random is the result of throwing dice? After all, the physical processes that underlie the toss of a die can be completely described in terms of the laws of mechanics. If one understood all aspects of the situation, then it should be possible in principle to calculate in advance the precise movement of the die. Is it the case, then, that with dice the randomness is also only subjective, that is, the result of a lack of information? The leading initiator of such a viewpoint was Laplace, who shaped the advances in mechanics and astronomy of his time into a deterministic world view. In 1783, he noted the following:[1]

> Thus the appearance and movement of comets, which today we recognize as subject to the same law that ensures the cycle of the seasons, were once seen, by those who considered stars to be meteors, as the result of chance. Therefore, the word chance expresses nothing more than our lack of knowledge about the causes of events that we see appear and follow one another in an order that is invisible to us.

Consequently, Laplace saw probability theory as a tool that provides a degree of orientation in the face of lack of knowledge of causal relationships, so that the resulting ignorance can be partially overcome. Thus, with the law of large numbers, rather precise statements about the results of a series of trials in a game of chance can be made, without the requirement that the physical course of the throw of the dice be analyzed.

If chance, then, is no more than a subjective phenomenon, then the decimal digits of π are also random, though perhaps not quite so random as the results of throwing a die. But is there no such thing as objective randomness? From today's viewpoint, the answer must be answered in the affirmative, on the basis of discoveries of physics and mathematics.

Probability theory obtained its first significant application in the natural sciences in kinetic gas theory, founded in 1859 by James Clerk Maxwell (1831–1879). The temperature of a gas is interpreted as a function of the motion of the molecules of the gas, and therefore, can be derived in principle from the laws of mechanics as they relate to motion. What is new, here, is the complexity resulting from the astronomical number of particles, which makes explicit calculation impossible. Nonetheless, mathematical results can be obtained if one views the velocity of a single particle as random. Macroscopically measurable quantities such as volume, pressure, temperature, and chemical composition can be explained by the average behavior of the molecules. In particular, the temperature can be described as the average kinetic energy of the molecules. As with any other model, kinetic gas theory must be tested against the results of experiment. And it passes the test. Here is an example involving the notion of entropy: a container is filled with two different gases, one after another, in an environment of weightlessness. Then soon, the gases combine with each other into a uniform mixture. Is the reverse direction possible? That is, can the mixed

---

[1]Ivo Schneider, *Die Entwicklung der Wahrscheinlichkeitsrechnung von den Anfängen bis 1933*, Darmstadt 1988, p. 71.

gases separate on their own? There is nothing in the laws of mechanics that prohibits this. The probabilistic model states that while such a separation can happen in principle, the probability is so small that it would almost never happen, and thus macroscopically, the outcome is determined. A simpler version of this situation can be seen in shuffling two large packs of cards, one of red cards, the other of black, so that they are mixed. The event that a later shuffle will separate the red from the black is possible, but extremely unlikely. That is the effect of the law of large numbers.

Does randomness have an objective character in the context of the kinetic theory of gases? The fact that the data of a system are in practice concentrated on a fundamental core of average values such as the temperature demonstrates an existing subjective state of ignorance, nothing more. However, such a system can in principle be calculated in advance based on a mechanistic interpretation and the deterministic laws of physics, even though the practical realization of such a project is just as hopeless as in the case of throwing dice. That is, one may nurture the hope of being able to explain apparently random processes of macrophysics through the deterministic laws of microphysics. That this hope must be dashed even theoretically has been known since the development of quantum physics in the 1920s by Werner Heisenberg (1901–1975) and Erwin Schrödinger (1887–1961). According to that theory, randomness is an insuperable obstacle to the observer of events, and the apparently deterministic laws of macrophysics are revealed as statistical laws of microphysics. This inability to predict outcomes is encapsulated in the Heisenberg "uncertainty principle," which asserts that the location and velocity of a particle can be determined only up to limited precision, since every measurement—though it be "only" that of the light being used—has itself an effect. In Heisenberg's words, this means that, "In the precise formulation of the law of causality—'If we know the present, we can calculate the future'—it is not the conclusion that is false, but the hypothesis." Since the entirety of the state parameters that determine the model can never be precisely measured, statements about the future development of a system are always subject to randomness. That is, randomness is objectively present.

In the realm of classical mechanics, we now return to our dice: the apparently simple toss of a die, with all its bouncing and spinning, is influenced by so many factors that it is practically impossible to predict its course. Even two tosses with apparently identical initial conditions can have completely different outcomes, since even the smallest differences can multiply, eventually leading to extremely divergent behaviors: "Small causes, large effects." One can thus see a special form of causal relationship in the randomness of a die, in which the occurrence or nonoccurrence of an event depends on unobservable tiny influences. It is therefore impossible to predict how these influences will affect the outcome of a concrete trial. Causal relations that are sensitive in this way to very small perturbations are called *chaotic*. They represent, as we now know, much more the normal state of affairs than the orderly, continuous relations that appear in the classical interpretations of natural law.

Randomness in chaotic relations is actually something purely mathematical. By this we mean that chaos can be completely explained on a mathematical level with the aid of the deterministic formulas of classical physics. The sort of chaos that even the simplest formulas can "cause" is known to all who have seen the popular images of fractal geometry. Although regularities are to be seen in the form of self-similar images, it is impossible to provide a simple description of the image without knowing the fundamental formula. Just as with the number π, and here we return whence we began, we may here speak of randomness, though only on a subjective level.

Just as in kinetic gas theory, it can be quite useful to employ methods of probability theory even when a random influence is not at all objectively guaranteed. Even in everyday life, there are many such situations whose courses are unforeseeable at a sufficiently early point in time:

- the make of the first car stopped at a red light;

- the amount of claims made against an insurance company in the course of a year;

- the amount of rain in the course of a day;

- the number of parking tickets that a habitual scofflaw received in the course of year;

- the number of workers in a firm who called in sick.

We have already seen in our examination of Buffon's needle problem that it can be reasonable to bring randomness into play in mathematics. It is not only π, but the area of arbitrary planar regions that can be determined through random experiments. Even when the subject turns to prime numbers, it can be advantageous to consider divisibilty properties from the point of view of randomness. For example, a randomly chosen integer is even with probability $1/2$. From this viewpoint, it is possible to make estimates of how frequently, within a specified range, integers appear with particular properties, such as prime numbers, twin primes (two primes, like 41 and 43, that differ by 2), or integers with a specified number of prime divisors.

We have up to now been able to discover phenomena that can be considered truly random only in quantum mechanics. Thus there is no a priori moment at which an atomic nucleus of a radioactive isotope should decay. But can there be true chance within mathematics, which appears to be so certain in all its aspects? In the historical development of chance and probability there has been considerable difficulty in stating its principles satisfactorily. Thus at the beginning of the twentieth century, probability theory consisted of a hodgepodge of methods for solving a variety of problems. But there was great uncertainty with regard to the foundations of the field: are the laws of probability laws of nature, such as those of physics, or is there an abstract theory whose objects are so idealized, as in the field of geometry, that they can be considered outside of the material world? In that case, the notions of randomness and probability must be able to be explained mathematically in the same way as length, area, and volume. Concrete trials and series of measurements have no place in such explanations, no more than in geometry would a volume be defined as the result of submerging the shape in question in a tub of water and measuring the displacement.

Since the certainty of mathematics, on the one hand, and the uncertainty of randomness and probability, on the other, appear to exclude each other, mathematical authorities on the subject tended at the beginning of the twentieth century to view probability theory as a physical disciplne.[2] In 1919, the mathematician Richard von Mises (1883–1953) sought a way to provide a mathematical foundation for probability theory. Mises studied relative freqencies in series of trials. However, he did not base his studies on real series of trials, but on sequences of results that could come from a series of trials. For this, Mises required that the sequence of results be irregular. But how does one formulate such a criterion precisely? How irregularly must a sequence of integers such as

4,   2,   1,   1,   6,   2,   3,   5,   5,   5,   6,   2,   3,   6,   2,   ...

---

[2] Here is a relevant quotation form the year 1900: "Through investigations into the foundations of geometry the task suggests itself to treat according to this model such physical (!) disciplines in which even today mathematics plays such a prominent role; these are primarily probability theory and mechanics." This quotation comes from a famous lecture given by David Hilbert (1862–1943) at the second International Congress of Mathematicians. The lecture presented a list of the important open mathematical problems of the time, and the sixth of these was Hilbert's just-quoted suggestion. Later, we shall have something to say about some of the other problems. See also *Ostwalds Klassiker der exakten Wissenschaften* **252**, Leipzig 1976 (Russian original 1969); Jean-Michel Kantor, Hilbert's problems and their sequels, *The Mathematical Intelligencer* **18/1** 1996, pp. 21–30.

be constructed for it to be considered random? Regular sequences such as

$$1, \quad 2, \quad 3, \quad 4, \quad 5, \quad 6, \quad 1, \quad 2, \quad 3, \quad 4, \quad 5, \quad 6, \quad 1, \quad 2, \quad 3, \quad \ldots$$

are assuredly not random. But for what sequences does the opposite hold?

In the end, von Mises's approach could not be deemed satisfactory. His approach of abstracting the measurement process for probability mathematically broke down. It was only decades later, namely, in the 1960s, that Gregory Chaitin and Andrei Kolmogorov (1903–1987) independently succeeded in providing a formal definition of randomness in sequences of integers.[3] What is crucial for the randomness of a sequence of integers is how large a computer program would have to be to generate the sequence. Thus random sequences in their infinite length could not be generated by a finite program; thus the sequence of integers in the decimal expansion of π is not random. And for finite beginnings of sequences, there must be no general method of generating the integers with programs that are "too" simple. In other words, it should be impossible to compress the sequence too greatly.[4]

As interesting as such a characterization of randomness may be, it holds, surprisingly, hardly any significance for mathematical probability theory. A mathematical model of probability can be constructed without reference to randomness.

Thus already in 1900, the mathematician Georg Bohlmann (1869–1928) was able to reduce the known laws and techniques of probability theory to a system of basic properties: in addition to establishing that probabilities are values between 0 and 1 associated with events, his system primarily encompasses the addition and multiplication laws. In and of itself, Bohlmann's approach is not at all unusual, just as Kepler's laws can be derived from the basic laws of mechanics and the theory of gravitation. What is crucial here is that Bohlmann changed the meaning of these fundamental properties by taking them as definitions. That is, whatever satisfies these properties is considered a probability. And even the independence of two events is no longer understood as the lack of "mutual influence." Two events are considered mutually independent in the sense of Bohlmann if they satisfy the multiplication law, that is, if the probability that both events occur

---

[3] Chaitin has written two popular accounts of his work: Randomness and mathematical proof, *Scientific American* **232:5** 1975, pp. 47–52, and Randomness in arithmetic, *Scientific American* **259:1** 1988, pp. 80–85.

[4] From an information-theoretic point of view, compression is possible for sequences—even those randomly generated—whenever the numbers that appear do so with differing frequencies. For example, a sequence that consists of 90% zeros and 10% ones can be compressed by storing the distance between the ones. Since the orignional sequence can be thereby reconstructed, no information is lost.

is equal to the product of the individual probabilities. Further properties of probabilities, such as the law of large numbers and the methods for measuring probabilities based thereon, are the result of, and only of, the logical conclusions that can be drawn from the fundamental properties—an axiomatic system, that is—which have here been elevated to a definition.

A weakness of Bohlmann's conception was that the notion of event was tacitly assumed. This difficulty was overcome in 1933, when Andrei Kolmogorov put forward a purely mathematical axiomatic system and an exposition of probability theory built upon it. Kolmogorov used exclusively mathematical objects, namely, numbers and sets. Probability theory thereby obtained a purely mathematical and thus universally applicable formulation, and was thereby transformed, like arithmetic, geometry, and analysis, into what is indisputably a mathematical discipline.

Kolmogorov's axiomatic system is based on the idea that every random experiment can be associated with a *set of possible outcomes*. In the case of dice, be they symmetric or not, the outcomes correspond to the integers from 1 to 6. One thus takes as the set of outcomes the set $\{1, 2, 3, 4, 5, 6\}$. Every *event* can now be understood as a subset of the set of outcomes, namely, that consisting of the "favorable" outcomes. For example, the event of throwing an even number is represented by the set containing all even numbers between 1 and 6, which is the set $\{2, 4, 6\}$. The certain event encompasses all possible outcomes and therefore corresponds to the set $\{1, 2, 3, 4, 5, 6\}$. The impossible event occurs for no outcome, and thus is represented by the empty set $\{\ \}$.

Every event is associated with a *probability*. In deference to the Latin *probabilitas* and the English *probability*, probabilities are traditionally denoted by the letter "P." The event whose probability is being considered is placed after the "P" in parentheses, and the expression $P(x)$ is read "P of x."[5] The statement that the certain event has probability 1 is abbreviated by the formula

$$P(\{1, 2, 3, 4, 5, 6\}) = 1.$$

The probability of rolling a 6 corresponds to $P(\{6\})$, while $P(\{2, 4, 6\})$ stands for the probability of rolling an even number. The concrete values of these probabilities are of no importance to the theory; in the end, physical formulas are also not dependent on concrete values. The values could, but need not, agree with those of the Laplacian model.

The approach of describing events by subsets of a fundamental set makes it possible to formalize, completely and mathematically, statements about

---

[5] Mathematically speaking, P is a mapping from the set of outcomes into the set of real numbers.

events and probabilities. Let us take as an example the formulation of the addition law.

Let A and B be two incompatible events. With respect to the sets A and B, this means that the two sets have no element in common, and therefore, their intersection $A \cap B$ is empty. For the two events can then never both occur in the same trial only if no event is "favorable" for both. Analogously, the result that one of the two events occurs is represented by the set union $A \cup B$, which contains both the set A and the set B. Formulated mathematically, the addition law can be stated thus:

for two events A and B with $A \cap B = \{\ \}$, one has

$$P(A \cup B) = P(A) + P(B).$$

For example, with a single die,

$$P(\{2, 4, 6\}) = P(\{2, 4\}) + P(\{6\}).$$

According to Kolmogorov, probabilities are defined by considering a system of subsets and their associated numbers P( ) as random experiments complete with events and probabilities when certain conditions are satisfied. These basically correspond to the axioms formulated by Bohlmann and well-known intuitively derived laws of probability:

- probabilities are numbers between 0 and 1.

- the certain event has probability 1.

- the addition law holds.

We are not going to look too closely here into the mathematical formulation, because then it would be necessary to go into some details that while important for the mathematics, are not so important for practical applications (see Note 1 at the end of the chapter). It remains to mention that the axioms are so universal that they encompass all practical applications, including, of course, the infinite event sets of "geometric probabilities."

As was already proposed by Bohlmann, the statement of the multiplication law for independent events achieves the status of a definition, in that the equality corresponding to the multiplication law is used: thus two events A and B are considered independent if the condition $P(A \cap B) = P(A) \times P(B)$ is satisfied, where the intersection set $P(A \cap B)$ corresponds to the event that assumes the occurrence of both events A and B. Three or more events are considered independent if the corresponding product formula holds for every choice of two or more events.[6]

In the Kolmogorov model, terms from probability theory such as *event, outcome, probability,* and *independence* have acquired new interpretations that apply exclusively to mathematical objects. Only these formal interpretations are of interest within mathematical probability theory when, for example, one is investigating entire classes of problems. Only when the results obtained are to be applied do the plain, everyday meanings again take center stage. Such a way of proceeding makes it possible to arrive at universal and provable conclusions, and is thus extremely efficient. However, one should not underestimate the danger of losing sight of the theory's connection to applications.

In practice, in which it is important to explain and predict events, one is happy to limit the scope of the mathematical formalism. Thus the interpretations and arguments of the Laplacian model, despite their inadequacy, again come into play. The knowledge that the model's formal deficits can be overcome provides peace of mind to those who require it.

Chapter Notes

1. To obtain a mathematical model that is both practical and maximally flexible, two particular factors must be taken into consideration, though they are of importance primarily only for infinite sets of outcomes:

   - all events form a closed subsystem of subsets of the fundamental set under the set operations. As set operations we include the intersection of two sets and the union of countably many sets.

   - the property corresponding to the addition law must hold for countably many mutually disjoint sets.

   It may happen that not all subsets can be thus obtained. But this is by no means a drawback; in fact, it is rather the opposite. Namely, there are subsets for which it makes no sense to associate a probability and that therefore are better not viewed as events. Here is an example from the domain of geometric probabilities.

   If one chooses a random point within a sphere, then every set of points in the sphere corresponds to an event. If all regions of the sphere are considered equally in the random selection, then the probability of an event is equal to the proportion of the volume within the sphere. Congruent point

---

[6]For example, for three independent events A, B, C the following equalities hold: $P(A \cap B \cap C) = P(A)P(B)P(C)$, $P(A \cap B) = P(A)P(B)$, $P(B \cap C) = P(B)P(C)$, and $P(A \cap C) = P(A)P(C)$.

sets that can be superposed on one another by translation and rotation therefore possess equal probabilities. However, that some rather strange sets can arise is demonstrated by the Banach–Tarski theorem, according to which a sphere can be divided into finitely many pieces that—after translation and rotation—can be put together without any holes into a sphere of double the diameter of the original sphere. Such a construction appears to be impossible to normal human reasoning. Some insight can be obtained from the explanation that the pieces are not connected regions, as one may have thought. Rather, they are so complicated that they cannot be visualized; by comparison, the sets of fractal geometry seem like equilateral triangles. Although mathematics frequently describes conditions that correspond to our experience, it does not always do so.

What is the volume, then, of the pieces produced by the sphere deconstructed according to Banach–Tarski? Their sum must equal the volume of the sphere, but also that of a sphere with eight times the volume! We would do well to restrict ourselves at the outset to subsets that can actually represent events.

Further Literature on the Axiomatization of Probability Theory

[1] B. L. van der Waerden, Der Begriff der Wahrscheinlichkeit, *Studium Generale* **4**, 1951, pp. 65–68.

[2] Ivo Schneider, *Die Entwicklung der Wahrscheinlichkeitsrechnung von den Anfängen bis 1933*, Darmstadt 1988, pp. 353–416.

[3] Ulrich Krengel, Wahrscheinlichkeitstheorie, in: *Ein Jahrhundert Mathematik, 1890–1990*, Braunschweig 1990, pp. 457–489, especially Chapters 1 to 4.

[4] Thomas Hochkirchen, *Die Axiomatisierung der Wahrscheinlichkeitsrechnung und ihre Kontexte*, Göttingen 1999.

9

In Quest of the Equiprobable

There is an American television show in which the contestant can win an automobile by guessing which of three doors conceals the car. Behind the other two doors, as a symbol of the loser, are a couple of goats. To spice things up, after the contestant has selected a door, the quizmaster opens up one of the two remaining doors, always one with a goat, since he knows which door hides the car. Now the contestant has the option of switching his choice from the door that he originally chose to the remaining third door. Should he make the switch?

A great debate raged in the years 1990–1991 when this question was discussed in the pages of the magazine *Skeptical Inquirer*. It even crossed the Atlantic, seeping into the letters columns of the German weeklies *Die Zeit* and *Der Spiegel*.[1] And all because the opinion was offered in the magazine that the contestant can increase his chances of winning by abandoning his original choice. This seems a rather dubious assertion. It seems much more plausible to argue as follows: the probability of winning the car begins at 1/3. But when the quizmaster opens one of the doors, then there are only two possible outcomes, both equiprobable. Thus the probability for both doors has increased from 1/3 to 1/2. And so the decision to switch doors makes no sense.

---

[1] *Der Spiegel* 1991/34, pp. 212–213 and also (intentionally and unintentionally) amusing letters to the editor in 1991/36, pp. 12–13; *Spektrum der Wissenschaft* 1991/11, pp. 12–16; Gero von Randow, *Das Ziegenproblem*, Hamburg 1992. The problem itself is not new. It appeared in another formulation in Martin Gardner, *Second Book of Mathematical Puzzles and Diversions from "Scientific American,"* New York 1961, Chapter 19.

On the other hand, supporters of switching doors argued that the probability for the door originally chosen has not changed: since the quizmaster always opens one of the other two doors, and always one that has a goat behind it, one obtains no additional information about the originally chosen door. However, after that door has been opened, there are but two doors remaining. Therefore, the probability of success for the third door has increased to 2/3, and so the switch is to be recommended.

Surprisingly, the contestant's decision is made intuitively easier when the number of doors is larger. Let us assume that there are 100 doors with 99 goats and a single automobile, and the contestant has pointed at the first door. Then it is clear that the chances of success are very small indeed, since the probability of having chosen the correct door is—at least at the outset—only 1/100. If the quizmaster now opens 98 of the remaining 99 doors, what might we suppose is behind the remaining door? That's right! It must be the car, unless the initial choice was correct, which is highly unlikely. In the case of 100 doors, it is clear that one should alter one's original choice.

It should be clear, then, that the original version with three doors represents only a quantitative difference: to solve the original problem with clarity, we need to identify the equiprobable outcomes. The question is only what time and what state of knowledge are relevant for the symmetries in question. If the quizmaster was to open a door with a goat at the outset, then there would be only two equiprobable outcomes. In reality, however, the contestant first chooses one of the three doors, and only then is one of the other two doors opened. This does not necessarily make the remaining two doors equiprobable. The only thing that is certain is the initial situation, in which the automobile can be behind any one of the three doors with equal probability. Therefore, the initial choice is correct with probability 1/3 and incorrect with probability 2/3. After the quizmaster has opened one of the doors, revealing a goat, it is clear what the effect of switching doors would be:

	Original Choice	Probability	Change Mind?
First Case	Correct	1/3	bad
Second Case	Incorrect	2/3	good

Since the contestant does not know whether the first or second case obtains the automobile, he can only make a global decision to change or not to change. The table shows clearly that the contestant should switch doors, since that is twice as likely to improve the outcome as to worsen it.

Another way of viewing the situation is this: suppose that the game is played differently, so that after the contestant has chosen one of the three doors, the quizmaster offers to let him change his choice to *both* of the two remaining doors. That is, by switching his bet, the contestant will win if the car is behind either one of the two remaining doors. At this point, when no doors have been opened, it is clear that the probability of the car being behind one of those two doors is 2/3. Since it is a certainty that there is a goat behind at least one of those two doors, and the quizmaster knows where the goats are hidden, no new information is provided by opening a door with a goat behind it, and so makes no difference whether that door is opened before or after the contestant decides whether to switch doors. Therefore, from the point of view of probabilities, this variant is equivalent to the original game.

There is an important principle of probability theory hiding behind the problem of guessing the correct door: probabilities of events are not necessarily absolute. They can be seen as depending on the occurrence of other events. For example, the probability of rolling at least an 11 with two dice is equal to 3/36. If the first die shows a 6, then 30 of the original 36 equiprobable outcomes are excluded. Among the six remaining possible outcomes, 6-1, 6-2, 6-3, 6-4, 6-5, 6-6, two of them, namely, 6-5, 6-6, lead to success. Therefore, the probability of rolling at least 11 on the assumption of the event that a 6 was rolled with the first die is equal to $2/6 = 1/3$.

In general, one can proceed analogously within the confines of the Laplacian model. We begin with two events, A and B. We then divide the number of outcomes that are favorable for both events A and B by the number of outcomes favorable for the event B, and thereby obtain the *conditional probability* of event A, on the assumption that event B has occurred. This probability is denoted by $P(A \mid B)$. Such probabilities can also be defined outside of the Laplacian model, where instead of numbers of favorable outcomes, the quotient of the corresponding probabilities is taken:

$$P(A \mid B) = \frac{P(A \cap B)}{P(B)}.$$

For our example above, A corresponds to the event that at least 11 dots will show up on the two dice, while B stands for the event that the first die comes up with a 6. The equation can be expressed in full detail as follows:

P(sum is at least 11, assuming that the first die shows a 6)

$$= \frac{P(\text{sum is at least 11 and first die shows a 6})}{P(\text{first die shows a 6})}$$

$$= \frac{\frac{2}{36}}{\frac{1}{6}} = \frac{2}{6} = \frac{1}{3}.$$

For two independent events A and B, which by definition satisfy the condition $P(A \cap B) = P(A)P(B)$, the definition of conditional probability yields the equality $P(A \mid B) = P(A)$, and analogously, $P(B \mid A) = P(B)$. The probabilities of the two events do not change when one considers each event as conditional on the other.

It is often possible to calculate the probability of an event more simply with the help of conditional probabilities. To do this, the event in question must be viewed as the intersection of two or more events so that the equality $P(A \cap B) = P(A \mid B) \times P(B)$, called the *generalized multiplication law*, can be used. It is fairly clear how this approach operates when the underlying random experiment breaks down into several steps, even if only in imagination, as in the above example of the dice. For example, if two cards are chosen at random from a deck of 52 cards, then

- the probability of the first card being an ace is 4/52, and

- the conditional probability of an additional ace is 3/51.

Therefore, the probability of drawing two aces is equal to $3/51 \times 4/52 = 1/221$.

The probability of drawing two aces is thus calculated by investigating the "intermediate" result of drawing an ace with the first card. However, this method does not always proceed so easily, for example, if the event of interest can be achieved in a number of ways. Let us consider, for example, the event that by drawing 2 of the 52 cards we get a "blackjack."[2] A blackjack consists of an ace together with a ten or one of the face cards (jack, queen, king). We again proceed by considering events characterized by the first card drawn. In contrast to the example of two aces, there are now two completely different ways in which the drawing can proceed, since a blackjack may arise from the first card being an ace, as well as from the first card being a ten or face card. It is only after we separate these two cases from each other that we will be able to use the multiplication law. If we then add together the two probabilities associated with the two ways of achieving a blackjack, we obtain the total probability of drawing a blackjack:

---

[2]A blackjack is the highest combination in the like-named card game. Blackjack, which is played in many casinos, is the subject of Chapter 17.

First Card	Probability	Conditional Probability of a Blackjack	Net Probability of a Blackjack
Ace	4/52	16/51	$4/52 \times 16/51 = 16/663$
10, J, Q, K	16/52	4/51	$16/52 \times 4/51 = 16/663$
2 to 9	32/52	0	$32/52 \times 0 = 0$

In general, the principle that we have presented is described by the *formula for total probability*: if the events $B_1, B_2, \ldots, B_m$ form a disjoint decomposition of the certain event, then for an arbitrary event A, the following equality holds:

$$P(A) = P(A \mid B_1) \times P(B_1) + P(A \mid B_2) \times P(B_2) + \cdots + P(A \mid B_m) \times P(B_m).$$

And now back to our goat problem. Although we have actually solved it already, we would like to solve it once more. For each of the two strategies, we simply use the formula for the total probability.

Original choice is changed:

First Choice Was	Probability	Conditional Probability of Winning	Net Probability
Correct	1/3	1	1/3
Incorrect	2/3	0	0
Total Probability of Winning (Sum)			1/3

Original choice is not changed:

First Choice Was	Probability	Conditional Probability of Winning	Net Probability
Correct	1/3	0	0
Incorrect	2/3	1	2/3
Total Probability of Winning (Sum)			2/3

What is so complicated? How was it possible for even professionals to be led so astray? There seem to be two reasons for this:

- in the case of three doors, the conditional probabilities are equal to 0 and 1. Such values keep us from having to deal with fractions, but on the other hand, they are scarcely seen as probabilities. Often, the correct approach is not found at all. In the case of four or more doors, the situation is simpler, since all the probabilities assume "genuine" values.

- where are the outcomes that we can consider equiprobable on the basis of symmetry? They are recognizable only at the outset, when each door is equivalent to every other one. Every attempt to find symmetries later becomes quickly entangled in pure speculation.

Is there no longer any doubt about the matter? What are we to make of the following argument, put forth in *Der Spiegel* by an individual with "Prof. Dr." in front of his name?[3]

When the quizmaster now opens an additional door and thereby removes the second door from consideration, then according to your calculation, the probability that the automobile is behind the first door remains $1/3$ (not 1!); that is, the probability that the goats have meanwhile eaten the car is $2/3$.

---

[3] *Der Spiegel* **36**, 1991, p. 12.

10

Winning the Game:
Probability and Value

In the game of chance called chuck-a-luck, three dice are thrown. One may bet on each of the six symbols that adorn the faces of each die. One loses if the symbol bet upon does not appear on any of the three dice. However, if the symbol does appear, the player receives, in addition to his stake, the amount bet for each of the dice displaying the symbol. (For example, if the number bet on appears twice, the player receives three times the amount wagered.) Does the bank have the advantage in this game? And if so, by how much?

Chuck-a-luck, which also goes by the name crown and anchor,[1] is a game that is relatively easy to analyze. However, the player's chances of winning are frequently overestimated. The appearance of the six symbols on three dice can lead to the deceptive conclusion that the probability of winning is at least 1/2. And since one can win not only double one's bet, but even three or four times that amount, it would seem that the odds favor the player over the bank.

---

[1]The only difference between the two games is the symbols shown on the dice. In chuck-a-luck, normal dice are used, while in crown and anchor, the four playing-card symbols club, diamond, heart, and spade are used, together with a crown and an anchor. Illustrations and more information on the two games can be found in David Pritchard, *The Family Book of Games*, London 1983, p. 174; Erwin Glonnegger, *Das Spiele-Buch*, Munich 1988, p. 61; John Scarne, *Complete Guide to Gambling*, New York 1974, pp. 505–507.

Score	Dice Combinations	Number	Probability
4	6-6-6	1	1/216
3	6-6-a, 6-a-6, a-6-6, with $a = 1, 2, 3, 4, 5$	15	15/216
2	6-a-b, a-6-b, a-b-6, with $a, b = 1, 2, 3, 4, 5$	75	75/216
0	a-b-c with $a, b, c = 1, 2, 3, 4, 5$	125	125/216
Total		216	1

Table 10.1. Probabilities of winning in chuck-a-luck (betting on 6).

In calculating the chances of winning, it is crucial to consider not simply the chances of winning versus losing, but the amount that is won as well. Therefore, one cannot simply compute the chances of winning and losing at chuck-a-luck. The mathematical model that we use must be extended to allow us to make calculations of the *value* of a win in addition to the probability of merely achieving such a win.

We begin by calculating the probabilities for the possible results of the game. There are 216 possible combinations of dice to check. The results are displayed in Table 10.1.

We now know the probabilities of attaining the different levels of winning. But how do we use this information to calculate our winning outlook? That is, we seek a measure for how our bankroll would fare over a large number of games. To put it concretely, we seek a relationship that in a long series of games would tell us how much we would expect to win on average for a given amount wagered. We can calculate the average amount won in a series of games if we know the relative frequencies of the various levels of winning: each amount that can be won is multiplied by its relative frequency, and then these products are added together. The sum is the average amount won.

As the number of rounds played increases, the law of large numbers kicks in. This means that the relative frequencies of the individual scores approach (with increasingly high probability) the associated probabilities. Consequently, the average amount won approaches a number that can be calculated by multiplying all the scores by their probabilities and summing the products.

In this way, one discovers that the long-term prospects for winning at chuck-a-luck are

$$\frac{1}{216} \times 4 + \frac{15}{216} \times 3 + \frac{75}{216} \times 2 + \frac{125}{216} \times 0 = \frac{199}{216} = 0.9213.$$

Thus one expects to win about 8% less than the amount bet. In other words, in the long run, one expects a net loss of about 8% of the total amount wagered.

In games of chance we are dealing with a number whose value is determined in a random experiment. Similar situations occur frequently:

- the number of squares in the game "Mensch ärgere dich nicht" that one is allowed to move. It is the direct result of a dice roll.

- the number of throws of the dice that a backgammon player requires to roll out his counters in the endgame, the so-called running game.

- the damages that an insurance company must settle in a year.

- the number of citizens in a randomly chosen collection of one thousand individuals who answer a polltaker's question a certain way.

- the number of radioactive decay processes observed during an experiment.

- the amount won in a particular win class in the lottery.

A numerical value that is determined randomly is called a *random variable*. In particular, a random variable consists of a random experment and data on how the results of the experiment determine the values of the random variable. Thus in the context of a random experiment, a random variable is associated with an order relation that assigns a number to each result of the experiment. Of course, completely different random variables can be defined for a random experiment.

Although a random variable is described by its relation to a random experiment, it is often of little interest to know concretely what this relation looks like. What is of greater interest is the probabilities with which the individual values are obtained. For a win in chuck-a-luck, we have already determined the probabilities: the probability of the win value 4 is $1/216$, for the value 3 it is $15/216$, for the value 2 it is $75/216$, and finally, for the value 0, the probability is $125/216$. If the random variable is denoted by X, then one can write the *probability distribution* compactly in the following form (see Note 1 at the end of the chapter):

$$P(X = 0) = \frac{125}{216},$$
$$P(X = 2) = \frac{75}{216},$$
$$P(X = 3) = \frac{15}{216},$$
$$P(X = 4) = \frac{1}{216}.$$

For chuck-a-luck, we calculated the odds directly from the various scores and their associated probabilities. In particular, each amount that could be won was multiplied by its probability, and then the products were summed. The formula makes sense because of the law of large numbers, since in long series of games, the average amount won approaches the number that was calculated arbitrarily closely (with ever increasing probability).

Needless to say, the underlying principle is not restricted to the game of chuck-a-luck. Thus for a random variable X that can assume only values from the finite set $x_1, x_2, \ldots, x_n$, we define the *expectation*, abbreviated $E(X)$ and read "E of X," by the formula (see Note 2 at the end of the chapter)

$$E(P) = P(X = x_1) \times x_1 + P(X = x_2) \times x_2 + \cdots + P(X = x_n) \times x_n.$$

We have already determined that the expectation in chuck-a-luck is $199/216$. For the number thrown on a die, we have the expectation

$$\frac{1}{6} \times 1 + \frac{1}{6} \times 2 + \frac{1}{6} \times 3 + \frac{1}{6} \times 4 + \frac{1}{6} \times 5 + \frac{1}{6} \times 6 = \frac{21}{6} = 3.5.$$

Since all six outcomes are equiprobable, this expectation is equal to the average of the six numbers on the die, which are the six possible outcomes. For the sum on two dice, one obtains the expectation

$$\frac{1}{36} \times 2 + \frac{2}{36} \times 3 + \frac{4}{36} \times 4 + \cdots + \frac{3}{36} \times 10 + \frac{2}{36} \times 11 + \frac{1}{36} \times 12 = 7.$$

There is a particular category of expectations that we should emphasize: if a random variable assumes only the values 0 and 1, then the expectation is equal to the probability that the random variable assumes the value 1. Thus expectation can be viewed as a generalization of the notion of probability.

Intuitively, it is clear that the expectation characterizes the relative weights of the values of random variable, and it does so with a single number. The value of the expectation is influenced by all the values that the random variable can assume, where values with higher probability should have a stronger influence than those whose probability is small. Just as for probabilities, there is also a law of large numbers for random variables: if the experiment underlying a random variable is repeated in a series of trials, and if the results of the individual experiments are independent of one another, then the average value of the random variable approaches the expectation arbitrarily closely, ignoring exceptional values whose probability of occurring becomes arbitrarily small. Thus in the analysis of a game, the concept of the expected winnings, or simply expectation, assumes central importance:

- the odds of a game are fair if the amount wagered is equal to the expectation.

- if a game admits a choice of strategies, then the player should behave so as to maximize the expected winnings. In the long run, such a strategy leads to maximal success.

In order to calculate expectations and to interpret the quantitative results, there is a rather broad palette of techniques and general laws that we should introduce, at least superficially. We are interested primarily in how to compute expectations. As an example, let us take the expectations that arise from two successive games of chuck-a-luck:

- $X_1, X_2, \ldots, X_6$ denote the amounts won if in the first round, a simple bet of one, two, $\ldots$, respectively, six is placed.

- $Y_1, Y_2, \ldots, Y_6$ denote the amount won if in the second round a simple bet of one, two, $\ldots$, respectively, six is placed.

All 12 random variables possess equal probability distributions; their expectations are equal to $199/216$. What is important is that the relations between the random variables are quite different: thus it is impossible that X_1 and X_6 simultaneously hold the maximal value of 4, since the required dice throws 1-1-1 and 6-6-6 are mutually exclusive. On the other hand, it can easily happen that one wins maximally in each of the two trials. This means, for example, that X_1 and Y_6 can simultaneously have the value 4. The events in which these two random variables take on particular values are always independent of one another; one speaks of *independent random variables*.

We can now calculate with random variables that relate to the same random experiment. With respect to the random experiment of two chuck-a-luck throws, expressions like

$$2X_6, \quad X_6 - 1, \quad X_1 + X_6, \quad X_6 + Y_6, \quad X_6 Y_6,$$

are not only mathematically reasonable,[2] but they also have practical application. Thus

- $2X_6$ is the amount won by betting twice a unit amount on 6 for the first round.

- $X_6 - 1$ is the possibly negative net profit when the amount bet is subtracted from the amount won (on a simple bet on the 6 in the first round).

---

[2]Mathematically, these expressions represent the addition, multiplication, etc., of maps that share a common domain of definition.

- $X_1 + X_6$ is the amount won by betting on both 1 and 6 in the first round.

- $X_6 + Y_6$ is the amount won if 6 is bet on in both rounds.

- $X_6 Y_6$ is the amount won if 6 is bet on in the first round, and the winnings from that round are bet on 6 in the second round.

What are the expectations for these five random variables? One can always calculate the probability distributions combinatorially, but that is much too complicated. A better approach is to relate the expectations to the underlying random variables. And indeed, that is possible. We have

$$\mathrm{E}(2X_6) = 2\mathrm{E}(X_6) = 1.843,$$
$$\mathrm{E}(X_6 - 1) = \mathrm{E}(X_6) - 1 = -0.079,$$
$$\mathrm{E}(X_1 + X_6) = \mathrm{E}(X_1) + \mathrm{E}(X_6) = 1.843,$$
$$\mathrm{E}(X_6 + Y_6) = \mathrm{E}(X_6) + \mathrm{E}(Y_6) = 1.843,$$
$$\mathrm{E}(X_6 Y_6) = \mathrm{E}(X_6) \times \mathrm{E}(Y_6) = 0.849.$$

Most of the properties that we have used are not particularly surprising, for it was to be expected that doubling the wager doubles the amount won, and so on. Of particular interest is the last equality, which is an application of the multiplication law for independent random variables. In the case that we are considering, the result is plausible: on the first throw, we expect on average to win $\mathrm{E}(X_6)$. Then in the second round, we bet that amount, and expect to win $\mathrm{E}(Y_6)$ times the amount wagered. Since the two events are independent, we expect on average for the two rounds to win $\mathrm{E}(X_6) \times \mathrm{E}(Y_6)$.

To summarize, for random variables X and Y and constants a, b, the following equalities hold:

$$\mathrm{E}(aX + b) = a\mathrm{E}(X) + b,$$
$$\mathrm{E}(X + Y) = \mathrm{E}(X) + \mathrm{E}(Y).$$

If the random variables X and Y are independent, then the *multiplication law*

$$\mathrm{E}(XY) = \mathrm{E}(X) \times \mathrm{E}(Y)$$

holds as well.

The random variables $2X_6$, $X_1 + X_6$, and $X_6 + Y_6$ all have the same expectation, but not the same probability distribution. In terms of the game itself, this difference manifests itself in that the three different betting strategies represent different levels of risk tolerance (see also Table 10.2):

- with $2X_6$ one is taking a great risk on a single number with a doubled bet. With probability $125/216 = 0.569$ one will win nothing, though one has the possibility of a return of eight times the amount wagered.

- with $X_1 + X_6$ one is playing cautiously, since bets are placed on two different numbers. Only for 64 of the 216 possible outcomes, that is, with probability 0.296, is nothing won. However, the most that one can win is five times the unit wager, which occurs when on the first throw either two 1s and a 6 or a single 1 and two 6s appear.

- the random variable $X_6 + Y_6$ represents a strategy whose risk tolerance is somewhere between those of the other two strategies. The probability of winning nothing is 0.335, and one can win up to eight times the amount wagered, though this is likely to happen with considerably less frequency than with the first strategy.

Qualitatively, these three random variables are distinguished by their degree of scattering, or *deviation*, that is, the degree and and probability of the deviation of the values of the random variable from the expectation. Mathematically, the deviation of the random variable X is described by the transformed random variable $|X - E(X)|$. It contains precisely the information as to what absolute differences from the expected value are possible and with what probabilities they occur. For example, if X is the result of the throw of a die, then $|X - E(X)| = |X - 3.5|$ is a random variable that takes on each of the values $1/2, 3/2, 5/2$ with probability $1/3$. A possible measure of deviation is represented by the average deviation, which in our example is equal to $3/2$ and in general is equal to the expectation $E(|X - E(X)|)$. The fact that one generally measures the deviation of a random variable with the *standard deviation*

$$\sigma_X = \sqrt{E\left((X - E(X))^2\right)}$$

is due exclusively to the fact that absolute values are inconvenient to deal with mathematically. The expression under the square root sign is called the *variance* and denoted by $\mathrm{Var}(X)$. An explicit formula for the variance—and thus for the standard deviation—can be easily derived for random variables X that take on only the finitely many values $x_1, x_2, \ldots, x_n$. If $m = E(X)$ is the expectation, then the variance is equal to

$$\mathrm{Var}(X) = P(X = x_1)(x_1 - m)^2 + P(X = x_2)(x_2 - m)^2 + \cdots$$
$$+ P(X = x_n)(x_n - m)^2.$$

The odds of winning as described for the three different random variables differ considerably in their deviations from their common expectation, as shown in Table 10.2.

t	$P(2X_6 = t)$	$P(X_1 + X_6 = t)$	$P(X_6 + Y_6 = t)$
0	0.57870	0.29630	0.33490
1			
2		0.44444	0.40188
3		0.11111	0.08038
4	0.34722	0.12037	0.12592
5		0.02778	0.04823
6	0.06944		0.00804
7			0.00064
8	0.00463		0.00002
$E(X)$	1.842593	1.842593	1.842593
$Var(X)$	4.956704	2.003001	2.478352
σ_X	2.226366	1.415274	1.574278

Table 10.2. Different ways of placing two bets in chuck-a-luck.

Here are some important rules for the variance and standard deviation: if X is a random variable, and a, b constants with $a \geq 0$, then

$$\sigma_{aX+b} = a \times \sigma_X.$$

For independent random variables X and Y, one has the relation[3]

$$\sigma_{X+Y} = \sqrt{\sigma_X^2 + \sigma_Y^2}.$$

The standard deviations for the random variables $2X_6$ and $X_6 + Y_6$ could also have been calculated from the standard deviation of X_6.

In order not to lose our overview of these matters in the thicket of the many formulas that have been unavoidable in this latter part of the chapter, let us sum things up: when values are determined in a random experiment, they can be described mathematically by random variables. In particular, a random variable can be used to describe the winnings in

---

[3]This equation is based on the multiplication law for random variables. The proof is a good exercise in calculating with random variables: we first observe that the variance formula can be easily reformulated, yielding the variance of a random variable X in the form

$$Var(X) = E\left((X - E(X))^2\right) = E\left(X^2 - 2E(X)X + E(X)^2\right)$$
$$= E\left(X^2\right) - 2E(X)^2 + E(X)^2 = E\left(X^2\right) - E(X)^2.$$

Then for independent random variables X and Y, we obtain

$$Var(X + Y) = E\left((X + Y)^2\right) - E(X + Y)^2 = E\left(X^2 + 2XY + Y^2\right) - (E(X) + E(Y))^2$$
$$= Var(X) + Var(Y) + 2E(XY) - 2E(X) \times E(Y) = Var(X) + Var(Y),$$

from which follows the corresponding formula for the standard deviation.

a game of chance. Since random variables are difficult to calculate with all their variety, their fundamental properties are described with the aid of two characteristic values:

- the expectation is a sort of average, whose value appears in practice when an experiment that determines the random variable is repeated in a large number of independent trials. Then according to the law of large numbers, the average value of the random variable measured approaches, eventually, and with increasing probability, the value of the expectation. In particular, a game of chance offers *fair* odds if the amount that one expects to win on average is equal to the amount wagered, that is, if one expects to "break even" in the long run.

- the standard deviation is a measure of how frequently and with how great a deviation the values of a random variable deviate from its expected value.

Frequently, it happens that only these two values, the expectation and standard deviation, are known about a random variable. Such can be the case, for example, when the random variable is derived from other random variables by arithmetic operations and the expectation and standard deviation can be determined directly from those of the original random variables. We shall see later that in such situations, a knowledge of these two values suffices to make adequate statements about the random variable itself.

Chapter Notes

1. Since a random variable X associates a number with every result, mathematically speaking, we are dealing with a mapping from the set of results into the real numbers. For the probability $P(X = 0)$, the expression $X = 0$ is the abbreviation for the event $\{\, w \mid X(w) = 0 \,\}$, which contains those 125 of the 216 dice outcomes in which the player wins nothing.

   From a purely mathematical point of view, there are almost no restrictions on how a random variable can associate numbers with the outcomes. It must only be ensured that every set that is a preimage of the form $\{\, w \mid X(w) < t \,\}$ corresponds to a probability. This is always achievable with finite sets of outcomes.

2. If a random variable can assume infinitely many values, then the expectation becomes an infinite series, or in the case of continuous values, an integral. For example, the number V of rolls of a die that one needs before

a 6 comes up has expectation

$$E(V) = P(V = 1) + 2 \times P(V = 2)$$
$$+ 3 \times P(V = 3) + 4 \times P(V = 4) + \cdots$$
$$= 1 \times \frac{1}{6} + 2 \times \frac{5}{36} + 3 \times \frac{25}{216} + 4 \times \frac{125}{1296} + \cdots.$$

The sum of this series is 6. That is, the number of rolls that one expects to take on average until a 6 is rolled is six, a quite plausible value.

The following game, known as the Petersburg paradox, does not have a finite expectation: a coin is tossed until "tails" appears. If this occurs on the first toss, then the player wins the value 1. If tails first occurs on the second toss, then the value is 2. For the third toss, it is 4, then 8 for the fourth toss, and so on. The expectation is therefore

$$1 \times \frac{1}{2} + 2 \times \frac{1}{4} + 4 \times \frac{1}{8} + 8 \times \frac{1}{16} + \cdots.$$

But the series does not converge; the expectation is infinitely large. A practical interpretation of this state of affairs is there is no amount large enough to be a reasonable wager in this game.

11

Which Die Is Best?

Two players, Jack and Jill, are playing a game of dice in which each player rolls a die, and whoever rolls the highest number wins. There are three dice available, all of them different from the standard model in that the first die has faces with the numbers 5,6,7,8,10,18; the second has the numbers 2,3,4,15,16,17; and the third has 1,6,11,12,13,14. The players each choose one of the three dice, taking turns choosing first. It is Jack's turn to choose first. Which die should he take?

The three dice represent three random variables whose values can be compared. In the case of everyday numbers, among any three of them there is always one that is not exceeded by any of the others. Does that hold for random variables as well? Which of the three dice produces the "largest" random variable?

Let us start out by comparing the first two dice with each other. If we look at all 36 equiprobable combinations, as listed in Table 11.1, we see that in 21 of the 36 cases, the value of the first die is greater than that of the second. Thus the probability of winning is 21/36 for the player who chooses the first of these two dice in playing against the second.

Table 11.2 shows that the third die is even worse than the second. In competition with the third die, the second wins with probability 21/36.

That would appear to settle the matter. The first die is better than the second, and the second is better than the third. Jack should choose the first die. But what if Jill then chooses the third die? Against all expectation, the first die is not at an advantage. In fact, it loses with probability 21/36, as can be seen in Table 11.3.

		Die I					
		5	7	8	9	10	18
Die II	2	I	I	I	I	I	I
	3	I	I	I	I	I	I
	4	I	I	I	I	I	I
	15	II	II	II	II	II	I
	16	II	II	II	II	II	I
	17	II	II	II	II	II	I

Table 11.1. Die I versus Die II: which one shows the larger number?

		Die II					
		2	3	4	15	16	17
Die III	1	II	II	II	II	II	II
	6	III	III	III	II	II	II
	11	III	III	III	II	II	II
	12	III	III	III	II	II	II
	13	III	III	III	II	II	II
	14	III	III	III	II	II	II

Table 11.2. Die II versus Die III: which one shows the larger number?

		Die III					
		1	6	11	12	13	14
Die I	5	I	III	III	III	III	III
	7	I	I	III	III	III	III
	8	I	I	III	III	III	III
	9	I	I	III	III	III	III
	10	I	I	III	III	III	III
	18	I	I	I	I	I	I

Table 11.3. Die III versus Die I: which one shows the larger number?

It turns out that Jack is at a great disadvantage in having the dubious privilege of choosing the first die. No one die is the best, since each of them has a die that is better: the first die is better than the second; the second is better than the third; and the third is better than the first. What at first glance seems to defy the laws of reason can be expressed quite clearly in the language of mathematics: the relation "better than" between two random variables is not transitive. That is, the familiar transitive property of numbers, that from $a > b$ and $b > c$ it always follows that $a > c$, is not satisfied for random variables. We note that the relation "better than" is indeed suitably defined for the game under consideration: a random variable X is considered to be better than a random variable Y if $P(X > Y) > 1/2$.

One might now ask whether the advantage to the second player increases if dice with different numbers are used. Let us formulate the question in greater generality. We seek independent random variables $X_1, X_2, \ldots, X_n$ for which the minimum of the probabilities $P(X_1 > X_2)$, $P(X_2 > X_2)$, $\ldots$, $pr(X_n > X_1)$ is as large as possible. For the case $n = 3$ random variables the maximum turns out to be 0.618, where the random variables satisfying that condition cannot be achieved with suitably numbered dice. Moreover, it is clear that the value $21/36 = 0.583$ determined for the game that we have been considering cannot be much improved. In contrast, for the case $n = 4$ independent random variables, the theoretical maximum $2/3$ can actually be achieved with real dice:[1]

Die I:	3	4	5	20	21	22;
Die II:	1	2	16	17	18	19;
Die III:	10	11	12	13	14	15;
Die IV:	6	7	8	9	23	24.

---

[1] The set of four dice is taken from Martin Gardner's *Wheels, Life, and Other Mathematical Amusements*, New York, 1983, Chapter 5. The labeling of the three sets is from G. J. Székely, *Paradoxa*, Frankfurt 1990, pp. 65 f.

12

A Die Is Tested

A die whose symmetry is to be tested is rolled ten thousand times. The sum of the numbers thrown in 37 241, which corresponds to an average of 3.7241. Is such a deviation from the ideal value of 3.5 possible if the die is fair? Or can the result be explained only by the assertion that the die is asymmetric?

Such questions are typical in the practice of applied statistics. Does the result of a series of trials lie within the range of what can be expected from random variations? Or can the assumptions held at the outset no longer be maintained? Must we conclude that the die thought to be symmetric is in fact asymmetric, the medicine thought to be worthless in fact effective, the politician whose popularity has been at a steady level now no longer so popular. Therefore, we might rephrase our original question more precisely thus: how likely is such a deviation from the ideal value of 3.5 if the die is fair? Can the result be much better explained by the assertion that the die is asymmetric?

Such problems are frequently investigated by starting with an assumption, generally called a *hypothesis*, and then deriving assertions about the possible behavior of the results of a series of trials to be made. Generally, such statements deal with a range within which a *test value*, also called a *statistic*, must lie with a certain high degree of probability. Thus in the series of dice throws, one determines limits within which the sum of the numbers obtained must almost certainly lie, say, with a probability of 0.99. If then in the trial series the sum falls below or exceeds this value by a large amount, the die is declared asymmetric. This decision is justified in large measure because it is entirely plausible that an otherwise very improbable deviation *can indeed* be caused by an asymmetry. On the other hand, with a less sensible plan of experiment, the occurrence of an a priori improbable result could not necessarily be taken as an indication that the hypothesis should be rejected. For example, in 10 000 throws of a die, it is highly improbable to obtain *exactly* the sum 35 000. But since there is no asymmetry that could plausibly cause such a result, it would not make much sense to conclude from this that an asymmetry existed.

But one can arrive at a false conclusion even with the "outrider" criterion:

- even if a die is absolutely symmetric, there is still a finite probability, 0.01 in our example, that it will be declared asymmetric.

- conversely, there may be no evidence adduced against the die. That is, an asymmetric die does not necessarily lead to conspicuous deviant behavior. In particular, if the die is only slightly asymmetric, one cannot realistically expect anything much different from normal behavior.[1] Furthermore, a test that considers only the average value of a die is not able to detect every sort of asymmetry.

Although mathematical statistics employing probability-theoretic test methods has existed only since about 1890, isolated examples of *hypothesis testing* can be found much earlier. Thus, for example, in 1710, the English mathematician John Arbuthnot (1667–1735) disproved the hypothesis that male and female human births are equiprobable. In the birth statistics available to him, in each of the 82 years, the number of males was greater than that of females. If the events were in principle equiprobable, the probability of such a result occurring by chance would be $(1/2)^{82}$, an unusually convincing refutation of the hypothesis.[2]

With our die, the situation is not so simple. Nevertheless, it is plausible in considering how to evaluate the results of our experiment to ask whether they violate the law of large numbers. Does the average value in 10 000 throws of a fair die have to lie closer to the expected value 3.5 than the

---

[1] In other cases, one takes the fact that the results of the experiment do not contradict the hypothesis as sufficient cause to declare the hypothesis satisfied. If the hypothesis is actually false, then one speaks of having made a type 2 error, in contrast to a type 1 error, in which a valid hypothesis is in fact rejected. A significant portion of the subject of mathematical statistics is devoted to questions such as how likely type 1 and type 2 errors are to occur and how tests can be designed to avoid them to the greatest possible extent.

[2] Arbuthnot's investigations are discussed at greater length, in addition to early methods in statistics by other scholars, in Robert Ineichen, Aus der Vorgeschichte der Mathematischen Statistik, *Elemente der Mathematik* **47**, 1992, pp. 93–107.

observed value of 3.7241? Since we have thus far discussed the law of
large numbers only qualitatively, we need to come up with a quantitative
formulation. In this, the rules for calculating expectation and standard
deviation will be most useful.

For the sake of generality, let us not limit ourselves to the special case
of dice throws, but instead, investigate an arbitrary sequence of identically
distributed and independent random variables $Y = Y_1, Y_2, \ldots, Y_n$. For the
total of the numbers from n throws of a die, which is the random variable
$Y_1 + Y_2 + \cdots + Y_n$, we first obtain

$$E(Y_1 + Y_2 + \cdots + Y_n) = nE(Y),$$
$$\sigma_{Y_1+Y_2+\cdots+Y_n} = \sqrt{N}\sigma_Y.$$

The average throw, which is given by the random variable

$$X = \frac{Y_1 + Y_2 + \cdots + Y_n}{n},$$

is thus characterized by the values

$$E(X) = E(Y),$$
$$\sigma_X = \frac{\sigma_Y}{\sqrt{n}}.$$

That is, the random variable X, which represents the average value of the
n random variables, possesses the same expectation as the original random
variable Y. On the other hand, the standard deviation is less than that
of the original value by a factor of $\sqrt{n}$. Therefore, with an unchanged
expectation, the deviation becomes smaller in the course of the series of
throws. But that is precisely the statement of the law of large numbers!
Note that the deviation from the average value of the random variable
becomes less, but not the deviation from the sum, whose standard deviation
actually becomes greater, namely, by a factor of $\sqrt{n}$ for n throws. There is
no law of equalization that levels out the results in the sum $Y_1+Y_2+\cdots+Y_n$,
but only a law of large numbers that pushes average values in the direction
of the expectation.

We know, then, that with sufficiently many trials, the measured average
value exhibits a reduced standard deviation. But what can we say about
the probabilities of the observed average values; that is, what numbers,
and with what probabilities, can be attained by the experimentally determined average value? A very coarse bound on how the random variable
X can be distributed is given by *Chebyshev's inequality* (Pafnuty L'vovich
Chebyshev, 1821–1894): the probability that an arbitrary random variable

deviates from its expectation by at least t times its standard deviation is at most $1/t^2$. Thus large deviations are relatively improbable, where statements that can be made are of interest only when the value of t is greater than 1. For example, for $t = 1.5$, Chebyshev's inequality tells us that the probability of a deviation of at least one and one-half times the standard deviation from the expectation is at most $1/1.5^2 = 0.444$. For $t = 10$, a deviation of more than ten times the standard deviation occurs with probability at most 0.01.

Let us see what that last statement means for our series of dice throws. For one throw, we have an expectation of 3.5 and a standard deviation of 1.708. For 10 000 throws, we obtain for the average value of the random variable X the standard deviation $\sigma_X = 0.01708$. According to Chebyshev's inequality, for 10 000 throws, the average of the numbers thrown should lie outside the interval $3.5 \pm 10\sigma_X$ with probability at most 0.01. And it is precisely this unlikely event that has occurred in our series of trials with the result 3.7241. If we are looking for a fair die, then this one should not be used, since we are forced to reject the hypothesis of symmetry.

Now that we have seen how useful Chebyshev's inequality can be, we would like to delve into its consequences a bit more deeply. In mathematical formulation, for a random variable X, we have

$$P(|X - E(X)| \geq t\sigma_X) \leq \frac{1}{t^2}.$$

If we read this inequality symbol by symbol, we have the following interpretation: the event that the random variable X deviates "significantly" from its expected value $E(X)$, namely, $|X - E(X)| \geq t\sigma_X$, can occur with probability at most $1/t^2$ (see Note 1 at the end of the chapter). That is, "significant" deviations cannot appear too "frequently."

The most important application of Chebyshev's inequality is the law of large numbers. If a random variable X assumes values from a series of independent trials as described above, then Chebyshev's inequality takes the form

$$P\left(|X - E(X)| \geq t\frac{\sigma_Y}{\sqrt{n}}\right) \leq \frac{1}{t^2}.$$

The effect of the law of large numbers is vividly recognizable in this inequality when the parameter t is gradually increased in the course of the series of trials. The tolerance interval and associated probability become smaller simultaneously, as shown, for example, for the dice experiment in Table 12.1.[3]

---

[3] For the examples in the table, the value $t = \sqrt[6]{n}$ was chosen.

Trials	Tolerance Interval		Probability That the Average Roll Is Outside
	From	To	
10	2.7073	4.2927	≤ 0.4642
100	3.1321	3.8679	≤ 0.2154
1000	3.3292	3.6708	≤ 0.1000
10 000	3.4207	3.5793	≤ 0.0464
100 000	3.4632	3.5368	≤ 0.0215
1 000 000	3.4829	3.5171	≤ 0.0100
10 000 000	3.4921	3.5079	≤ 0.0046

Table 12.1. The course of average values of a die in a series of dice throws.

We note finally that the estimates of the probabilities are very generous. That is, the probability is generally much smaller than the upper bound provided by Chebyshev's inequality. A much more precise statement about such probabilities is made by the *central limit theorem*, which will be discussed in the next chapter. However, the formulas of the central limit theorem are significantly more complex, for which reason Chebyshev's inequality retains its significance: that the law of large numbers can be derived in a relatively elementary manner from the axioms of probability theory.

The classical law of large numbers, as we have remarked repeatedly in the first chapters of this book, applies only to probabilities and relative frequencies. Of course, this special case can also be made quantitatively more precise with Chebyshev's inequality: from a random experiment and event A with probability $p = P(A)$, a random variable is constructed that has the value 1 when the event occurs, and is otherwise equal to 0. If the experiment is now repeated n times in a series of independent trials, one obtains uniformly distributed independent random variables $Y_1, Y_2, \ldots, Y_n$. They satisfy the following equalities:

$$P(Y_i = 1) = p,$$
$$P(Y_i = 0) = 1 - p,$$

and therefore,

$$E(Y_i) = p$$

and

$$\text{Var}(Y_i) = (1 - p)p^2 + p(1 - p)^2 = p(1 - p) \leq \frac{1}{4}.$$

Trials	Tolerance Interval	Probability That the Relative Frequency Is Outside
10	$p \pm 0.2321$	≤ 0.4642
100	$p \pm 0.1077$	≤ 0.2154
1000	$p \pm 0.0500$	≤ 0.1000
10 000	$p \pm 0.0232$	≤ 0.0464
100 000	$p \pm 0.0108$	≤ 0.0215
1 000 000	$p \pm 0.0050$	≤ 0.0100
10 000 000	$p \pm 0.0023$	≤ 0.0046

Table 12.2. The empirical measurement of an unknown probability p.

The average X of these random variables $Y_1, Y_2, \ldots, Y_n$ is now nothing but the relative frequency of the event A in these n trials. Chebyshev's inequality now has something to say about the probability of deviations of this relative frequency X from the probability p, namely,

$$P\left(|X - p| \geq \frac{t}{2\sqrt{n}}\right) \leq \frac{1}{t^2}.$$

If we again let the deviation determined by the parameter t gradually increase along with the number of trials, the consequences of the inequality can be seen in the results displayed in Table 12.2.[4]

In comparison with the version of the law of large numbers formulated in Chapter 5, the version formulated here is significantly weaker. It is therefore known as the *weak law of large numbers*. The difference is that here, the probability always refers only to the deviation in a particular number of trials. In contrast, the strong law of large numbers, as described in Chapter 5, also makes assertions about the deviations in the further course of the series of experiments.

The formal derivation of the law of large numbers gives probability theory an important confirmation. The empirical discovery that relative frequencies in series of trials approach a limit over time was the original motivation for introducing an abstract number: the probability of an event. At the same time, one obtains a method of measuring this number and information about how precise this method is. Thus the transition from uncertainty to near certainty in long series of trials has become quantifiable.

---

[4]We have again used $t = \sqrt[6]{n}$.

Chapter Notes

1. From a mathematical point of view, Chebyshev's inequality is relatively elementary. For example, if a random variable X can assume only the finite values $x_1, x_2, \ldots, x_n$, then every value x_i that deviates by at least u from the expectation, that is, that satisfies the inequality $|x_i - \mathrm{E}(X)| \geq u$, results in a summand

$$\mathrm{P}(X = x_i)(x_i - \mathrm{E}(X))^2 \geq \mathrm{P}(X = x_i)u^2$$

in the variance formula, so that the variance attains at least the value

$$\mathrm{P}(|X - \mathrm{E}(X)| \geq u)u^2$$

from these "outliers" alone. By algebraic manipulation, the inequality becomes

$$\mathrm{P}(|X - \mathrm{E}(X)| \geq u) \leq \frac{\mathrm{Var}(X)}{u^2},$$

which for $u = t\sigma_Y$ yields the desired version of Chebyshev's inequality.

Further Literature on Statistics

[1] Karl Bosch, *Elementare Einführung in die Statistik*, Braunschweig 1976.

[2] Ulrich Krengel, *Einführung in die Wahrscheinlichkeitsrechnung und Statistik*, Braunschweig 1988.

[3] Marek Fisz, *Wahrscheinlichkeitsrechnung und mathematische Statistik*, Berlin (East) 1970 (Polish original, 1967).

[4] Helmut Swoboda, *Knaurs Buch der modernen Statistik*, Munich 1971.

[5] Hermann Witting, Mathematische Statistik, in: *Ein Jahrhundert Mathematik, 1890–1990*, Braunschweig 1990, pp. 781–815.

13

The Normal Distribution: A Race to the Finish!

In a racing game whose goal is to be first to move a counter across the goal line, player Jill has 76 squares to traverse. On each move, she is permitted to move her counter a number of squares equal to the result of throwing a pair of dice. What is the probability of Jill's reaching the goal in fewer than ten moves?

Such racing games with dice have a long tradition, and they exist in great variety. Among the games in which the players do not try to block one another (in contrast to backgammon and pachisi) can be found in such classics as the game of goose and that of snakes and ladders. A modern variant is the very successful game Railway Rivals, invented in 1970 by the Englishman Dave Watts.[1]

It is frequently of great importance in this game to determine probabilities for particular routes, namely, when one must decide whether one

---

[1] In this game, played on a honeycomb-patterned board showing a simplified map with bodies of water, mountains, and cities, stretches of railway track are laid and then traveled. In each round, two towns are selected with a dice roll as starting and ending locations. Each player who wants to travel must decide on a route where only one's own track can be traversed free of charge; one is charged a tariff to ride on an opponent's track. After a player has decided on a route, the actual race begins, where moves are governed by the roll of the dice. The first two players to reach the goal get points. More information on the game can be found in Erwin Glonnegger, *Das Spiele-Buch*, Munich 1988, p. 75; Jury "Spiel des Jahres," *Spiel des Jahres*, Munich 1988, pp. 62 f.

should pay a tariff to be able to take a shorter route. For more on this problem, see "Assymetric Roulette."

In principle, the problem stated at the beginning of the chapter can be solved along the same lines as the division problem introduced in Chapter 4: for longer and longer routes, the probabilities are determined for traversing them via various rolls of the dice. Here we shall fall back on the results that we obtained earlier. Let $Y_1, Y_2, \ldots$ be the numbers rolled with the dice. Then the desired probability of obtaining the sum 76 is given by $P(Y_1 + \cdots + Y_9 \geq 76)$, which can be calculated with the formula

$$
\begin{aligned}
P(Y_1 + \cdots + Y_9 \geq 76) = {} & \frac{1}{12}P(Y_1 + \cdots + Y_8 \geq 74) \\
& + \frac{2}{12}P(Y_1 + \cdots + Y_8 \geq 73) \\
& + \frac{3}{12}P(Y_1 + \cdots + Y_8 \geq 72) \\
& + \cdots \\
& + \frac{1}{12}P(Y_1 + \cdots + Y_8 \geq 64),
\end{aligned}
$$

on the assumption that we know the probabilities on the right-hand side of the equation.[2] Every situation is thus conditioned on the possible results of the dice roll just analyzed. Step by step, one thus obtains the desired probability 0.042138. One requires either a computer[3] or sufficient patience, since several hundred intermediate values have to be calculated. It would therefore be desirable to have a simpler way of computing an approximate result. Indeed, that is possible, and it is done with the help of the central limit theorem. This theorem makes assertions regarding independent and identically distributed random variables $Y = Y_1, Y_2, \ldots, Y_n$ as they appear in series of trials: assertions about the sum $Y_1 + Y_2 + \cdots + Y_n$ for

---

[2] This formula is an application of the formula for total probability (see Chapter 9). The event $Y_1 + \cdots + Y_9 \geq 76$ is investigated conditioned on the possible results Y_9 of the ninth throw of the dice. Here we have

$$P(Y_1 + \cdots + Y_9 \geq 76 \mid Y_9 = k) = P(Y_1 + \cdots + Y_8 \geq 76 - k).$$

[3] Using a spreadsheet calculation is simpler than the obvious method of writing a program in a programming language such as Pascal, C, Basic, or Fortran. A table is made of the probabilities of achieving the goal for the squares yet to be traversed, from 76 down to -10 (allowing for the possibility that the goal is exceeded by 10 squares) and the remaining number of rolls from 0 to 9. Aside from the initial values 0 and 1, which correspond to the probabilities at the end of the game, only a single formula needs to be entered. The balance can be taken care of with table calculations such as "fill down" and "fill right."

n sufficiently large. It is known that the sum has expectation $n\mathrm{E}(Y)$ and standard deviation $\sqrt{n}\sigma_Y$. What the random variable looks like exactly is seen from the probabilities

$$\mathrm{P}(Y_1 + Y_2 + \cdots + Y_n \leq u).$$

But how do these probabilities behave as the parameter u is varied? The central limit theorem asserts that for a sufficient number of trials, a good approximation can be calculated from the quotient

$$t = \frac{u - n\mathrm{E}(Y)}{\sqrt{n}\sigma_Y}.$$

Of importance is how far the parameter u is from the expectation $n\mathrm{E}(Y)$, where this distance is measured relative to the standard deviation $\sqrt{n}\sigma_Y$. The actual statement about the limiting value can best be understood by choosing a fixed value of t: if one fits, for fixed t and increasing number of trials, the value of the parameter u continuously to the number of trials n according to the formula

$$u = u_n(t) = n\mathrm{E}(Y) + t\sqrt{n}\sigma_Y,$$

then the probability $\mathrm{P}(Y_1 + Y_2 + \cdots + Y_n \leq u)$ approaches a number that is independent of the original random variable Y. That is, the limiting value of the named probabilities is always the same, regardless of which random variable you start with. There are differences only for various values of the parameter t, so that the limiting values, which are generally denoted by $\phi(t)$, can be tabulated. To provide an overview, let us content ourselves here with a small selection of values of t (see Table 13.1). More complete data can be found in any book of mathematical tables under the heading "normal distribution."

The tabulated numbers $\phi(t)$ describe in their entirety a special random variable, which can take arbitrary values in the set of real numbers. Here $\phi(t)$ is the probability that this random variable is less than or equal to t. The random variable called the *standardized normal distribution* has expectation 0 and standard deviation 1 (see Note 1 at the end of the chapter).

The central limit theorem is applied in practice to approximate probabilities of the form $\mathrm{P}(Y_1 + Y_2 + \cdots + Y_n \leq u_n(t))$ with the associated probabilities $\phi(t)$ of the normal distribution. The error involved becomes smaller for larger values of the number of trials n.

In our example of a series of dice rolls, the approximation already returns good results for small numbers of trials such as $n = 9$: starting with the characteristic data of the random variable Y, namely, $\mathrm{E}(Y) = 7$ and

t	$\phi(-t)$	$\phi(t)$
0.0	0.50000	0.50000
0.2	0.42074	0.57926
0.4	0.34458	0.65542
0.6	0.27425	0.72575
0.8	0.21186	0.78814
1.0	0.15866	0.84134
1.2	0.11507	0.88493
1.4	0.08076	0.91924
1.6	0.05480	0.94520
1.8	0.03593	0.96407
2.0	0.02275	0.97725
2.2	0.01390	0.98610
2.4	0.00820	0.99180
2.6	0.00466	0.99534
2.8	0.00256	0.99744
3.0	0.00135	0.99865

Table 13.1. Values of the normal distribution.

$\sigma_Y = 2.4152$, one has to choose the parameter t such that the equation $u_7(t) = 9 \times 7 + t \times 3 \times 2.4152 = 75.5$ is satisfied, which is the case for $t = 1.7232$. The reason for choosing the number 75.5 instead of 76 is that the sum of the dice can assume only integer values, while the normal distribution extends over the whole real line. The values between 75 and 76 of the normally distributed random variable are divided approximately evenly between the two neighboring dice sums. We thus obtain the result

$$P(Y_1 + Y_2 + \cdots + Y_9 \leq 75) \approx \phi(1.7252) = 0.9578.$$

The desired probability $P(Y_1 + Y_2 + \cdots + Y_9 \geq 76)$ is therefore approximately 0.0422, which is a good approximation to the exact value 0.042138.

    There are many applications of the central limit theorem and the normal distribution. To give an impression of their scope, we look back at some of the topics discussed in previous chapters:

- with the central limit theorem, the law of large numbers can be improved over what is attainable with Chebyshev's inequality. One can describe how in long series of trials the average X formed of the independent and identically distributed events $Y = Y_1, Y_2, \ldots$ is distributed about the expectation $E(Y)$ via the approximation

$$P\left(|X - E(Y)| \geq t\frac{\sigma_Y}{\sqrt{n}}\right) \approx 1 - \phi(t) + \phi(-t) = 2\phi(t).$$

- even for the value $t = t$, about which Chebyshev's inequality has nothing to say, the central limit theorem tells us that the probability is actually significantly smaller, namely, about 0.317. With the value $t = 2$, Chebyshev's inequality provides the generous upper bound 0.25, which the central limit theorem pares down to 0.046.

- with the dice test discussed in the previous chapter, the normal distribution can provide more precise information about the sum of the dice rolls. Based on the assumption that the die is completely symmetric, one assumes an expectation $E(Y) = 3.5$ and standard deviation $\sigma_Y = 1.708$. Then the value $t = 2.58$ corresponding to the rolled average must lie between 3.456 and 3.544 with probability 0.99. Thus asymmetric dice can be more easily identified than was possible with the Chebyshev inequality.

- even the binomial distribution can be approximated with the normal distribution. We again start with a series of trials, in which an experiment is repeated n times, each trial independent of the others, where the relative frequency X of a particular event is measured. If the probability of that event in an individual experiment is p, then the central limit theorem provides the following approximation for long series of trials:

$$P\left(X \le p + t\sqrt{\frac{p(1-p)}{n}}\right) \approx \phi(t).$$

- let us take up an example that we introduced in Chapter 4. We are to determine the probability of obtaining at least 900, and at most 1100, 6s in 6000 rolls of a die. With respect to relative frequencies, this corresponds to the interval from $1/6 - 0.01675$ to $1/6 + 0.01675$. Using the values ± 3.4814 of t, one then obtains the approximation $\phi(3.4814) - \phi(-3.4814) = 0.0005$ for the desired probability. To be sure, the result could have been obtained using the formulas for the binomial distribution, but that would have been computationally impractical.

Asymmetric Roulette

The roulette wheels in gambling casinos are manufactured with great precision, and moreover, they are regularly checked for symmetry. The reason for this is obvious. If some asymmetry in the wheel was to lead to the probability of any particular number exceeding $1/36$, that would open to the public a clear winning strategy. That is, the tiniest asymmetry could lead to great losses for the casino if a player was to discover it, which appears to have happened a number of times in the past. Thus in the 1960s, Dr. Jarecki's streak of wins, due apparently to a defect in the wheel, received considerable press coverage.[4]

If one wishes to establish the existence of an asymmetry based on statistics, then one must first take note of the fact that such an asymmetry will not reveal itself all the time. For example, the ball is tossed into the spinning wheel now from the left, now from the right into the oppositely spinning wheel. If one of the bridges is slightly higher than the others, its effect will depend on the direction of motion. Furthermore, it is technically feasible to rotate the ring of numbers so that they lie over different slots, which would allow possibly favorable slots to be associated with different numbers over time.

In order to refute the hypothesis that a roulette wheel is fair, one should perform a series of experiments under uniform conditions. One must doubt, though, whether one would actually be able to perform such experiments under the required conditions in a real-world casino.

The first person to recognize a roulette permanence as highly improbable was Karl Pearson (1857–1937), one of the founders of mathematical statistics. He wrote about a two-week permanence in Monte Carlo that even if the Monte Carlo casino had been operating since the beginning of geological time, such a permanence would not have been expected if one were dealing with an unbiased wheel. It turned out later, as Edward Thorp related (see Note 2 at the end of the chapter), that the actual runs had apparently been invented by journalists, so that they could avoid the tedious job of writing everything down.

The great significance of the normal distribution in probability theory is not, of course, that one can use it for calculating probabilities of dice

---

[4]See, for example, the *Stuttgarter Zeitung* of 7.7.1973, also reproduced in Max Woitschach, *Logik des Fortschritts*, Stuttgart 1977, p. 75.

rolls. As has already been suggested, the normal distribution is a distribution of probabilities and frequencies that occurs frequently in nature, technology, and economics. It comes into play when a process is determined by a large number of independent random factors. If an overall result is obtained as the sum of independent random variables, each of which influences the result only slightly, then, like the sum of the dice rolls, its distribution is approximately normal. Thus, for example, the heights of adult human beings are approximately normally distributed, and not only the entire population, but subpopulations as well, such as those of men and of women, though of course with different expectations and standard deviations. Quasi-random processes that are subject to the normal distribution can even be found in the divisibility properties of the integers. Thus, for example, in 1940, the mathematicians Paul Erdős (1913–1997) and Marc Kac (1914–1984) proved that the number of primes by which a number is divisible behaves approximately like a normal random variable. Thus it is possible to approximate the frequencies with which such numbers of divisors occur in a range of numbers bounded above by a large integer n using a normal distribution with expectation and standard deviation both equal to $\ln \ln n$.[5]

Railway Rivals: A Race Between Two Players

What are the odds when two players engage in a race in the game Railway Rivals discussed earlier in this chapter? Information is provided by the random variable that corresponds to the difference in progress of the two players. We must compare this difference in the dice sums of the two players with the original lead, which can occur in approximation to the central limit theorem for every fixed number of throws of the dice. However, in a race, the necessary number of turns is not fixed; rather,

---

[5]Let $\nu(m)$ denote the number of primes by which the integer m is divisible. Then for an arbitrary fixed value of t, one has the relation

$$\lim_{n \to \infty} \frac{1}{n} \# \left\{ m \mid 1 \leq m \leq n \text{ with } \nu(m) \geq \ln \ln n + t\sqrt{\ln \ln n} \right\} = \phi(t).$$

The relative frequency of numbers between 1 and n that have more than $\ln \ln n + t\sqrt{\ln \ln n}$ prime divisors thus converges to $\phi(t)$. One can find a proof of a greatly simplified version of this theorem based on Chebyshev's inequality and the prime number theorem in Noga Alon, J. H. Spencer, *The Probabilistic Method*, New York, 1992, Theorem 2.1.

it is itself a random variable. It therefore makes sense to begin by approximating the number of turns by determining its expectation.

Let us investigate a race in which player Jack has w squares to traverse, while his rival Jill has $w + d$ squares to the goal. Since both players throw the dice the same number of times (if both players cross the finish line, then the player who exceeded the goal by the most squares is the winner), it makes no difference whether one or two dice are thrown (except for the frequency of tie games). We therefore assume that only one die is thrown. Since the difference between two rolls of a die has standard deviation 2.415, the standard deviation for $w/3.5$ turns is $1.291\sqrt{w}$, so that the probability of victory for the leading player is approximately

$$\phi\left(0.775\frac{d}{\sqrt{w}}\right)$$

(we count a draw as half a win for each player). With route lengths of 25 squares we obtain usable approximations to the probabilities; thus, for example, for the races 25 : 20, 25 : 30, 65 : 60, and 65 : 70, we obtain the values 0.193 (exact value 0.210), 0.781 (exact value 0.768), 0.308 (exact value 0.313), and 0.685 (exact value 0.681). In particular, the approximation formula allows us to estimate how profitable it is to take routes over an opponent's track. Since the winner of the race gets 20 points, and the runner-up gets 10, for the two players Jack and Jill, the expectation for Jack is

$$10 + 10\mathrm{P}(\text{Jack beats Jill}).$$

If there are three players—Jack, Jill, and Mira—then taking shorter routes is even more advantageous, since the expectation for Jack is now

$$10\mathrm{P}(\text{Jack beats Jill}) + 10\mathrm{P}(\text{Jack beats Mira}).$$

Backgammon

If in the endgame of a backgammon match no more checkers can be hit, then the game takes on the character of a race, called the running game. Such endgame situations are generally evaluated by adding up the remaining fields through which all a player's checkers must pass. If we imagine a hypothetical situation in which each player has a single checker that is the distance of this sum from the goal, then this corresponds to a simple model that can be used to estimate the odds in an actual situation. The power of this model can be improved if one slightly alters the positions of both checkers depending on the details of the original situation as well as the number and distribution of the checkers to compensate for the points lost in the removal of pieces from the board.

In practice, one is often interested in estimating the odds of winning in order to decide whether one should double the stakes. This special problem will be dealt with in detail in Chapter 31.

For our backgammon model we would now like to estimate the odds of winning in the situation in which the leading player, Jill, has w fields to traverse, and her opponent Jack has $w + d$. Except for two details, all is the same as in Railway Rivals:

- if a player reaches the goal, the game ends at once, so that the player whose turn it is has an advantage. Since the players alternate in their turns, the advantage of first move should be calculated at half a turn; that is, the player whose turn it is receives a bonus in the number of fields equal to half the expectation of a single turn.

- one plays with two dice, and if doubles are thrown, then one moves twice the number shown. For the number of fields traversed per turn, this gives an expectation of 8.167 and standard deviation of 4.298.

The game situation of w to $w + d$ fields corresponds to an expected number of turns of $w/8.167$ and standard deviation $2.127\sqrt{w}$. The probability of winning for the leading player is

therefore approximately

$$\phi \left(0.470 \frac{d \pm 4.083}{\sqrt{w}} \right),$$

where the distance between the players is increased or decreased by 4.083 depending on whether or not the leading player goes first. The approximations are relatively exact for numbers of fields that are not too small. Thus for the situations 20 : 25, 25 : 20, 65 : 55, 65 : 65, 65 : 75, and 65 : 85, the approximate probabilities are 0.830 (exact value 0.829), 0.462 (exact value 0.451), 0.354 (exact value 0.358), 0.594 (exact value 0.595), 0.794 (exact value 0.787), and 0.920 (exact value 0.906). If we consider a realistic situation, that is, one with several checkers, then additional imprecision arises in using our model.

Risk

A race of quite another sort takes place in the game of Risk, invented by the Frenchman Albert Lamorisse. It made its first appearance in 1957. The Risk game board shows a stylized map of the world. The fields on the board correspond to imaginary countries, which—depending on the version—are to be either conquered or liberated.[6] In one variant, the winner is the one who frees the entire world by capturing all the enemy pieces, which represent armies.

A turn consists in carrying out one or more attacks, in which up to three armies occupying one country attack those of a neighboring country. The defender can put up as many as two (in the old variant three) armies in defense. The result of an attack is determined by rolling the dice, where one die is rolled for each army involved. The results of the roll are then, individually for attacker and defender, sorted in ascending order,

---

[6]Further description and illustrations can be in Erhard Gorys, *Das grosse Buch der Spiele*, Hanau ca. 1987, pp. 283–286; David Pritchard, *The Family Book of Games*, Brockhampton Press 1983; *Spielbox* **3**, 1983, p. 22; Roberto Convenevole, Francesco Bottone, *La storia di risiko*, Rome 2002.

so that they can be compared, to the extent possible, with one another. Each pair then decides a duel between an attacking army and a defending army, where the attacker wins if his die is higher. For example, a 3 : 2 attack with a roll of 6-4-2 against 4-4 results in both attacker and defender having to remove a single army from the board: 6 wins against 4, while 4 against 4 represents a loss for the attacker.

So much for an overview of the rules. What the odds are in a single battle are shown in the following table, which reflects the combinatorial situation:

Attacker: Defender	Defender's Loss					
	0	1	2	3	Expectation	Standard Deviation
1 : 1	21	15			0.42	0.49
1 : 2	161	55			0.25	0.44
1 : 3	1071	225			0.17	0.38
2 : 1	91	125			0.58	0.49
2 : 2	581	420	295		0.78	0.79
2 : 3	4816	1981	979		0.51	0.71
3 : 1	441	855			0.66	0.47
3 : 2	2275	2611	2890		1.08	0.81
3 : 3	17871	12348	10017	6420	1.11	1.07

In the later stages of a game of Risk, the number of armies on the board generally increases dramatically. The odds for longer duels between two strong armies can be estimated with the aid of the central limit theorem. To simplify matters, let us assume that the entire battle consists of a number of identical attacks in the relation 3 : 2 and 3 : 3. Since after each attack the number of armies on the board is reduced by two or three, we may proceed as follows: if a is the number of attacking armies, and d the number of defenders, then an attack is successful when the total loss by the defender after $a+d/2$, respectively $a+d/3$, attacks is at least d. In that case, the attacker still has at least one army, while the defender theoretically is in the minus column, that is, has actually already lost the duel. Using the numbers $a + d/2$ and $a+d/3$, as well as the tabulated expectations and standard deviations, we finally obtain the following approximations for the attacker's probability of victory: for a $3 : 2$ attack, we have

$$\phi\left(\frac{1.08\frac{a+d}{2} - d}{0.81\sqrt{\frac{a+d}{2}}}\right) = \phi\left(0.94\frac{a - 0.85d}{\sqrt{a+d}}\right),$$

while for $3 : 3$, we obtain

$$\phi\left(\frac{1.11\frac{a+d}{3} - d}{1.07\sqrt{\frac{a+d}{3}}}\right) = \phi\left(0.60\frac{a - 1.71d}{\sqrt{a+d}}\right).$$

Chapter Notes

1. By converting to the form $X' = aX + b$, one can obtain the other normally distributed random variables. An identifying feature of normally distributed random variables is that the sum of independent normally distributed random variables is again normally distributed. As can be seen in Table 13.1, the standardized normal distribution is concentrated in a narrow region around zero. The values of ϕ can be calculated either with the integral

$$\phi(t) = \frac{1}{2} + \frac{1}{\sqrt{2\pi}} \int_0^t e^{-x^2/2}\, dx$$

or with the rapidly converging power series

$$\phi(t) = \frac{1}{2} + \frac{1}{\sqrt{2\pi}} \sum_{n=0}^{\infty} (-1)^n \frac{t^{2n+1}}{2^n n!(2n+1)}.$$

The function to be integrated in the first formula, together with the normalization factor, namely,

$$\frac{1}{\sqrt{2\pi}} e^{-x^2/2},$$

is called the density of the normal distribution. The graph of this function is the well-known Gaussian bell curve, where each value $\phi(t)$ can be interpreted as an area under the curve. This curve was to be found, together with a portrait of its discoverer, Carl Friedrich Gauss (1777–1855), on the German ten-mark note. The symmetry of the curve is a consequence of the relation $\phi(t) + \phi(-t) = 1$.

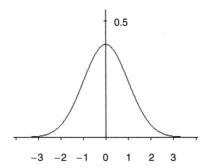

2. E. O. Thorp, Optimal gambling systems for favorable games, *Revue de l'institute international de statistique/Review of the International Statistical Institute* **37**, 1969, pp. 273–293; particularly p. 276.

14

And Not Only at Roulette: The Poisson Distribution

It is hardly to be expected that in 37 spins of the roulette wheel, all 37 numbers appear exactly once each. How many different numbers will appear on average?

Although a naive interpretation of the laws of equal probability might suppose that all 37 numbers should appear once each, in fact, such an event is almost impossible. For among the 37^{37} possible events, there are only 37! that correspond to the possible permutations of the different numbers. Therefore, the probability that all 37 numbers appear in 37 spins of the wheel is $37!/37^{37} = 1.304 \times 10^{-15} = 0.000000000000001304$. It is much more likely to win the lottery twice in a row with all six numbers correct.

We know now that we can hardly expect to obtain all 37 numbers in 37 roulette spins. But how many different numbers should we expect to see on average? That is, what is the expectation of the random variable that represents the number of different numbers obtained? If we choose a number, then the binomial distribution tells us how probable the different possible numbers of "hits" are for this particular number. For example, if X is the target number, then the probability $P(X = k)$ that in n trials there are k hits on this target is given by

$$P(X = k) = \binom{n}{k} p^k (1 - p)^{n-k}.$$

In our concrete example, the number of trials is $n = 37$, and the probability is $p = 1/37$. It was indicated in our study of the normal distribution

Wheaton College : Norton, MA

        #210 04-05-2010 12:24PM
   Item(s) checked out to p14203807.

TITLE: Luck, logic, and white lies : the
AUTHOR: Bewersdorff, J rg
CALL #: QA269 .B39413 2005
BARCODE: 36307101440359
DUE DATE: 05-03-10
LOCATION: Wheaton Stacks
COPY #: 1

   Madeleine Clark Wallace Library
    Wheaton College : Norton, MA

that the formulas of the binomial distribution are cumbersome to work with. In addition to the option of using the normal distribution as an approximation, in our present example, a much simpler approximation is offered by the *Poisson distribution*. It is named for the mathematician Siméon Denis Poisson (1781–1840). The Poisson distribution is based on the observation that the probability of getting a particular one of the 37 numbers k times in 37 roulette spins remains practically unchanged when the number of numbers and the number of spins are enlarged by the same amount, that is, with 100 numbers and 100 spins, the probabilities $P(X = k)$ are about the same as before. We can see the underlying principle behind this phenomenon by examining the formulas of the binomial distribution. We assume that the product of the number of trials n and the probability p in a single trial, namely, $\lambda = np$, has a fixed value; in our example, we have $\lambda = 1$. If the probability p is now replaced by the expression λ/n, then one obtains the chain of equalities

$$P(X = k) = \binom{n}{k} p^k (1 - p)^{n-k}$$

$$= \frac{1}{k!} \times \frac{n}{n} \times \frac{n-1}{n} \times \cdots \times \frac{n-k+1}{n} \times \lambda^k \left(1 - \frac{\lambda}{n}\right)^n \left(1 - \frac{\lambda}{n}\right)^{-k}$$

$$\approx \frac{\lambda^k}{k!} e^{-\lambda}.$$

At the end of the chain, the exact value is approximated with the help of its limiting value, which is obtained as the number of trials n increases and the probability $p = \lambda/n$ becomes correspondingly smaller, say, in passing to a 370-number roulette wheel and ten times as many trials, and so on. The error due to this approximation remains small if the probability p is relatively small. Indeed, it can be shown that the sum of all deviations is at most $2np^2$.[1] Regardless of how exact the approximations are in a concrete case, the limiting values should be viewed as a probability distribution of a random variable Y. The range of values encompasses the natural numbers $k = 0, 1, 2, \ldots$, and the probability distribution, namely, the Poisson distribution, is given by the formula

$$P(Y = k) = \frac{\lambda^k}{k!} e^{-\lambda}.$$

In our example, that is, for the parameter $\lambda = 1$, one obtains the values shown in Table 14.1. They show approximately how likely it is that in 37

---

[1]This theorem and more detailed discussion can be found in standard books on probability theory, such as Ulrich Krengel, *Einführung in die Wahrscheinlichkeitstheorie und Statistik*, Braunschweig 1988; particularly pp. 88 ff., Theorem 5.9.

k	Poisson Distribution $P(Y = k)$	Binomial Distribution $P(X = k)$	Error (Difference)
0	0.36788	0.36285	0.00503
1	0.36788	0.37293	0.00505
2	0.18394	0.18647	0.00253
3	0.06131	0.06043	0.00088
4	0.01533	0.01427	0.00106
5	0.00307	0.00262	0.00045
6	0.00051	0.00039	0.00012
7	0.00007	0.00005	0.00003
...	...	...	...

Table 14.1. Probabilties for multiple hits in 37 roulette spins.

roulette spins, a paricular number will appear k times. In particular, the probability that a particular number will not appear at all in 37 spins is more than one-third. As a comparison, we give the exact values of the binomial distribution and the resulting errors.

For individual numbers, we now know the probabilities of various numbers of hits. But how does this relate to the totality of numbers? How many different numbers are to be expected in 37 spins? A little trick allows us to obtain the answer with the data at hand: we define, based on the 37 spins, the random variables $Z_0, Z_1, \ldots, Z_{36}$, where each variable takes on the value 0 or 1 according to whether the corresponding number occurs exactly once. Based on the results already obtained, we have

$$E(Z_0) = E(Z_1) = \cdots = E(Z_{36}) = 0.37293.$$

It follows that the number of numbers $Z_0 + Z_1 + \cdots + Z_{36}$ hit exactly once has the expectation

$$E(Z_0) + E(Z_1) + \cdots + E(Z_{36}) = 37 \times 0.373 = 13.8,$$

a value that in a long series of 37 spins should give approximately the average value, according to the law of large numbers. This is the same number that one would expect for numbers that do not appear at all. In the roulette literature, this state of affairs is known as the "two-thirds law": in a series of 37 spins, called a *rotation*, approximately two-thirds of the 37 numbers should appear.

In everyday practice, the Poisson distribution comes into play above all in determining the frequency of rare events. This could involve insured accidents, repair service contracts, or events of atomic decay. In each case, the event is a rare one in relation to a particular object, be it an insurance-policy holder or a particular atom, while due to the large number of objects, the event itself occurs frequently, whence its probability distribution is given by the Poisson distribution. The first statistical observation of the Poisson distribution occurred toward the end of the 19th century in the study of cases of death in the German army due to a horse's kick. The example can be found in the book *The Law of Small Numbers*, by the mathematician Ladislaus von Bortkiewicz (1868–1931), which appeared in 1898. The title of Bortkiewicz's book relates to the rarity of the underlying events. The title suggests a negation of the law of large numbers, though that is not at all the case, and is thus more misleading than helpful. Nonetheless, it is still occasionally used.

Finally, we should mention that the approximation formula derived in Chapter 2 in relation to de Méré's problem is also a special case of the Poisson distribution. There we asked for the number of trials necessary for the probability that an event occurs at least once to reach at least $1/2$. If we seek an approximate answer with the help of the Poisson distribution, then the number n of trials must satisfy the condition $P(Y = 0) \leq 1/2$. On account of the formula $P(Y = 0) = e^{-np}$, this inequality holds precisely when $n \geq \ln 2/p$.

15

When Formulas Become Too Complex: The Monte Carlo Method

Two players, Jill and Jack, play a series of games of chance. They bet a set amount each round. The probability that Jill wins any one round is 0.52. If she loses a round, her stake goes to Jack and vice versa. At the start, Jill has 5 chips, while Jack has 50. They play until one player is broke. What is the probability the Jill wins, and how many rounds are played on average before one player is bankrupt?

Even good odds of winning are no guarantee against bad luck. If one's starting capital is too small, it can happen that premature financial ruin can prevent a player from profiting from the law of large numbers. But how can this risk of ruin be calculated? Although there are formulas for this classical problem, which goes back at least to Christian Huygens (1629–1695), we would like to approach the problem from a different point of view. Instead of looking for a formula, we shall simply set up a series of trials and see what happens. But since that could become a bit boring after a while, we are not going to play the game ourselves. Instead, we shall entrust the play and evaluation of the results to a computer.

But how is the computer going to determine the results of a game? It does not have any built-in dice, does it? There are two ways to proceed:

- one could carry out a random experiment apart from the computer, and then input the results into the computer. To save effort, one could use roulette permanences obtained from casinos. The list of *random numbers* thus obtained could then be used for a variety of purposes, including the one that we are presently considering.

- the computer could generate random numbers itself. As we mentioned in Chapter 8, computation and randomness are mutually exclusive. However, there are calculational procedures whose results behave statistically like random numbers. We therefore speak of *pseudorandom numbers*. To all appearances, according to the results of empirical investigations, the digits in the decimal expansion of π behave like a sequence of pseudorandom numbers: at each place, each of the ten digits appears equally likely to appear.

In practice, today, in general, only the second method is used, since it involves considerably less effort. However, one does not need to calculate the digits of π, since there are much simpler algorithms that serve the purpose, and that furthermore have the advantage that one can make precise statements about the "quality" of the randomness that is produced. To program a sequence of random numbers does not require a great deal of thought, for every compiler or interpreter of a programming language has available a pseudorandom number generator. For example, the expressions INT(100*RND(a))+1 in Basic and Random(99) + 1 in Pascal produce uniformly distributed pseudorandom numbers between 1 and 100. The result of a single game as described at the beginning of this chapter can thereby be simulated by comparing the randomly obtained number to 52. If it is less than or equal to 52, then we chalk up a win for Jill, which will occur with probability 0.52. The remainder of the program keeps track of the amounts of money in possession of the players, and performs a statistical evaluation of the results. Table 15.1 shows the results of a computer simulation.

In spite of her slight advantage of winning a single game, Jill's long-term prospects are bleak. The precision of these experimentally obtained

Number of Series	Average Game Length	Jill's Winnings	Jack's Winnings
10	567.40	0.4000	0.6000
100	323.54	0.3600	0.6400
1000	338.16	0.3430	0.6570
10 000	326.70	0.3347	0.6653
100 000	333.89	0.3344	0.6656

Table 15.1. Results of a simulation of the game between Jack and Jill.

results can be seen from considerations of the law of large numbers and the central limit theorem. These tell us that if a series of trials is long enough, then a large deviation between the relative frequency and probability of an event is highly unlikely. However, to achieve a high degree of accuracy, long series of trials are necessary, since for a given level of confidence, the precision is doubled only when the length of the series is quadrupled. In general, as we said in Chapter 13, with n attempts, the probability of a deviation of more than $2.58/2\sqrt{n}$ is at most $2\phi(-2.58) = 0.01$. So with $100\,000$ trials, the error is less than 0.004 with 99% certainty.

An experiment of the type that we have just described is said to be carried out by a *Monte Carlo method*. Its advantage is that approximate results can be obtained quickly and simply using a universal approach whose precision is generally acceptable in practice. The simplicity of the method makes it possible in particular to carry out simulations under a variety of conditions and to compare the various results. In this way, one can determine numerically, for example, the dependency of the result on the intitial parameters. It is also possible to optimize a decision process, for example, in a game in which chance plays a role. However, caution is advised against trying to measure something, such as an infinite expectation, that cannot actually be measured.

Since Monte Carlo methods are impracticable to carry out without the aid of a computer, it is not surprising that such methods have existed only about as long as the computer has. Although the theoretical underpinnings, in particular the law of large numbers, have long been known, it was not until 1949 that there appeared a publication on Monte Carlo methods.[1]

In fact, the Monte Carlo method was established three years prior to that date, namely, in 1946, by Stanislaw Ulam (1909–1984). Important contributions to the subject go back to John von Neumann (1903–1957), who used them in connection with calculating nuclear reactions.

Perhaps the most truly magnificent idea regarding Monte Carlo methods was to use them in domains that in principle are not subject to the influences of randomness. The first, though historically isolated, example is Buffon's needle problem, on the basis of which we demonstrated in Chapter 7 how the decimal expansion of π can be experimentally determined. That procedure can be generalized to the determination of arbitrary areas and volumes. For example, if the three-dimensional coordinates (x, y, z) are generated by three independent uniformly distributed random variables in the interval from -1 to 1, then using the inequality $x^2 + y^2 + z^2 \leq 1$ one can ask whether the randomly generated point lies within the sphere of

---

[1]N. Metropolis und S. Ulam, The Monte Carlo method, *Journal of the American Statistical Association* **44**, 1949, pp. 335–341.

radius 1 and center at the origin. The probability that the inequality is satisfied is the ratio of the volume of the sphere to that of the cube of side length 2. Even higher-dimensional hypervolumes can be approximated in this way. For example, one could attempt to calculate the volume $\pi^2/2$ of the hyphersphere of radius 1.

When we talked about the ruin problem, we did not go into the question of how long the game is expected to last. To use the central limit theorem to say something about this, we must know the standard deviation of the length of the game. However, in practice, it suffices to approximate the standard deviation by means of a simulation. This can be done by using, instead of the unknown probability distribution, the approximately equal distribution of the relative frequencies. In the simulation results recorded in Table 15.1, for one million series of games there was a standard deviation of 439. With 99% certainty, there is then an error for the average length of a game that is at most $2.58 \times 439/1000 \approx 1.1$.

The Generation of Random Numbers

We have referred to the program libraries available for the various programming languages that have functions for the generation of random numbers. But how do these random number generators work? How can one generate one's own random numbers? This is an important question, because with long Monte Carlo simulations, the precision of the results is greatly dependent on the quality of the random numbers used. The random number generators implemented in the various programming languages are not always adequate.

In the period in which the computer was not generally available, one simply used tables of random numbers. One of the most extensive such tables, from the year 1955, contained one million random digits. If we had not already seen applications of this in the examples presented, we might well have shaken our heads in disbelief at such a monumental opus. Furthermore, such random numbers were not generated with a roulette wheel or with dice, but instead, sources were the likes of the middle digits from statistical census tables. Moreover, in order to generate random numbers more quickly, modified "wheels of fortune" were constructed, whereby a number was selected using an electronically controlled flash. To exclude errors due to conceptual shortcomings, the random numbers were then subjected to statistical tests.

For computer simulations, such permanent generated lists of random numbers are not very suitable, since it is uneconomical to store long sequences of random numbers. Therefore, mathematicians who participated in the development of Monte Carlo methods came up with other techniques. They sought programs that would generate sequences of random numbers as needed. That seems simpler than it actually is. The main difficulty is that computers are entirely deterministic in their calculations. That is, a given input will always produce the same output, and thus that output cannot be random. However, it is possible to generate numbers that "act as if" they were random. What is meant is sequences of numbers that, like the decimal digits of π, do not reveal any regularities when subjected to statistical tests. Such sequences are called *pseudorandom*.

The method used most frequently for generating pseudorandom numbers depends on a theorem about prime numbers. The following table demonstrates the principle in a simple case, one that is not suitable for practical use. It shows 100 numbers between 1 and 99 generated in random order:

15	24	99	17	7	92	26	82	10	16
66	45	72	95	51	21	74	78	44	30
48	97	34	14	83	52	63	20	32	31
90	43	89	1	42	47	55	88	60	96
93	68	28	65	3	25	40	64	62	79
86	77	2	84	94	9	75	19	91	85
35	56	29	6	50	80	27	23	57	71
53	4	67	87	18	49	38	81	69	70
11	58	12	100	59	54	46	13	41	5
8	33	73	36	98	76	61	37	39	22

Let us check the randomness of these numbers without any knowledge of how they were generated. The first thing that we observe is that each of the numbers between 1 and 100 appears exactly once, which in fact, is highly unlikely to be the result of a random process. Indeed, the formula that produced this list generates each of the numbers from 1 to 100 exactly once. Each number is generated from the one before it, and after all

of them have been run through, the cycle begins again. That
is, after 22 comes 15. The formula for this sequence of numbers
$x_1, x_2, x_3, \ldots$ is

$$x_{n+1} = 42x_n \quad (\text{mod } 101).$$

The expression "mod (101)" is short for "modulo 101," which
means that after the product $42x_n$ is calculated, 101 is repeatedly subtracted from the result until an integer in the range 0
to 100 is obtained. For example, since the first number is 15,
the next number is calculated by taking $42 \times 15 = 630$ and then
subtracting 101 six times, leaving the remainder 24. Therefore,
the next number in the sequence is 24. Since the number of
times 101 is subtracted varies erratically, it is difficult to see
the regularities in the sequence.

The reason that this formula generates all the numbers between
1 and 101 is, as we have mentioned, a theorem about prime
numbers. That theorem tells us that for each prime number p,
there is at least one number a between 1 and $p - 1$ such that
the numbers $1, a, a^2, a^3, \ldots, a^{p-1}$, when taken modulo p, yield
a complete set of the integers from 1 to $p - 1$ (see Note 1 at the
end of the chapter). In our example, $p = 101$ and $a = 42$. To
generate pseudorandom numbers, it is necessary to use much
larger prime numbers; usually, numbers of order of magnitude
at least 10^9 are used. Once a prime number p is fixed, there
are quite a few choices for the number a that will generate a
sequence of all the integers from 1 to $p - 1$ in pseudorandom order. One is limited to those choices that will give the sequence
a random character. In particular, each number should be followed by numbers of a variety of orders of magnitude, and that
excludes relatively small values of a.

To obtain random numbers that are universally applicable, the
sequence that is generated is usually transformed to fall in the
range between 0 and 1. This is done via division by p. From
such a random number y, one could use, say, INT(6*y+1) to
obtain the integer part of the real number $6y + 1$, which is less
than 7 and greater than 0, thus obtaining an integer in the
range from 1 to 6 that could simulate the result of throwing a
die.

Many random number generators in use employ a generalized
algorithm. In some cases, the remainder taken is not based

on a prime number, while in others, a more complex recursion formula is used. Generally, one obtains a new random number x_{n+k} from the k previous numbers with a formula of the type

$$x_{n+k} = a_0 x_n + a_1 x_{n+1} + \cdots + a_{k-1} x_{x+k-1} + b \pmod{m}.$$

Here $a_0, a_1, \ldots, a_{k-1}, b, m$, as well as the starting values $x_1, x_2,$ $\ldots, x_k$, are suitably chosen integers. All the numbers generated are integers between 0 and $m - 1$, where the period of the sequence can be at most m^k.

For the simply constructed random number generators, that is, those with $k = 1$ and $b = 0$, the following parameters are usual:

a_0	m
5^{13}	2^{39}
5^{15}	2^{35}
5^{17}	2^{40}
23	100 000 001
100 003	10 000 000 000

In practice, one also sees combinations of various random sequences (see Note 2 at the end of the chapter).

Snakes and Ladders

A nice example of a Monte Carlo method is the 1960 investigation[2] of a game of snakes and ladders.[3] The original version

---

[2] N. W. Bazley, P. J .Davis, Accuracy of Monte Carlo methods in computing finite Markov chains, *Journal of Research of the National Bureau of Standards, Mathematics and Mathematical Physics* **B64**, 1960, pp. 211–215. See also S. C. Althoen, L. King, K. Schilling, How long is a game of snakes and ladders? *The Mathematical Gazette* **77**, 1993, pp. 71–76.

[3] Directions and more information about the game can be found in Erwin Glonnegger, *Das Spiele-Buch*, Munich 1988, pp. 54–55; R. C. Bell, *Board and Table Games from Many Civilizations*, New York 1979, volume 2, pp. 10–11; Frederic V. Grunfeld, *Spiele der Welt*, Frankfurt 1985 (Dutch original 1975), volume I, pp. 74 f.

of snakes and ladders involves a race in which each player tries to be the first to move a playing piece to the winning square, where moves are made according to a throw of the dice. In the version that was investigated in the research article, the board contained 100 squares. The game starts at square 1, and the winning square is 100. A single die is thrown, and the player moves his or her piece that many squares forward. The player has no choice as to how to move. One must reach square 100 exactly: if the number on the die would take the piece beyond square 100, the piece stays where it is. There is no influence of one piece by another. That is, players' pieces do not block or capture each other. What adds spice to the game is the ladders and snakes (or chutes, in one version) depicted on the board, as shown in the figure.

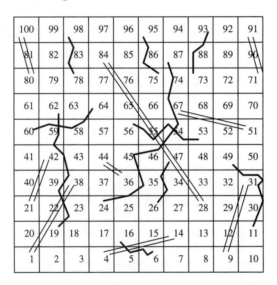

If a piece lands at the bottom of a ladder, it ascends to the top, and if it lands at the top of a snake, it slides down to the bottom.[4]

In the investigation under discussion, the researchers determined the average number of turns it would take for a player to arrive at the winning square. A Monte Carlo method turned

---

[4]In a particularly dramatic version of the game the board depicts a mountain that pops up when the board is opened. The game pieces represent mountain climbers who are held to the mountain by invisible magnets fastened from behind. If a player arrives at a square without a magnet, the climber falls down a distance on the mountain.

out to be quite simple, in comparison to working out the probabilities mathematically, as we shall do in Chapter 16. The result was an approximation of 39.225 for the expected number of turns.

Statistics: Sample Functions and Their Distribution

In Chapter 12, a die was judged asymmetric because the theoretical expectation was not matched by the experimental result: beginning with the hypothesis that the die was in fact symmetric, particular results of a series of trials were predicted with almost certain probability. Because the results of the experiment did not correspond to the predictions, the starting hypothesis had to be rejected. In our concrete case, the sum of the numbers from the various throws of the die was taken as the test variable. Of course, the underlying principle is independent of the details of how the test variable is calculated.

As an example, let us again take a die, which is thrown n times. In principle, one may take for the test variable, which in statistics is generally called a sample function, any random variable determined by the series of trials. Particularly suitable are those random variables that react strongly to deviations in the property being tested. For finite distributions, this is achieved by the the χ^2 function, which was devised in 1900 by Karl Pearson, which in the case of a die is calculated from the absolute frequencies $h_1, h_2, \ldots, h_6$ with the formula

$$\chi^2 = \frac{\left(h_1 - \frac{1}{6}n\right)^2}{\frac{1}{6}n} + \frac{\left(h_2 - \frac{1}{6}n\right)^2}{\frac{1}{6}n} + \cdots + \frac{\left(h_6 - \frac{1}{6}n\right)^2}{\frac{1}{6}n}.$$

In general, within a series of trials, the χ^2 function measures all deviations between the measured frequencies and their expectations, which in our example is $n/6$ in each case. One can derive, with deep mathematics, how the values of the χ^2 function are distributed. It turns out that the distribution of the χ^2 function depends almost exclusively on the number of possible results, and not on the number of trials and the probabilities of the individual results; however, the expectations resulting from these data cannot be too small for the individual frequencies. The theoretical considerations and the not-so-simple calculations can be avoided if one uses a Monte Carlo method. That is, before one actually carries out a series of trials with a die that is under investigation, one should simulate a large number of series of trials on the computer. For example, one can run 999 series of trials. If the result with the die is significantly different in comparison to these 999 results, then the hypothesis that the die is symmetric can be rejected. But what is meant by "significantly different"? Since asymmetric dice produce larger χ^2 values than symmetric dice, one simply considers the greatest ten of the 999 simulation results as outriders. If the test result lies above those of the 989 lowest simulation results, then there are two possible causes:

- the die is asymmetric; the hypothesis is correctly rejected.
- the die is symmetric, so that the result of the experiment is an outlier; rejecting the hypothesis in this case is an error.

In a particular situation, one cannot say which is the case. However, the a priori probability that such a procedure will lead to an incorrect rejection of the hypothesis, that is, before the beginning of the simulation and the dice experiment, is at most 0.01. The reason for this is clear: if the die is actually symmetric, then the simulated and actual series of trials produce identically distributed random variables. If at the end of the experiment one sorts all one thousand results according to size, then every level from 1 to 1000 is equiprobable for the χ^2 value of a series of trials, and for exactly ten of them, the initial hypothesis is rejected, although it was actually correct.

The possible objection that such an elaborately constructed χ^2 sample function was unnecessary is only partially justified. Even with other, less cleverly constructed, functions, correct hypotheses are rejected with the same probability. However, the quality of the χ^2 function is shown elsewhere: since it reacts strongly to deviant probabilities, the danger that the hypothesis of a symmetric die will not be rejected in the case of a strongly asymmetric die is relatively small.

Incidentally, the 999 series of trials with 1200 rolls each yielded a rejection region of 14.67 up; in statistical tables, in the section

on the χ^2 distribution with five degrees of freedom, one finds
the associated expectation of 15.09.

Chapter Notes

1. The arrangement of the integers from 1 to 16 obtained in this way for the
   case $p = 17$ was used by Carl Friedrich Gauss to prove that the regular
   17-gon is constructible with straightedge and compass. He made his discovery,
   as he later noted, "by strenuous thinking... in the mornings... (even
   before I arose)," and he thereby solved a general construction problem
   that had been formulated in antiquity and left unsettled. With the entry,
   "Foundations on which the division of the circle is based, and indeed, its
   geometric divisibility into seventeen parts, etc., Braunschweig, March 30,"
   the 18-year-old Gauss began a diary in which throughout the period of his
   creative work he set down many remarkable results. The notebook, in the
   original Latin together with a German translation, is available in the series
   Ostwalds Klassiker (number 226, Leipzig 1976). The first quotation above
   comes from one of Gauss's letters (*Carl Friedrich Gauss, the Prince of
   Mathematics in Letters and Talks*, published by Kurt-R. Biermann, Leipzig
   1990, p. 54). The methods of constructing the regular 17-gon—the coordinates
   of the vertices are calculated with the aid of the complex solutions
   of the equation $x^{17} - 1 = 0$ expressed in terms of quadratic equations—
   can be found in B. L. van der Waerden, *Algebra*, New York 1991, and
   Jörg Bewersdorff, *Algebra für Einsteiger: Von der Gleichungsauflösung zur
   Galois-Theorie*, Braunschweig 2002, pp. 67–71. An explicit construction is
   described by Ian Stewart in his article Gauss, *Scientific American* **237:1**,
   1977, pp. 122–131, as well as in the ninth lecture in Heinrich Tietze's book,
   Famous Problems of Mathematics, New York 1965.

2. Among the decisive criteria for the quality of a random sequence are the
   uniformity of the distribution in the interval from 0 to 1 and the general
   independence of the numbers as they follow in sequence. The latter is important,
   for example, when several random numbers in sequence determine
   an event. In a controlled setting, the independence of a sequence of random
   numbers can be achieved if several sequences of pseudorandom number
   sequences are interleaved into a single sequence. If the periods of these sequences
   are mutually coprime (no common factor), then the total sequence
   contains all combinations in which the numbers follow one another. This
   can be done, for example, with the *Sophie Germain primes* $p_1, p_2, \ldots$, such
   as $999521, 999611, 999623, 999653, 999671, 999749, 1000151$, which have the
   property that all the numbers $2p_i + 1$ are prime. With an arbitrary factor

$a \neq -1, 0, 1 \pmod{2p_i + 1}$, one first obtains individual sequences with period p_i or $2p_i$. If one mixes these sequences together, the result is a sequence with period $p_1 p_2 \cdots$ or $2p_1 p_2 \cdots$. If one then transforms the normalized pseudorandom number y in the interval $(0, 1)$ by $y' = 2\min(t, 1 - y)$, one then obtains a uniformly distributed sequence with period $p_1 p_2 \cdots$, where there are as many independent consecutive numbers as distinct primes p_i used.

Further Literature on Monte Carlo Methods

[1] I. M. Sobol, *Die Monte-Carlo-Methode*, Frankfurt 1985.

[2] S. M. Ermakow, *Die Monte-Carlo-Methode und verwandte Fragen*, Munich 1975.

16

Markov Chains and the Game Monopoly

In the game Monopoly, one wishes to evaluate the various properties according to the expected income from rent. How should one proceed?

Among games that are protected by copyright, Monopoly,[1] with over 185 million copies sold, is one of the most successful. Since its invention by the American Charles Darrow in the year 1931 it has influenced the development of many other games based on economics, none of which, however, has achieved the worldwide popularity of Monopoly. We should mention, though, that Monopoly did not materialize out of nowhere; it had forebears whose similarity to Monopoly suggests that one or more of them may have inspired its creator. For example, there exists a 1904 patent application for a "Landlord's Game."[2] Not only did this game employ a game board with 40 squares around the perimeter, it also had special corner fields and

---

[1] For more information on Monopoly, see Erhard Gorys, *Das Buch der Spiele*, Hanau, ca. 1987, pp. 357–359; Werner Fuchs, *Spieleführer 1*, Herford, 1980, pp. 75 f.; David Pritchard, *The Family Book of Games*, Brockhampton Press, 1983, pp. 186 f.; David Pritchard (ed.), *Modern Board Games*, London 1975, pp. 85–91 (with contributions by David Parlett); Mit grossen Scheinen und kleinen Steinen, *Spielbox* **4** 1983, pp. 8–14, 40–43. Maxine Brady's *Monopoly*, New York 1974, is devoted exclusively to Monopoly.

[2] For information on the Landlord's Game, see Sid Sackson, *A Gamut of Games*, New York 1969; Erwin Glonnegger, *Das Spiele-Buch*, Munich 1988, p. 114; Dan Glimme, Barbara Weber, Monopoly: die internationale Geschichte, *Spielbox* **4** 1995, pp. 10–14 and 1995/5, pp. 4–8; Willard Allphin, Who invented Monopoly? *Games and Puzzles* **34** 1975, pp. 4–7; Philip Orbanes, *The Monopoly Companion*, Boston 1988, pp. 25 ff.

railroads at the midpoints of the four sides. The location of the utilities differs from that of Monopoly by a single square. Moreover, the Landlord's Game has as its theme the buying and renting of 22 pieces of property.

In the opening phase of a game of Monopoly, each player attempts to acquire various properties. A player may buy only those unsold properties on which his game piece has landed. A player whose piece lands on a property owned by an opponent must pay the owner rent. At first, these rents are rather low. However, once a player owns a matched set, a "monopoly," of two or three properties, then the rent is doubled. With additional investment, the owner of a monopoly may build houses or a hotel, and the rents rise steeply.

Any serious analysis of Monopoly must provide the player with well-founded advice on making the choices that arise in the game, which have mostly to do with the purchase and sale of properties, including to and from the other players, as well as the building of houses and hotels. Just as in real-world real estate transactions, the costs of investment must be weighed against the prospective increased return. Since Monopoly contains a large measure of chance, the prognosis of returns can be made only on the basis of probabilities and expectations. Thus the amount of income that can be expected from a monopoly on which buildings have been erected depends on the amount of rent that will be collected per "visit," as well as the probability that such a visit will occur. What are the probabilities for the 40 fields? They are certainly not equiprobable, for the symmetry is greatly disturbed by the "Go to Jail" square, the "Chance" and "Community Chest" cards, and the rule that if one rolls doubles three times, one is sent directly to jail.

It is certain that computer simulations are the best route to determining the probabilities of landing on the various squares of the game board. However, the probabilities can be calculated. In order to see how that can be done, let us consider first a simpler example, which can be seen in Figure 16.1.

The figure shows a circuit of four squares in which a game piece is moved the number of squares determined by the roll of a single die. The game begins at square 1, labeled "Go." If the piece ends up on square 4,

Figure 16.1. A dice circuit with four fields.

		End Field			
		1	2	3	4
Start Field	1	1/6	1/2	1/3	0
	2	1/6	1/2	1/3	0
	3	1/3	1/2	1/6	0
	4	1/3	1/2	1/6	0

Table 16.1. Transition probability matrix for the dice circuit.

it is moved immediately to square 2. As we asked in Monopoly, we can ask here for the probabilities of the game piece landing on the various squares. There is no point looking for a situation of equiprobability. We need to look at the various throws of the die, and how they translate into positions on the game board. These determine the *transition probabilities*, which specify the probabilities of moving between two given squares. These probabilities never change. Thus one always goes from square 1 to square 3 with probability 2/6, namely, the probability of rolling a 2 or a 6. The probability of starting on square 2 and remaining there is 3/6, namely, the probability of rolling a 2, a 4, or a 6. The transition probabilities are shown in Table 16.1.

If we now wish to calculate the probability of ending up on a particular square after a certain number of moves, we can use the transition probabilities. If we let $p_n(1), p_n(2), p_n(3), p_n(4)$ denote the probabilities of landing on the indicated square after n turns, then we note the following:

- the starting situation, in which the game piece is standing on the first field, can be represented by the values

$$p_0(1) = 1, \qquad p_0(2) = p_0(3) = p_0(4) = 0.$$

- a move can be described by the transition equations[3]

$$p_{n+1}(1) = \frac{1}{6}\big(p_n(1) + p_n(2) + 2p_n(3) + 2p_n(4)\big),$$

$$p_{n+1}(2) = \frac{1}{2}\big(p_n(1) + p_n(2) + 2p_n(3) + 2p_n(4)\big),$$

$$p_{n+1}(3) = \frac{1}{6}\big(2p_n(1) + 2p_n(2) + p_n(3) + p_n(4)\big),$$

$$p_{n+1}(4) = 0.$$

---

[3] This result comes from the formula for total probability (see Chapter 9), since transition probabilities are a form of conditional probabilities.

n	$p_n(1)$	$p_n(2)$	$p_n(3)$	$p_n(4)$
0	1.0000000	0.0	0.0000000	0.0
1	0.1666667	0.5	0.3333333	0.0
2	0.2222222	0.5	0.2777778	0.0
3	0.2129630	0.5	0.2870370	0.0
4	0.2145062	0.5	0.2854938	0.0
5	0.2142490	0.5	0.2857510	0.0
6	0.2142918	0.5	0.2857082	0.0
7	0.2142847	0.5	0.2857153	0.0
8	0.2142859	0.5	0.2857141	0.0
9	0.2142857	0.5	0.2857143	0.0
...	...	...	...	...

Table 16.2. Development of the four state probabilities of the dice circuit.

Thus, after the first throw, the probabilities are

$$p_1(1) = \frac{1}{6}, \quad p_1(2) = \frac{1}{2}, \quad p_1(3) = \frac{1}{3}, \quad p_1(4) = 0,$$

while after two throws, they are

$$p_2(1) = \frac{2}{9}, \quad p_2(2) = \frac{1}{2}, \quad p_2(3) = \frac{5}{18}, \quad p_2(4) = 0.$$

The further development is recorded in Table 16.2.

The table indicates that as the number of turns increases, the state probabilities rapidly approach a *stationary probability distribution*. Although it was not obvious a priori, we note that when we asked for the four probabilities, we were implicitly assuming such a stability. The way in which this stationary probability distribution was demonstrated was rather complicated, and we might ask whether there is a simpler approach, and indeed, that is the case. Clearly, if there is a stationary limiting distribution $p(1), p(2), p(3), p(4)$, then it should repeat itself when the transition probabilities are applied. That is, they must satisfy the system of equalities

$$p(1) = \frac{1}{6}\big(p(1) + p(2) + 2p(3)\big),$$
$$p(2) = \frac{1}{2}\big(p(1) + p(2) + p(3)\big),$$
$$p(3) = \frac{1}{6}\big(p(1) + 2p(2) + p(3)\big),$$
$$p(4) = 0,$$

with the condition

$$p(1) + p(2) + p(3) + p(4) = 1.$$

Without difficulty, one obtains the desired probability distribution

$$p(1) = \frac{3}{14}, \quad p(2) = \frac{1}{2}, \quad p(3) = \frac{2}{7}, \quad p(4) = 0.$$

The example that we just analyzed, as well as our actual topic, the game of Monopoly, suggests that the phenomena that we have been observing are special cases of general principles that are applicable to other games as well. Before we return our attention to Monopoly, we would like first to discuss some of the foundational ideas that go back to the Russian mathematician Andrei Andreyevich Markov (1856–1922), namely, the theory of *Markov chains*.

Up to now, when we have investigated random sequences, it has been for the most part sequences of independent events, such as those obtained in a sequence of dice rolls. As we have put it, dice do not have a "memory." The situation is quite different when we consider the situation of a piece on the game board, say, on our dice circuit. Here the events of being on one or another of the four fields after n rolls of the die are not independent of where the piece was sitting after m rolls. However, and this is the fact that we wish to emphasize, it is only the current field on which the piece is standing that has any bearing on where it will be after the current turn. That is, the past history of the game piece's movement plays no further role. Therefore, the dependence relationships within the random sequence are limited, namely, by a "memory" that is only one turn long. A general model for such situations is given by Markov chains.

A *Markov chain* is a sequence of random trials in which precisely one of a finite set of events occurs. Furthermore, the probability that a particular event occurs on the $(n+1)$th trial depends only on the event of the nth trial, and not on any of the prior events. That is, the conditional probabilities for the event occurring on the $(n+1)$th trial are the same whether the condition is based on the event of the nth trial or on the nth event together with some of the previously occurring events.

A special vocabulary has developed for Markov chains, based on the terminology of physics. Thus the occurrence of an event is interpreted as a *sojourn* in a *state*. One thus obtains a system that always finds itself in one of a finite number of states and whose changes of state take place at fixed intervals in a random fashion. In every case, the probability that the system will move from one state to another is governed by these two states only, not by the time at which the state change occurs or any of the previous history. Mathematically, a Markov chain amounts to a square array of transition probabilities, called the *transition matrix*. For more details, see "A Brief Primer on Markov Chains."

In our example of the dice circuit, the Markov chain comprises four states, corresponding to the four fields, where the current state of the Markov chain is determined by the location of the game piece. We have already written down the transition matrix. Another example of a Markov chain is the game snakes and ladders, introduced in the previous chapter. And even the ruin problem discussed earlier can be seen as a Markov chain if the current distribution of wealth is viewed as a state (see "The Ruin Problem as a Markov Chain").

In order to investigate the development of a Markov chain, the sojourn probabilities are calculated, that is, the probabilities that the system is in a particular state at the nth trial. However, it is frequently sufficient to ascertain the trend of the sojourn probabilities. Thus for the Markov chain of the dice circuit, a stationary state distribution arrived at over time can be derived from the transition probabilities. With the ruin problem and with snakes and ladders, there are other issues to deal with.

Snakes and Ladders as a Markov Chain

In addition to the 100 squares, the start situation is also considered a state, so that one obtains a Markov chain with 101 states. The transition matrix consists of $101 \times 101 = 10\,201$ probabilities, so that we are able to reproduce only a part of it here. Note that as with all transition matrices, the probabilities along a row always sum to 1:

		To										
		0	1	2	3	4	5	6	7	8	...	100
From	0	0	0	1/6	1/6	0	1/6	1/6	0	0	...	0
	1	0	0	1/6	1/6	0	1/6	1/6	1/6	0	...	0
	2	0	0	0	1/6	0	1/6	1/6	1/6	1/6	...	0
	⋮	⋮	⋮	⋮	⋮	⋮	⋮	⋮	⋮	⋮		⋮
	100	0	0	0	0	0	0	0	0	0	...	1

Using these data, beginning with the initial probabilities $p_0(0) = 1$ and $p_0(1) = p_0(2) = \cdots = p_0(100) = 0$, we can determine the further development of the probability distribution, just as we did in the dice circuit problem. In the limit, as stationary distribution we have $p(0) = p(1) = \cdots = p(99) = 0$ and $p(100) = 1$; that is, eventually the player reaches the goal square. What is more important, though, is that from the development of the sojourn probabilities one can also obtain the probability distribution of the length of the game, together with its expectation of 39.224.

The Ruin Problem as a Markov Chain

The ruin problem that we studied in the previous chapter using a simulation is considered here in a generalized version. We begin with a total capital of n units and the probability p that Jill wins a round against Jack. We shall denote Jill's probability of losing by $q = 1 - p$. This problem may be represented by a Markov chain with $n + 1$ states, where the current state is indicated by the amount of capital in Jill's possession: at state 0, she is ruined, while at state n, it is Jack who is ruined. The transition matrix looks as follows:

		State After							
		0	1	2	3	$\ldots$	$n-2$	$n-1$	n
State	0	1	0	0	0	$\ldots$	0	0	0
Before	1	q	0	p	0	$\ldots$	0	0	0
	2	0	q	0	p	$\ldots$	0	0	0
	$\vdots$	$\vdots$	$\vdots$	$\vdots$	$\vdots$		$\vdots$	$\vdots$	$\vdots$
	$n-2$	0	0	0	0	$\ldots$	0	p	0
	$n-1$	0	0	0	0	$\ldots$	q	0	p
	n	0	0	0	0	$\ldots$	0	0	1

In contrast to our other examples, here there is no unique limiting distribution, since the relation between the two probabilities of ruin depends on the initial distribution of the capital n. To calculate these probabilities is not particularly difficult: if $r(k)$

is the probability that Jill loses her initial capital of k in the course of the game, then we have

$$r(0) = 1, \quad r(n) = 0,$$

since in these two cases the game is already over. For $0 < k < n$, we can determine the probabilities $r(k)$ from $r(k-1)$ and $r(k+1)$ if one assumes knowledge of the next round:

$$r(k) = q\,r(k-1) + p\,r(k+1).$$

For $q > 0$, setting $s = p/q$, one thereby obtains the general formula (see Note 1 at the end of the chapter)

$$r(k) = \frac{1 + s + \cdots + s^{n-k+1}}{1 + s + \cdots + s^{n-1}}.$$

One may also obtain a general formula for the expected length $\ell(k)$ of a game. Clearly, we have $\ell(0) = 0$ and $\ell(n) = 0$. In the case $0 < k < n$, we again consider the course of the subsequent round:

$$\ell(k) = p\,\ell(k+1) + q\,\ell(k-1) + 1.$$

For $p \neq q$, we then obtain the formula (see Note 2 at the end of the chapter)

$$\ell(k) = \frac{1}{q(s-1)}\left(n - k - \frac{n\left(s^{n-k} - 1\right)}{s^n - 1}\right).$$

In the case $p = q = 1/2$, we have simply $\ell(k) = n(n-k)$.

For the example of the previous chapter, that is, with $p = 0.52$ and $n = 55$, we obtain $r(5) = 0.6661$ and $\ell(5) = 334.1304$.

We once again turn our attention to Monopoly. We first need to figure out just what the states are that we need to distinguish. There is the complication that if doubles are thrown, the player gets to throw the dice again. The same applies to a second throw of doubles. However, a third set of doubles does not result in that number of squares being traversed, but in the player being sent at once to Jail. Thus on a single turn, a player can land on one, two, or three squares, with all rights and responsibilities

Square	Property German Edition	Property US Edition	Prob. (German)	Prob. (US)	Maximal Rent (US) Abs.	Exp.	Group
0	Los	Go	0.02889	0.02914			
1	Badstr.	Mediterranean Avenue	0.02436	0.02007	250	5	
2	Gemeinschaftsfeld	Community Chest	0.01763	0.01775			
3	Turmstr.	Baltic Avenue	0.02040	0.02037	450	9	14
4	Einkommenssteuer	Income Tax	0.02210	0.02193			
5	Südbahnhof	Reading Railroad	0.02686	0.02801	200	6	
6	Chausseestr.	Oriental Avenue	0.02169	0.02132	550	12	
7	Ereignisfeld	Chance	0.00972	0.00815			
8	Elisenstr.	Vermont Avenue	0.02246	0.02187	550	12	
9	Poststr.	Connecticut Avenue	0.02217	0.02168	600	13	37
10	Nur zu Besuch	Just Visiting (Jail)	0.02184	0.02139			
11	Seestr.	St. Charles Place	0.02596	0.02556	750	19	
12	Elektrizitätswerk	Electric Company	0.02378	0.02614	70	2	
13	Hafenstr.	States Avenue	0.02213	0.02174	760	16	
14	Neue Str.	Virginia Avenue	0.02457	0.02426	900	22	57
15	Westbahnhof	Pennsylvania Railroad	0.02531	0.02635	200	5	
16	Münchener Str.	St. James Place	0.02703	0.02680	950	25	
17	Gemeinschaftsfeld	Community Chest	0.02306	0.02296			
18	Wiener Str.	Tennessee Avenue	0.02821	0.02821	950	27	
19	Berliner Str.	New York Avenue	0.02794	0.02812	1000	28	80
20	Frei parken	Free Parking	0.02806	0.02825			
21	Theaterstr.	Kentucky Avenue	0.02594	0.02614	1050	27	
22	Ereignisfeld	Chance	0.01209	0.01045			
23	Museumsstr.	Indiana Avenue	0.02549	0.02567	1050	27	
24	Opernplatz	Illinois Avenue	0.02983	0.02993	1100	33	87
25	Nordbahnhof	B & O Railroad	0.02718	0.02893	200	6	
26	Lessingstr.	Atlantic Avenue	0.02540	0.02537	1150	29	
27	Schillerstr.	Ventnor Avenue	0.02521	0.02519	1150	29	
28	Wasserwerk	Water Works	0.02480	0.02651	70	2	4
29	Goethestr.	Marvin Gardens	0.02441	0.02438	1200	29	87
30	Gefängnis	Go to Jail	0.09422	0.09457			
31	Rathausplatz	Pacific Avenue	0.02501	0.02524	1275	32	
32	Hauptstr.	North Carolina Ave.	0.02438	0.02472	1275	32	
33	Gemeinschaftsfeld	Community Chest	0.02193	0.02228			
34	Bahnhofstr.	Pennsylvania Ave.	0.02312	0.02353	1400	33	97
35	Hauptbahnhof	Short Line Railroad	0.02243	0.02291	200	5	21
36	Ereignisfeld	Chance	0.00934	0.00816			
37	Parkstr.	Park Place	0.02023	0.02060	1500	31	
38	Zusatzsteuer	Luxury Tax	0.02023	0.02052			
39	Schloßallee	Boardwalk	0.02457	0.02483	2000	50	81

Table 16.3. Sojourn probabilities and rents in Monopoly. The maximum rents relate to a monopoly, with hotels for regular properties and a roll of 7 for the utilities. The total expectations are given to the right of the last property of each group.

that appertain to each square. The player can thus perhaps purchase two properties or have to pay rent twice. For this reason, one constructs a Markov chain in which a transition represents the effect of a single throw. Any intermediate stops on a Chance or Community Chest square do not have to be explicitly accounted for. Thus if the player draws a card such as "Take a walk on the Boardwalk," the transfer to that square can be viewed as a transition in conjunction with the actual dice throw, without changing the rent expectations.

If a transition within a Markov chain always comprises exactly the effect of a single roll of the dice, then the current state must contain the information necessary for the next move to be taken according to the rules of the game. Thus in addition to the current location, a state should tell whether this square was reached by doubles, or indeed two sets of doubles, having been thrown. Thus each square corresponds to three states: reached without doubles, reached on one throw of doubles, reached on two throws of doubles. There are also three states associated with "in Jail," since one has three chances to roll out of Jail with doubles.[4]

Strictly speaking, the states that we have described need to be further subdivided. The reason for this is the several "transfer" cards among the Chance and Community Chest cards that require the game piece to move to a particular location. If any of these have already been removed from the deck, then the transition probabilities for the affected squares are slightly altered. However, without introducing too large an error, we may assume that cards are always drawn from a complete, well-shuffled deck.

In order to calculate the sojourn probabilities for the individual squares, we must investigate a Markov chain with $3 \times 40 = 120$ states, which would be a hopeless enterprise without a computer. The probabilities of the state transitions depend on the probabilities of the various dice rolls and the special cases such as rolling doubles and the instructions of the transfer cards (see Note 3 at the end of the chapter). The natural division of a transition into the actual dice roll and the subsequent transfer are best folded into the calculations. To accelerate the iteration, one may start with a probability distribution that approximates the expected outcome, for example, $3/42$ for Jail, and $1/42$ for each of the other squares. The resulting probabilities for the individual squares are collected in Table 16.3. Since the composition of the Chance and Community Chest cards differs in different versions of the game, we offer two variants, one from the German edition, the other from the American.[5]

The sojourn probabilities for the various squares exhibit a rather large range. In the "street-name" properties in the American edition, they range

---

[4]Since one can collect rent while one is in Jail, but is safe from having to pay any, in the end phase of the game it is worthwhile to remain in Jail as long as possible. Thus one should not immediately purchase one's freedom.

[5]The results for the American edition have been published in several venues: Robert B. Ash, Richard L. Bishop, Monopoly as a Markov process, *Mathematics Magazine* **45**, 1972, pp. 26–29. Bishop compares, among other things, how the two Jail strategies—remain as long as possible or get out at once—affect a player's finances. An extensive version of this article has appeared in which minor errors in the journal article have been corrected. Irvin R. Hentzel, How to win at Monopoly, *Saturday Review of Sciences* April 1973, pp. 44–48. Dr. Crypton, How to win at Monopoly, *Science Digest* September 1985, pp. 66–71. Hentzel's results also appear in the book by Maxine Brady cited earlier. The probabilities for the individual squares are given in the book by Orbanes cited earlier.

It is reasonable to consider the probabilities of three doubles in a row as approximately equal for each square, with the exception of Jail. This corresponds to a Markov chain of 42 states, whose results are only slightly imprecise. See Steve Abott, Matt Richey, Take a walk on the Boardwalk, *The College Mathematical Journal* **28**, 1997, pp. 162–171. Even greater simplifications are made by Ian Stewart in his two articles, How fair is Monopoly? *Scientific American* **274**, 1996, pp. 86–87; Monopoly revisted, *Scientific American* **275**, 1996, pp. 92–95; see also Feedback, *Scientific American* **277**, 1997, p. 104.

from 0.02060 for Park Place to 0.02993 for Illinois Avenue, which represents a relative difference of 45%. A hotel on Illinois Avenue thus has a higher rent expectation than a hotel on Park Place, namely, 656 German marks versus 607 marks. In general, the squares between Jail and the "Go to Jail" square have a relatively high sojourn probability. There are particularly high probabilities for squares that can be reached from Jail by a throw of doubles or one of the more likely dice throws. In the special case of Illinois Avenue, which is 14 squares beyond Jail and therefore frequently reached in two turns from Jail, there is also the Chance card that advances the player to Illinois Avenue.

How are the calculated rent expectations to be interpreted? How can they be used to optimize decsions such as whether to buy, sell, or build on a property? Of course, due to the complexity of the game as played with a number of players, and given the goal of the financial ruin of one's opponents, we can make only general statements. In this we must distinguish the benefits that accrue from an investment in a property according to the phase of the game:

- in the early phases, when the first houses are being built, the amount of capital available to the players is generally small. A high priority of each player is to maintain a degree of liquidity. A potential investment must be weighed in light of how the long-term rent expectation can be maximized given the present budget and the amount of capital expected to be available in the near term. For example, building a house is evaluated according to its expected return, that is, according to how quickly the expected receipts of rent will amortize the building costs.

- in the later stages of the game, when more money is in play, it is of greater urgency to try to ruin one's opponents financially. One-time costs, such as those paid to the bank for the construction of houses, are of small significance compared to the continual receipt of income from rent. Therefore, investment is evaluated on the basis of projected income, that is, the absolute expectation of rents. In particular, hotels are built whenever possible.[6]

Table 16.4 contains, for eight groups of properties, both the absolute rent expectations and the percentage return for additional houses. All values relate to a single turn and therefore to 1.1869 dice throws on average.[7]

---

[6]An exception is the possibility of keeping four houses on a property to limit the number of available houses, thereby blocking an opponent from building.

[7]Within the Markov chain, the first throws of a turn can be localized as the transitions that begin at the 39 states arrived at without throwing doubles or to one of the three Jail states. These represent a portion of 0.8425 in the stationary state distribution.

	Rent Expectation: Hotel	Profit for Each Additional House: Percent Per Turn				
		1st	2nd	3rd	4th	5th
Purple	17	0.4	1.4	4.3	5.1	5.3
Light Blue	44	1.0	3.1	9.6	7.0	7.7
Maroon	68	0.9	3.0	8.7	5.2	4.3
Orange	95	1.4	4.4	11.8	6.6	6.6
Red	104	1.2	3.8	9.7	3.8	3.8
Yellow	104	1.3	4.5	9.4	3.5	3.5
Green	115	1.2	4.0	7.6	2.9	2.7
Dark Blue	96	1.4	4.9	9.6	3.4	3.4

Table 16.4. The rent expectation for various property groups and the percentage profit in building an additional house.

If one has the good fortune to be choosing among several building opportunities, then the ordering of profit margins in Table 16.5 might be of use.[8]

A Brief Primer on Markov Chains

The mathematical properties of Markov chains can be formulated most concisely in the language of matrices. We start with a square matrix A containing all the transition probabilities. The ith row contains the probabilities of how the system will develop from the ith state. In particular, the sum of the coefficients of each row is 1. If the state distributions of the row vectors are written as p, then the system of transition equations can be written

$$p' = pA.$$

---

[8]That the cited book by Maxine Brady and the article in *Spielbox* **4**, 1983, pp. 40 ff., give different orderings is due principally to the fact that there, the purchase price of the property is considered in the investment costs. Since in building a house, rights of possession are already available, rights are sometimes auctioned off at variable prices, and at times may be acquired strictly for strategic purposes, such as blocking another player, we have not followed that approach here. For example, if a mortgage has to be obtained in order to build, then those costs are to be included in the calculation.

Furthermore, Brady figures the average rent for the properties instead of the sum of the rents in relation to the total costs. This makes the groups of two properties, namely, purple and dark blue, seem more favorable than they really are.

| Investment | | Profit |
Color	Houses	(% per Turn)
Orange	1 to 5	6.2
Light Blue	1 to 5	5.7
Dark Blue	1 to 3	5.3
Yellow	1 to 3	5.1
Red	1 to 3	4.9
Maroon	1 to 5	4.4
Green	1 to 3	4.3
Red	4 to 5	3.8
Yellow	4 to 5	3.8
Dark Blue	4 to 5	3.4
Purple	1 to 5	3.3
Green	4 to 5	2.8

Table 16.5. Comparison of profits: where one should build first.

In further steps, the Markov chain develops to the state distributions $(pA)A = pA^2$, pA^3, and so on.

An important object in the study of Markov chains is the search for stationary state distributions, that is, distributions p with $pA = p$. Stationary state distributions always exist, though they are not always uniquely determined, as in the case of the ruin problem. Nevertheless, if there is a certain number of steps in which each state can be reached from every other state—such a Markov chain is called *regular*—then there exists precisely one stationary distribution, and it is the limiting distribution from any initial distribution. The Markov chain for Monopoly is regular, since every state can be reached from every other in three rolls. It is not hard to believe that this is so. But anyone who doubts may compute the 120×120 matrix A^3 and check that each entry is greater than zero.

Markov chains in which each state is reachable from every other state in some finite number of steps are called *irreducible*. Irreducible Markov chains are not necessarily regular, as the example

$$\begin{pmatrix} 0 & 1 \\ 1 & 0 \end{pmatrix}$$

demonstrates: the first state can be reached only in an even number of steps, while the second can be reached only in an odd number. The *period* of a state is defined to be the greatest common divisor of all numbers of steps in which one can leave the state and then return to it. In the case of irreducible Markov chains, the periods of all states are equal. If they are equal to 1, then the Markov chain is regular.

A state of a Markov chain is said to be *absorbing* if it cannot be exited. An example is the last square in the snakes and ladders game. If a Markov chain has at least one absorbing state that can eventually be reached from every nonabsorbing state, then the entire Markov chain is said to be absorbing. For example, the ruin problem can be modeled as an absorbing Markov chain whose two ruination states are absorbing. If one lists the absorbing states first, then the transition matrix assumes the following block form, where I is the identity matrix:

$$A = \begin{pmatrix} I & 0 \\ R & Q \end{pmatrix}.$$

One obtains information about the long-term behavior of an absorbing Markov chain by multiplying the block form of the matrix A repeatedly by itself:

$$A^n = \begin{pmatrix} I & 0 \\ \left(I + Q + \cdots + Q^{n-1}\right) R & Q^n \end{pmatrix} \to \begin{pmatrix} I & 0 \\ (I - Q)^{-1} R & 0 \end{pmatrix}.$$

In particular, using the limiting value of A^n, which in the ruin problem we obtained by other means, one can determine which absorbing states will be reached from a particular start state with what probabilities. How long this will take on average can be determined from the matrix $(I - Q)^{-1}$. If ℓ is the column vector whose coordinates equal the expected number of steps until an absorbing state is reached, then just as in the ruin problem, one obtains the equality

$$\ell = Q\ell + \begin{pmatrix} 1 \\ \vdots \\ 1 \end{pmatrix},$$

which can be transformed into

$$\ell = (I - Q)^{-1} + \begin{pmatrix} 1 \\ \vdots \\ 1 \end{pmatrix}.$$

Chapter Notes

1. Since $p + q = 1$, we first obtain

$$q\big(r(k-1) - r(k)\big) = p\big(r(k) - r(k+1)\big),$$

so that one may calculate the difference of two successive ruin probabilities recursively from $r(n-1) - r(n) = r(n-1)$:

$$r(k-1) - r(k) = s^{n-k} r(n-1).$$

If one now replaces k by $k+1, k+2, \ldots, n$ and then adds both sides of the corresponding equations, one obtains

$$r(k) - r(n) = \left(1 + s + s^2 + \cdots + s^{n-k-1}\right) r(n-1).$$

Since $r(n) = 1$ and $r(0) = 0$, one obtains the given formula, which can be simplified by distinguishing two cases:

$$r(k) = \begin{cases} \dfrac{n-k}{n} & \text{if } s = 1, \\ \dfrac{s^{n-k}-1}{s-1} & \text{if } s \neq 0. \end{cases}$$

2. For $p \neq q$, the equations of the system can be transformed to yield

$$\ell(k-1) - \ell(k) + \frac{1}{q-p} = s\left(\ell(k) - \ell(k+1) + \frac{1}{q-p}\right).$$

As in the case of the ruin probabilities, the values $\ell(1), \ldots, \ell(n-2)$ can be calculated from $\ell(n-1)$. Using the equation

$$p\,\ell(1) = q\,\ell(n-1) = n-1,$$

which is obtained by summing all the equations, one finally obtains the formula given in the sidebar. The special case $p \neq q$ can be solved by taking limits, say, by using L'Hôpital's rule twice.

3. The transfer cards are different for the different German editions. For the American edition they are as follows: there are Community Chest cards sending the player to Go and to Jail, and Chance cards that transfer the player to Go, Jail, Reading Railroad, St. Charles Place, Illinois Avenue, Boardwalk, the next railroad ($2\times$), the next utility, and three squares backward.

Further Literature on Markov Chains

[1] John T. Baldwin, On Markov chains in elementary mathematics courses, *American Mathematical Monthly* **96**, 1989, pp. 147–153.

[2] J. G. Kemeny, J. L. Snell, *Finite Markov Chains*, New York 1960.

17

Blackjack: A Las Vegas Fairy Tale

In gambling casinos, blackjack is considered the game with the best odds. It is even maintained that there exist strategies that make it possible for the player to beat the house. Is that possible?

The goal of blackjack is to draw cards until one has as high a sum of card points as possible that does not exceed 21. If 21 points are achieved with only two cards, then one has obtained a *blackjack*. This combination, of ace and either a 10 or a picture card, trumps any other combination of 21 points. In American casinos, in which blackjack has been played since 1920, the game enjoys an enormous popularity. Blackjack is offered in most European casinos as well.

In casinos, blackjack is a game played against the bank. That is, one plays against a casino employee, the dealer. In general, up to seven players can attempt simultaneously to draw a combination that beats the bank. Here is how the game is played: first, each player places a bet within a set limit. In the game itself, the dealer draws cards from a deck and turns them over: first one card for each player, then one for the dealer, and then again one for each player.[1] The players can ask for further cards if they wish, according to how they judge their own chances against those of the bank, as indicated by the single visible card. A player who "goes bust," that

---

[1] In American casinos it is usual that the dealer also takes a second card. This card is provisionally hidden from the players unless the dealer has obtained a blackjack. In comparison to the European variant, the players thereby obtain a bit more information: if the dealer, having an uncovered 10 or ace, does not at once show his cards, then the players know that the dealer has not obtained a blackjack.

is, exceeds 21, loses at once. When no players want additional cards, the dealer draws for the bank. The bank has no choice when to stop drawing cards. It must draw at least to 16, and must stop upon reaching 17 or more. If an ace is drawn, it must be counted as 11 if the result would be in excess of 21. Once the bank has finished drawing, each player who has not gone over 21 computes his or her score. A player whose total exceeds that of the bank gets his bet plus the amount of his bet. If the win occurs with a blackjack, then he gets his bet plus one and one-half times his bet. If it is a tie, the player gets back his bet. If the bank has the better hand, the bet is lost.

In contrast to roulette, in blackjack, the players have considerable strategic influence over the course of the game in deciding whether to take additional cards. To make the game more interesting, there are several additional rules.

- **Insurance.** If the bank's first card is an ace, then the bank will win with relatively high probability, namely, by obtaining a blackjack with any of 4 out of the 13 card values. To protect himself against the impending catastrophe, players can take out insurance against a blackjack. By placing an additional bet, half the size of the original amount, a player receives his original bet plus the supplement in the event that the bank draws a blackjack. If there is no blackjack, the insurance passes to the bank, while the original bet is calculated as usual.

- **Doubling down.** If the first two cards yield the sum of 9, 10, or 11 points, where an ace if present can be counted as 1, then the player is permitted to double his bet, with the proviso that he can draw only one more card.

- **Splitting.** If the first two cards have the same value, the player can break his hand in two, where an additional bet must be placed for the new hand. That is, the player draws independently for the two hands. However, a blackjack is counted only as a normal 21, and only one further card may be dealt to a split ace. Moreover, in many casinos, multiple splitting or doubling after splitting is not permitted.

The way the cards are dealt is of great importance. The deck is not reshuffled for each game, but instead, a number of decks, usually six, of 52 cards each are shuffled together and about one-fifth of the pack is divided off by a blank card. The game is played with this stack until the blank card is encountered. After the end of the game in which that card appears, the pack is reshuffled.

Blackjack is practically symmetric for the players and the bank, and is therefore more or less a fair game. Furthermore, the asymmetric aspects of blackjack tend to favor the player, which explains the popularity of the game:

- the player wins one and one-half times the original bet in the case of a blackjack.

- the bank must follow a prescribed course in dealing itself cards.

- the player knows the bank's first card.

- the bank is permitted neither to split nor to double.

The only, at first almost unnoticeable but therefore the more important, advantage to the bank is that the bank wins whenever a player goes bust, even if the bank also goes bust. It would seem, then, that a player should play more defensively than the bank. A good strategy should be oriented to the bank's first card, since it contains significant information about the course that the bank's hand will likely take.

A mathematical analysis of blackjack should begin with the bank, whose results can be determined in the form of a probability distribution. In the simplest case, one assumes that the probabilities for the individual card values are a constant $1/13$, except for the cards of value 10, whose probability is $4/13$. Such a supposition is of course true only for a stack of cards of infinite size, since the probabilities change as the cards are dealt out. However, the supposition will do for developing a fixed strategy that will approximate the optimal strategy on average.

Although it can happen that the bank will have to draw 12 cards, namely, six aces, a 6, and then five aces, the number is usually much smaller, seldom more than four. One can model the progression of cards drawn as a Markov chain, where the states correspond to the intermediate results. In addition to the special case of a blackjack, it is also necessary to consider *soft hands*, which are hands with an ace counted as 11, which can be viewed as a special case. Table 17.1 shows the end distribution for the bank.

It can be seen at once the great risk that the bank is taking, since in more than one in every four hands it draws over 21 points. The results of this table suffice for calculating the odds of a player who copies the bank's strategy: if a player plays like the bank, without ever splitting or doubling, until at least 17 is reached, then that player's probabilities are the same as those in the table. If the rules for winning were the same for the bank as for the player, then the player's expectation would be 0: to

Result	Probability
17	0.1451
18	0.1395
19	0.1335
20	0.1803
21	0.0727
blackjack	0.0473
bust	0.2816

Table 17.1. Probabilities for the bank in blackjack.

Situation	Advantage	Probability	Expectation
Player has blackjack, bank does not	0.5	0.0451	0.0225
Both player and bank go bust	−1.0	0.0793	−0.0793

Table 17.2. Asymmetries in winning and their effects when the player copies the bank's strategy.

break even. However, since the winning rules are somewhat different, some small correction needs to be made, and these are tabulated in Table 17.2, showing that in sum, the player who follows the bank's strategy has an average loss of 5.68% of the amount bet.

We have already mentioned that it hardly makes sense to copy the bank's strategy. In particular, in deciding how to play, a player should have a look at the bank's first card. In order to calculate his win expectation, the player should next determine the probability distributions that arise for the bank conditioned on that first card. This can best be done by calculating these conditional probabilities iteratively for arbitrary intermediate states. Only a single iteration is necessary, one that is realizable with a spreadsheet calculation.[2] The results are presented in Table 17.3.

The second step consists in investigating the player's profit and loss expectation if he decides not to draw upon reaching a certain number of points. Again, the probabilities are conditioned on the bank's first card. The random variables can assume only the values −1, 0, 1, and 3/2, corresponding to what the player can win, where their probability distribution comes directly from Table 17.3. This leads to the winning expectations tabulated in Table 17.4.

---

[2]If one places the special cases such as blackjack, soft hand, double, and split hands cleverly into the table, one can actually fairly easily incorporate the entire blackjack calculation presented here in a single spreadsheet.

Player	Bank 2	3	4	5	6	7	8	9	10	Ace
17	0.1398	0.1350	0.1305	0.1223	0.1654	0.3686	0.1286	0.1200	0.1114	0.1308
18	0.1349	0.1305	0.1259	0.1223	0.1063	0.1378	0.3593	0.1200	0.1114	0.1308
19	0.1297	0.1256	0.1214	0.1177	0.1063	0.0786	0.1286	0.3508	0.1114	0.1308
20	0.1240	0.1203	0.1165	0.1131	0.1017	0.0786	0.0694	0.1200	0.3422	0.1308
21	0.1180	0.1147	0.1112	0.1082	0.0972	0.0741	0.0694	0.0608	0.0345	0.0539
BJ	0.0000	0.0000	0.0000	0.0000	0.0000	0.0000	0.0000	0.0000	0.0769	0.3077
Bust	0.3536	0.3739	0.3945	0.4164	0.4232	0.2623	0.2447	0.2284	0.2121	0.1153

Table 17.3. Probabilities for the bank's results conditioned on the bank's first card.

Player	Bank 2	3	4	5	6	7	8	9	10	Ace
BJ	1.5000	1.5000	1.5000	1.5000	1.5000	1.5000	1.5000	1.5000	1.3846	1.0385
21	0.8820	0.8853	0.8888	0.8918	0.9028	0.9259	0.9306	0.9392	0.8117	0.3307
20	0.6400	0.6503	0.6610	0.6704	0.7040	0.7732	0.7918	0.7584	0.4350	0.1461
19	0.3863	0.4044	0.4232	0.4395	0.4960	0.6160	0.5939	0.2876	−0.0187	−0.1155
18	0.1217	0.1483	0.1759	0.1996	0.2834	0.3996	0.1060	−0.1832	−0.2415	−0.3771
17	−0.1530	−0.1172	−0.0806	−0.0449	0.0117	−0.1068	−0.3820	−0.4232	−0.4644	−0.6386
≤ 16	−0.2928	−0.2523	−0.2111	−0.1672	−0.1537	−0.4754	−0.5105	−0.5431	−0.5758	−0.7694

Table 17.4. Expectations, conditioned on the bank's first card, for the player's winnings (minus the amount bet), when the player takes no more cards.

With the aid of Table 17.4, a player can estimate his chances when he must decide whether to draw another card. The drawing strategy is optimized in the reverse direction of the chronological order of play; that is, one begins with the high-valued hands and then optimizes the strategy recursively step by step. The already optimized expectations then enter the calculation when the expectation of drawing to a lower-valued hand is calculated. This is compared to the expectation from not drawing a card. The higher of the two values is the maximum achievable expectation: the horizontal lines in Table 17.5 indicate the limits on drawing: in the upper region, that is, for the situations in which one does not draw, the winning expectations are the same as those in Table 17.4.

An examination of Table 17.5 shows that the player must play relatively defensively:

- against the bank's 4, 5, or 6, the player should draw only to 11; that is, he should draw only as long as drawing poses absolutely no risk.

- against the bank's 2 or 3, the player should pass at 13 and above.

- against higher cards, from 7 to ace, the player should draw to 17.

This optimal defensive strategy becomes plausible when one considers that if the bank starts with 6, it will frequently draw to 16 and then go bust. One can see the details in Table 17.3, where the conditional bank distributions are given.

Soft hands, that is, hands containing an ace valued at 11, must be investigated separately. Clearly, such hands provide greater flexibility in

Player	Bank 2	3	4	5	6	7	8	9	10	Ace
19	0.3863	0.4044	0.4232	0.4395	0.4960	0.6160	0.5939	0.2876	−0.0187	−0.1155
18	0.1217	0.1483	0.1759	0.1996	0.2834	0.3996	0.1060	−0.1832	−0.2415	−0.3771
17	−0.1530	−0.1172	−0.0806	−0.0449	0.0117	−0.1068	−0.3820	−0.4232	−0.4644	−0.6386
16	−0.2928	−0.2523	−0.2111	−0.1672	−0.1537	−0.4148	−0.4584	−0.5093	−0.5752	−0.6657
15	−0.2928	−0.2523	−0.2111	−0.1672	−0.1537	−0.3698	−0.4168	−0.4716	−0.5425	−0.6400
14	−0.2928	−0.2523	−0.2111	−0.1672	−0.1537	−0.3213	−0.3719	−0.4309	−0.5074	−0.6123
13	−0.2928	−0.2523	−0.2111	−0.1672	−0.1537	−0.2691	−0.3236	−0.3872	−0.4695	−0.5825
12	−0.2534	−0.2337	−0.2111	−0.1672	−0.1537	−0.2128	−0.2716	−0.3400	−0.4287	−0.5504
11	0.2384	0.2603	0.2830	0.3073	0.3337	0.2921	0.2300	0.1583	0.0334	−0.2087

Table 17.5. Expectations, conditioned on the bank's first card, for the player's winnings (minus the amount bet), when the player draws optimally.

Player	Bank 2	3	4	5	6	7	8	9	10	Ace
19s	0.3863	0.4044	0.4232	0.4395	0.4960	0.6160	0.5939	0.2876	−0.0187	−0.1155
18s	0.1217	0.1483	0.1759	0.1996	0.2834	0.3996	0.1060	−0.1007	−0.2097	−0.3720
17s	−0.0005	0.0290	0.0593	0.0912	0.1281	0.0538	−0.0729	−0.1498	−0.2586	−0.4320
16s	−0.0210	0.0091	0.0400	0.0734	0.0988	−0.0049	−0.0668	−0.1486	−0.2684	−0.4224
15s	−0.0001	0.0292	0.0593	0.0920	0.1182	0.0370	−0.0271	−0.1122	−0.2373	−0.3977
14s	0.0224	0.0508	0.0801	0.1119	0.1392	0.0795	0.0133	−0.0752	−0.2057	−0.3727
13s	0.0466	0.0741	0.1025	0.1344	0.1617	0.1224	0.0541	−0.0377	−0.1737	−0.3474
12s	0.0818	0.1035	0.1266	0.1565	0.1860	0.1655	0.0951	0.0001	−0.1415	−0.3219

Table 17.6. Expectations, conditioned on the bank's first card, for the player's winnings (minus the amount bet), when the player draws optimally from a soft hand.

deciding whether to draw another card, since one cannot go bust from the next card. Table 17.6 contains the relevant data for soft hands. Since additional cards can turn a soft hand into a normal hand, the table is in part identical to the previous table.

Continuing in reverse chronological order, one can calculate the expectations for hands down to 10 points. For hands that have other special features such as those consisting of a pair of like-valued cards or a single card, these values constitute only a lower bound for the winning expectations (see Table 17.7).

If we now consider the possibility of splitting or doubling, we finally obtain by the same methods the associated total expectation, which indicates an average loss of 2.42%. In addition, in an intermediate step not shown, it is necessary to determine the expectations of hands containing a single card; however, without doubling and splitting, it suffices to consider only 10 and ace separately (see Table 17.8).

The loss of about 2.42% is greater than that in roulette when one bets on a color or makes some other simple wager. However, the average loss at blackjack can be further reduced if one splits and doubles. We begin with an optimization of doubling. Again, there are two win expectations to compare: with and without doubling. Without doubling, we have the tabulated expectations for hands of 9, 10, and 11, as well the soft hands 19s and 20s. With doubling, we need to determine the expectation when these hands draw exactly one card, and this value is then multiplied by 2.

Player	Bank 2	3	4	5	6	7	8	9	10	Ace
10	0.1825	0.2061	0.2035	0.2563	0.2878	0.2569	0.1980	0.1165	−0.0536	−0.2513
9	0.0744	0.1013	0.1290	0.1580	0.1960	0.1719	0.0984	−0.0522	−0.2181	−0.3532
8	−0.0218	0.0080	0.0388	0.0708	0.1150	0.0822	−0.0599	−0.2102	−0.3071	−0.4441
7	−0.1092	−0.0766	−0.0430	−0.0073	0.0292	−0.0688	−0.2106	−0.2854	−0.3714	−0.5224
6	−0.1408	−0.1073	−0.0729	−0.0349	−0.0130	−0.1519	−0.2172	−0.2926	−0.3887	−0.5183
5	−0.1282	−0.0953	−0.0615	−0.0240	−0.0012	−0.1194	−0.1881	−0.2666	−0.3662	−0.5006
4	−0.1149	−0.0826	−0.0494	−0.0124	0.0111	−0.0883	−0.1593	−0.2407	−0.3439	−0.4829

Table 17.7. Expectations, conditioned on the bank's first card, for the player's winnings (minus the amount bet), when the player plays optimally.

Bank	2	3	4	5	6	7	8	9	10	Ace	Total
Expectation	0.0664	0.0938	0.1221	0.1530	0.1827	0.1215	0.0440	−0.0477	−0.1779	−0.3389	−0.0242

Table 17.8. Expectations, conditioned on the bank's first card and absolute, for the player's winnings (minus the amount bet), when the player plays optimally (without doubling or splitting).

Doubling should then be done in those situations that lie above the line in Table 17.9, since it is those for which the tabulated expectation exceeds the ordinary expectation, namely, that obtained by normal drawing. However, this does not hold for soft hands, which with normal drawing always have a greater expectation than that afforded by doubling.

It remains to optimize the splitting strategy. That is, for the case of two like-valued initial cards, we must determine under what conditions it is advantageous to split the hand. Since splitting rules vary from house to house, we are forced to consider several variants:

- after a split, the player may neither double nor split.

- splitting may be done several times, but split hands may not later be doubled.

- split hands may be doubled and further split.

Clearly, the last of these cases is the most liberal for the player, since it offers the greatest number of options. And in fact, it will turn out that there are cases in which splitting is advantageous only if the hand can be later doubled. On the other hand, the first two cases differ only in the win expectations, not in the optimal strategy. That is, if it is advantageous to split, then it remains so if later the same option is available.

The actual calculation proceeds in three phases, starting with results already obtained. First, the winning expectation is calculated for each pair of cards and each bank card for the case of splitting. Depending on the rule in force, it may be necessary to consider the case in which the next card can also be split or the case in which doubling is permitted. Then these results are compared with the win expectations for the optimal drawing strategy.

Player	Bank 2	3	4	5	6	7	8	9	10	Ace
11	0.4706	0.5178	0.5660	0.6147	0.6674	0.4629	0.3507	0.2278	0.0120	−0.5399
10	0.3589	0.4093	0.4609	0.5125	0.5756	0.3924	0.2866	0.1443	−0.1618	−0.6251
9	0.0622	0.1208	0.1819	0.2431	0.3171	0.1043	−0.0264	−0.3010	−0.5847	−0.9151

Table 17.9. Expectations, conditioned on the bank's first card, for the player's winnings (minus the amount bet), when the player draws exactly one more card.

Bank	2	3	4	5	6	7	8	9	10	Ace	Total
Only once	0.0891	0.1201	0.1529	0.1884	0.2232	0.1419	0.0574	−0.0409	−0.1770	−0.3389	−0.00883
Repeated	0.0903	0.1214	0.1543	0.1900	0.2252	0.1438	0.0588	−0.0399	−0.1763	−0.3389	−0.00772
Unrestricted	0.0918	0.1240	0.1576	0.1938	0.2295	0.1451	0.0592	−0.0397	−0.1763	−0.3389	−0.00639

Table 17.10. Player's expectation (minus the amount bet), for optimal play, depending on the splitting rule in force.

This yields the optimal splitting strategy. If one then wishes to calculate the total win expectation (see Table 17.10), then that is best done by way of the intermediate steps of hands consisting of a single card.

The optimal splitting strategy can be read from Table 17.11. The letter "S" indicates that splitting is advantageous, while "(S)" means that splitting is advantageous only if the next card is permitted to be doubled.

Depending on the variant of the rules, then, the average loss at blackjack can be brought down to the range 0.64% to 0.88% of the initial wager. This is much less than the 2.42% without splitting or doubling. Moreover, the loss is less than that for simple bets in roulette, where the number is $1/74 = 1.35\%$. However, and this is an important distinction, unlike roulette, blackjack offers these odds only to a skillful player. Bad play is

	2	3	4	5	6	7	8	9	10	Ace
Ace, Ace	S	S	S	S	S	S	S	S	S	
10, 10										
9, 9	S	S	S	S	S		S	S		
8, 8	S	S	S	S	S	S	S	S		
7, 7	S	S	S	S	S	S				
6, 6	(S)	S	S	S	S					
5, 5										
4, 4				(S)	(S)					
3, 3	(S)	(S)	S	S	S	S				
2, 2	(S)	(S)	S	S	S	S				

Table 17.11. When splitting is advantageous.

costly, and that includes insurance against blackjack, which is never good strategy. And even that obviously unfavorable option is observed over and over again in casinos.

The first extensive mathematical analysis of blackjack was carried out in 1956 by the Americans R. Baldwin, W. Cantey, H. Maisel, and J. Mc-Dermott.[3] They investigated the version of the game then current in the United States, which differs from that discussed here in a few details. These researchers came up with an average loss of 0.6%. Their optimal strategy was much more defensive than that previously extolled by gaming experts. In their publication, they refer, for example, to a recommendation of Culbertson and others to draw up to 13 or 15, according to whether or not the bank's card is in the range from 2 to 6.

Baldwin and his colleagues began, as we did, with the assumption of constant probabilities for the individual card values, namely, 1/13 and 4/13. They did not consider the information available to the player about the cards already played from the deck. We may ask, then, whether that knowledge could significantly increase the player's winning expectation. It was just this idea that motivated the young mathematician Edward Thorp after he read Baldwin's work. In fact, Thorp discovered certain situations, for example, when all 5s have been played, that are extremely advantageous to the player who follows a particular strategy. Indeed, the player can expect about a 3.3% profit. Since in those days, the game was played with a single deck, such a situation obtained in about 3.5 to 10 percent of hands. If one were to raise one's bet in such situations, one could reverse the bank's advantage. In theory, the bank could be beaten.

What happened then is something that occurs seldom in the world of mathematics: after Thorp made his results known at a meeting of the American Mathematical Society[4] a media frenzy ensued,[5] whose echo reached across the ocean to Europe. Of course, the mathematical theorem had to be put to the test at the casino. Thorp made a killing on his best-selling book, which sold over 500,000 copies to those who wanted to profit from his discoveries.[6] There, strategies were described that with much practice could give the player a significant advantage over the bank. The basic idea, which turns out to be a usable strategy in practice, is to use a

[3]R. Baldwin, W. Cantey, H. Maisel und J. Mc Dermott, The optimum strategy in blackjack, *Journal of the American Statistical Association* **51**, 1956, pp. 429–439.

[4]E. Thorp, A favorable strategy for twenty-one, *Proceedings of the National Academy of Sciences of the USA* **47**, 1961, pp. 110–112.

[5]E. Thorp, A prof beats the gamblers, *The Atlantic Monthly* June 1962, pp. 41–45; How to beat the game, *Scientific American* 1961/4, p. 84; P. O. Niel, A professor who breaks the bank, *Life* April 20, 1964, pp. 66–72; 17+4, Formel des Glücks, *Der Spiegel* **18**, 1964, pp. 127–131.

[6]E. Thorp, *Beat the Dealer*, New York 1962 (revised edition, 1966).

counting system to come up with an approximate, but sufficient, overview of the cards yet to be played.[7] The method associates a *weight* with each card value, which for Thorp's high–low system is

- +1 for 2 through 6;

- −1 for 10, face cards, and ace;

- 0 for the remainder.

After the cards are shuffled, one begins to add the values of all cards played. Depending on the total weight, generally called the *count*, and the number of cards remaining, one can vary one's strategy advantageously.

Blackjack is the first and only casino game to be "broken." And today? Blackjack continues to be played in casinos. Thanks to Thorp and other blackjack experts, as well as an inexhaustible supply of publications on the subject,[8] blackjack is more popular than ever. There are two reasons that the casino owners are not too worried about losses at the blackjack tables: first of all, blackjack is now played with several, usually six, decks of cards, from which about 80 cards are exempted by means of the blank card. Thus great imbalances among the various card values are to a large extent avoided. Second, counting strategies require a significant investment in practice and concentration, and indeed, each incorrect decision in the course of the rapid play goes against the player. Only those who continually count correctly and adapt their strategies accordingly can hope to turn their tiny advantage against the bank to account. Successful card counters are easily lost in the sea of other average players using an optimal strategy. That the minimal advantage is only an expectation and can come to nothing due to simple bad luck need hardly be mentioned.

We shall demonstrate, using the high–low system, how counting systems can be derived. We start with a blackjack calculation that must be generalized from what we have calculated thus far. The assumed equal distribution of the individual card values will be replaced with an arbitrary probability distribution. We shall still not take into account that the probabilities change in the course of the game; the error thus involved can be corrected later. What we require first is a sensitivity analysis; that is, we investigate how greatly the results change if the card probabilities are slightly increased or decreased. This is particularly quick and easy using a spreadsheet calculation for the case of a blackjack. Corresponding

---

[7]If the use of a computer was allowed in casinos, then one could simply input the values of the cards played. Indeed, the results of a spreadsheet calculation would suffice to present the current expectation as well as the associated strategy.

[8]See the list of literature at the end of the chapter.

Bank	2	3	4	5	6	7	8	9	10	Ace
± Expectation	0.0036	0.0044	0.0057	0.0073	0.0043	0.0027	−0.0002	−0.0018	−0.0044	−0.0059

Table 17.12. Change in winning expectation when one card is removed from a 52-card deck.

to a change in probability from 4/52 to 3/51, which results from the removal of the first card of a single complete 52-card deck, one obtains the changes in winning expectations (for a single permitted splitting) shown in Table 17.12.

The changes indicated, for which the strategy will have to be altered in some cases, are of the same order of magnitude as the expectation of loss from using the optimal fixed strategy. That is, in the style of play, today out of fashion, of a single deck, a player can quickly, namely, after only two cards have been played, find him- or herself in an advantageous position. And as play progresses, the winning expectation fluctuates dynamically up and down. If the game is played with six decks, then the effect of individual cards is much smaller, only about one-sixth of the tabulated values at the start of the game.

The table also shows that the play of card values 2 through 6 is positive for the player, while for the other cards, particularly the 10 and ace, the opposite is true. The high–low counting system reflects this situation exactly. Therefore, the high–low system is well suited for characterizing the changing winning expectation. For example, if a single card with count 1 is removed from the deck, then there are five equiprobable cases for the card value, namely, the cards 2 through 6. If one averages the results of these cases, one obtains an expectation improved by 0.0051.

One proceeds analogously to optimize the strategy based on the current count. First, the supposed composition of the remaining cards is determined; that is, conditioned on the current count C and the number n of remaining cards, the conditional probabilities of the individual card values in the rest of the deck are determined. Since the cards 7, 8, and 9 have no influence on the high–low count, their conditional probabilities are equal to 1/13 regardless of the count. On the other hand, the probabilities P_{high} and P_{low} for the low (2 through 6) and high (10, picture, ace) cards depend on the count. One has the equality

$$P_{\text{low}} - P_{\text{high}} = -\frac{C}{n}.$$

Together with the identity

$$P_{\text{low}} + P_{\text{high}} = \frac{10}{13},$$

we can determine the probabilities P_{low} and P_{high} at once. Moreover, on the basis of the count, no distinction can be made within the high and low card values. Thus, the probabilities conditioned on the relative count C/n for the individual card values are as follows:

$$P(2) = \cdots = P(6) = \frac{1}{13} - \frac{C}{10n},$$

$$P(10) = \cdots = P(\text{Ace}) = \frac{1}{13} - \frac{C}{10n},$$

$$P(7) = P(8) = P(9) = \frac{1}{13}.$$

With these conditional probabilities, assumed constant throughout the game, the strategy can be optimized depending on the quotient C/n. The results are summarized in "Optimal Strategy Based on the High–Low Count." The winning expectation from such a strategy depends on various factors, namely,

- the relation between the lowest and highest wagers,

- the number of decks of cards used,

- the number of cards removed from play after the cards are shuffled,

- the number of players; with more players, more cards will be played on average after the blank card is encountered.

How high the expectation grows in a given case can be determined with a Monte Carlo simulation that takes into account the special conditions. One can use it to test the efficacy of simplified strategies, such as determining whether to draw a card based on the high–low count.

Optimal Strategy Based on the High–Low Count

On the basis of the current high–low count C and the number n of cards remaining, one plays as follows:

Bet. When $100C/n \geq 3.6$, one raises one's bet; otherwise, the bet is kept to the minimum.

Drawing. Draw at "D" or when $100C/n$ is less than or equal to the given value:

Bank	2	3	4	5	6	7	8	9	10	11
19										
18										
17							−44.2	−40.2	−43.3	−12.0
16	−18.3	−20.4	−22.5	−24.9	−27.3	14.7	12.0	8.6	0.1	16.8
15	−11.9	−14.0	−16.1	−18.3	−19.8	18.3	17.9	15.3	8.2	18.8
14	−7.9	−10.3	−12.6	−15.1	−16.2	34.0	38.8	D	D	26.9
13	−2.0	−4.8	−7.4	−10.1	−10.5	D	D	D	D	38.7
12	5.8	2.5	−0.4	−3.3	−2.5	D	D	D	D	D
11	D	D	D	D	D	D	D	D	D	D

Draw at "D" or when $100C/n$ is less than or equal to the given value:

Bank	2	3	4	5	6	7	8	9	10	11
19s										
18s	−29.4	−29.5	−30.1	−30.5	−35.9		−29.7	D	D	2.7
17s	D	D	D	D	D	D	D	D	D	D

Double. Double when $100C/n$ is greater than or equal to the given value:

Bank	2	3	4	5	6	7	8	9	10	11
11	−23.3	−25.1	−26.7	−28.2	−31.5	−19.1	−14.8	−9.9	5.9	
10	−17.5	−19.5	−21.3	−22.8	−26.1	−12.6	−9.0	−3.4		
9	1.8	−2.2	−5.6	−8.5	−12.5	6.6	14.5			

Split. Split on "S" or when $100C/n$ is greater than or equal to the given value. The following table does not allow for split hands to be doubled:

Bank	2	3	4	5	6	7	8	9	10	11
11, 11	−22.2	−23.4	−24.5	−25.6	−27.7	−18.2	−16.0	−14.6	−11.8	
10, 10	20.0	15.9	12.4	9.4	8.4	25.0	38.6			
9, 9	−2.9	−5.6	−7.8	−10.1	−10.7	12.1	−15.6	−18.7		
8, 8	S	S	S	S	S	S	S	S		
7, 7	−16.8	−20.2	−23.3	−24.9	−35.5	S				
6, 6	3.6	−2.0	−6.4	−10.3	−14.9					
5, 5										
4, 4		36.6	24.6	15.7	34.0					
3, 3	14.8	6.3	0.0	−5.3	−18.9	S				
2, 2	12.6	5.0	−1.9	−8.7	−20.4	S	38.9			

Further Literature on Blackjack

The first six items are directed primarily at blackjack players, the remainder to the more mathematically sophisticated reader.

[1] Michael Rüsenberg, Andreas Hohlfeld, Geplantes Glück, *Bild der Wissenschaft* **10** 1985, pp. 60–71.

[2] Michael Rüsenberg, Andreas Hohlfeld, *Black Jack*, Düsseldorf 1985.

[3] Konrad Kelbratowski, *Black Jack*, Niedernhausen 1984.

[4] Bernd Katzenstein, Black Jack, *Capital* 1982/3, pp. 264–271.

[5] Virginia Graham, Ionescu Tulcea, *A Book on Casino Gambling*, New York 1978.

[6] Charles Cordonnier, *Black Jack*, Munich 1985.

[7] R. A. Epstein, *The Theory of Gambling and Statistical Logic*, New York, 1967 (second edition 1977).

[8] Edward Thorp, Optimal gambling systems for favorable games, *Revue de l'institut international de statistique/Review of the International Statistical Institute* **37**, 1969, pp. 273–293.

[9] Edward Thorp, William Walden, The fundamental theorem of card counting with applications to trente-et-quarante and baccarat, *International Journal of Game Theory* **2**, 1973, pp. 109–119.

[10] Edward Thorp, *The Mathematics of Gambling*, Hollywood 1984, pp. 11–28.

[11] Ulrich Abel, Black Jack mit der Fünf-Karten-Regel, *Der Mathematikunterricht* **28**, 1982, pp. 62–73.

[12] Martin Millman, A statistical analysis of casino blackjack, *American Mathematical Monthly* **90**, 1983, pp. 431–436.

[13] Gary Gottlieb, An analytic derivation of blackjack win rates, *Operations Research* **33**, 1985, pp. 971–988.

[14] Olaf Vancura, Judy A. Cornelius, William R. Eadington (eds.), *Finding the Edge: Mathematical Analysis of Casino Games*, Reno 2000, pp. 71–160.

Part II

Combinatorial Games

18

Which Move Is Best?

In the game of chess, a common opening move for white is the pawn move e2–e4. Among black's replies are e7–e5, e7–e6, c7–c5, and Ng8–f6 (see Figure 18.1). Are there two out of these four moves that are absolutely equivalent with respect to the chances of each side winning the game?

This question differs dramatically in kind from those that we have posed thus far. We recall the typical sources of uncertainty for games that we mentioned in the introduction. Chess is a purely combinatorial game. That is, the difficulty in determining how a game should develop from a given position lies wholly in the astronomical number of possible combinations of moves. Chance or hidden aspects of the game do not figure at all into the equation.

If it is our turn in a game of chess, we must consider future moves. There is a great difference between our moves and those of our opponent: with respect to the opponent, we must always reckon with his or her best move, that is, the one that is worst for us. In particular, each opponent's

Figure 18.1. Do two of these positions offer identical chances?

move that we overlook represents a danger that we have missed a move that will be bad for us. In contrast, when it is our turn, we need only find a good move. We do not have to consider other moves.

A move is good for us if it leads to the desired goal, that is, to a checkmate of the opponent's king or—if we are less ambitious—to a stalemate. However, such long-range goals are of little help to a player in the midst of a game, who needs criteria for the objective and immediate evaluation of a move without regard to the opponent's further play. To do this seems possible in chess—in contrast to roulette and the game rock–paper–scissors, in which no move can be absolutely characterized as good or bad—since all depends on what happens later. That is why there are no "good" players of those games, that is, players who almost always win. On the other hand, a chess player has practically no chance against a significantly better player. And the same holds for chess computer programs, against which an average player will almost never triumph.

Moves and positions in chess are generally characterized in the literature as "excellent," "advantageous," "somewhat better," "equal," "approximately equal," or "with equal prospects." In contrast, chess programs measure the player's chances of victory with a single number. It seems, then, that good moves can be calculated and quantified. Thoughts in this direction were expressed by Edgar Allan Poe on the occasion of a presentation of the famous chess automaton constructed in 1769 by Baron von Kempelen (1734–1804). In an article in the *Southern Literary Messenger*, Poe attempted to prove that this automaton was a hoax, operated by a chess-playing Turk. After seeing the calculating machine of the English mathematician Charles Babbage (1792–1871), Poe compared it with the chess automaton:

> Arithmetical or algebraical calculations are, from their very nature, fixed and determinate. Certain *data* being given, certain results necessarily and inevitably follow. These results have dependence upon nothing, and are influenced by nothing but the *data* originally given. And the question to be solved proceeds, or should proceed, to its final determination, by a succession of unerring steps liable to no change, and subject to no modification. This being the case, we can without difficulty conceive the *possibility* of so arranging a piece of mechanism, that upon starting it in accordance with the *data* of the question to be solved, it should continue its movements regularly, progressively, and undeviatingly towards the required solution, since these movements, however complex, are never imagined to be otherwise than finite and determinate. But the case is

widely different with the Chess-Player. With him there is no determinate progression. No one move in chess necessarily follows upon any one other. From no particular disposition of the men at one period of a game can we predicate their disposition at a different period... But in proportion to the progress made in a game of chess, is the *uncertainty* of each ensuing move. A few moves having been made, *no* step is certain. Different spectators of the game would advise different moves. All is then dependent upon the variable judgment of the players. Now even granting (what should not be granted) that the movements of the Automaton Chess-Player were in themselves determinate, they would be necessarily interrupted and disarranged by the indeterminate will of his antagonist. There is then no analogy whatever between the operations of the Chess-Player, and those of the calculating machine of Mr. Babbage.

Is it possible for the calculations that a machine is capable of carrying out according to a fixed algorithm truly to be upset by the opponent's play? That is certainly the case for games like rock–paper–scissors, but is it true for chess? Can moves be evaluated on the basis of an opposing strategy? Or is trying to figure out the opponent's psychology superfluous to optimal play? Such holds for any number of chess problems and endgames, in which the player can force a win no matter how the opponent plays. In other endgame situations, each player can prevent a loss, so that if no player makes an error, the game will end in a draw. But are there other positions, comparable to the starting position in rock–paper–scissors, that cannot be categorized in this scheme?

This question was raised and then answered in 1912 by the German mathematician Enrst Zermelo (1871–1953). This is how he began a talk presented to the Fifth International Congress of Mathematicians:[1]

> The following considerations are independent of the particular rules of the game of chess and indeed are valid in principle for all similar games of skill in which two opponents play against each other without the intervention of chance events. However, for definiteness, they will be exemplified by chess, the best known game of this type. Furthermore, we are not going to consider

---

[1] E. Zermelo, Über eine Anwendung der Mengenlehre auf die Theorie des Schachspiels, *Proceedings of the Fifth Congress of Mathematics*, volume II, Cambridge 1913, pp. 501–504. Zermelo, whose mathematical work concentrated on the axiomatic foundations of mathematics, later offered a suggestion for the ranking system in chess tournaments: Die Berechnung der Turnier-Ergebnisse als ein Maximierungsproblem der Wahrscheinlichkeitsrechnung, *Mathematische Zeitschrift* **29**, 1929, pp. 436–460.

a practical method of play, but will concern ourselves with the question whether the value of an arbitrary position that can occur in a game as well as a player's best possible move can be determined or at least defined in a mathematically objective manner, without recourse to such subjective, psychological notions such as the "perfect player." That this is possible at least in certain special cases is shown by "chess problems," that is, examples of positions in which it can be proved that the player whose turn it is to move can force checkmate in a prescribed number of moves. However, it seems to me worth considering whether such an evaluation of a position is at least theoretically conceivable and whether it makes any sense at all in other cases as well, where carrying out the analysis encounters the complication of an enormous number of possible continuations, and only such validation would provide a firm foundation for the practical theory of endgames and openings as we find them in books on chess. The methods that we use in what follows are taken from set theory and the logical calculus, demonstrating the fertility of this mathematical discipline in a case that concerns *finite* quantities almost exclusively.

Zermelo then proves with relatively little effort a theorem that states that the positions of chess and comparable games are completely determined; that is, they always satisfy one of the following three conditions:

- white can win regardless of how black plays.

- black can win regardless of how white plays.

- each player can achieve a draw, regardless of how the opponent plays.

If neither player makes an error that causes the loss of what was otherwise attainable, then the result holds for every position, including, of course, the opening position. And now we have arrived at the border between theory and practice. The fact that every position can be placed in one of three categories says nothing about how such a determination can actually be made. And that is precisely the most significant open problem in chess. If it were solved, then one would know the category of the opening position, and then, as Zermelo observed, chess "would of course lose completely the character of a game." Thus the fact that today, and certainly in the future, games between the world's best chess players and programs do not always end the same is evidence that the complexity of chess is too great to be overcome. To be sure, there is reason to believe

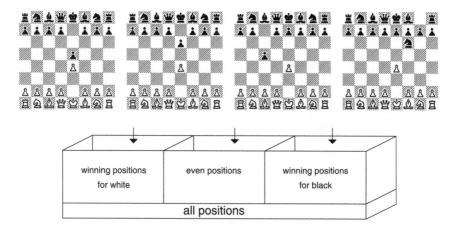

Figure 18.2. At least one "pigeonhole" contains two positions.

that black does not have an advantage by going second, although the large number of draws suggests that chess is essentially an even game. But all of this is pure speculation.

However, there is no speculation at all in the answer to the question raised at the beginning of the chapter. Since there are only three classes of different positions with respect to the prospects of winning, there must be at least two among the four given positions that belong in the same category (see Figure 18.2).[2] Which two positions are equivalent, and whether there are more than two that are equivalent, is simply not known.

Does it make any sense to state such vague results? It certainly offers no help to a chess player. However, such results create a foundation on which further results can be derived. How that can be done will be the topic of the next chapter. Here is a brief preview:

- for some games, much simpler than chess, optimal strategies and their accompanying prospects for winning can be explicitly determined. In other cases, one can determine the prospects for winning without being able to derive an optimal strategy.

- the principle underlying Zermelo's theorem is used, in modified form, in chess programs.

- games like chess and go are considered games of pure skill. Zermelo's theorem supports that view. In other words, the way a game is played

---

[2]This principle, which goes back to an argument of Lejeune Dirichlet (1805–1859), is known in number theory as Dirichlet's pigeonhole principle.

is closely related to it formal properties. It is thus of interest to check the conditions under which the theorem holds for other games.

This last point is worth going into a bit more deeply. The following five properties suffice to make Zermelo's theorem applicable to a particular game:[3]

1. the game is played by two players.

2. one player's win is equivalent to the other player's defeat.

3. the game ends after a finite number of moves, and at each stage, a player has only a finite number of moves from which to choose.

4. the game exhibits *perfect information*; that is, all information about the state of the game is always available to both players.

5. there is no influence from random processes.[4]

The second condition ensures that the outcome of the game is one of the pairs $(1, -1)$, $(0, 0)$, $(-1, 1)$, as games like chess are scored. If a loss is interpreted as a negative win, then the sum of winning values is always zero, and one therefore speaks of a *zero-sum game*. In a zero-sum game, the two players always have absolutely opposing interests. The third condition is satisfied in chess by the rule that a game is considered a draw if 50 moves pass in which no pawn is moved and no piece taken. Thus an infinite shifting of the pieces on the board is ruled out. It is essential for Zermelo's theorem that the players alternate moves. Simultaneous games like rock–paper–scissors are excluded by the fourth point.[5] Finally, results other than the three allowed in chess are possible, such as a double win $(2, -2)$ for white.

If a game satisfies all five conditions, then Zermelo's theorem holds, and the game is completely determined in the sense that the outcome is determined if both players play optimally. That is, with the game position is associated a unique result such as $(1, -1)$, $(0, 0)$, or $(-2, 2)$ that will always be achieved by error-free play. That is, at least one of the players

---

[3]We shall not state precisely at this point just what is meant by a "game" in the mathematical sense of the word. The notion of games that is available from the examples that we present will suffice for now.

[4]This condition can be removed if the notion of a win is replaced by its expectation. Then Zermelo's theorem holds as well for games such as backgammon.

[5]If both players play simultaneously, as in rock–paper–scissors, the game can be modified to have the moves made sequentially without changing the game substantively. In that case, the first player's move is not revealed until the second player has made a move. The result is that there is no longer perfect information.

can play in such a way that regardless of the opponent's play, the result is achieved. And conversely, a player cannot improve his result if his opponent plays optimally.

We may formulate Zermelo's theorem more precisely by introducing the notion of a *strategy*. A strategy represents for a player a complete set of instructions on how to play; that is, for every situation in which the player must move, the strategy offers an optimal move. Of course, such a strategy might encompass a great deal of information and may be enormously difficult to describe, but that is not going to inhibit our theoretical examination of the subject. At this level, we might even think about altering the rules of an arbitrary two-person game so that both players are required to announce their strategies before play begins. There are several variants to this scheme that we might consider; they differ in the amount of knowledge of the opponent's strategy possessed by a player before announcing his or her own strategy:

- both players come up with their strategies in secret and announce them simultaneously.

- Ms. White (who plays first) must announce her strategy before black is required to announce his.

- Mr. Black (who plays second) must announce his strategy before white is required to announce hers.

Let us now examine how these three variants of the rules affect the prospects of the players in a two-person zero-sum game.

If we look beyond the somewhat involved wording, we see that the first variant is a reformulation of the original game. The fact that the players have to announce their strategies has no effect on their chances of winning, since for any situation in which a decision has to be made, it makes no difference regarding strategy whether the situation has already arisen. What is important is simply that each decision to be made is based precisely on the state of information that would be available to a player in a real game.

On the other hand, the other two variants alter the information available to the players, since one of them will be able to examine the opponent's strategy and make use of possible weaknesses. Such a player therefore has more information than was available in the original game; namely, the player knows how the opponent will behave in any situation. In games like rock–paper–scissors, such an advantage is enormous, since it makes possible a certain win.

Maximin Value ≤ Minimax Value

The maximin is the value that white can be assured of if black plays optimally with respect to the strategy announced by white.

The minimax is the smallest value to which black can limit white's game value if white plays optimally with respect to the strategy announced by black.

Figure 18.3. The two values of a two-person zero-sum game.

If White has to announce her strategy first, she will attempt to find a strategy that optimizes her chances of winning regardless of black's reply to her moves. White thus considers the worst-case scenario, in which black's reply will minimize her chances of winning. She thus will choose a strategy that maximally cancels this minimum. The *game value* of such a strategy is quantified as the *maximin value*.

In the converse case, in which black must announce his strategy first, he must come up with moves that counter those of white to cancel white's chances as much as possible. Black must therefore play in such a way as to minimize white's attempt to maximize her chances. The game value of black's strategy is called the *minimax value*.

In sum, white pursues a strategy that achieves at least the maximin value of the game, while black's strategy seeks to prevent white from achieving more than the minimax value. In particular, the maximin value is less than or equal to the minimax value, and so one speaks of the lower and upper values of a game (see also Figure 18.3).

In the case of the game rock–paper–scissors, these two values are different, namely, they are −1 and +1. The situation is different for games like chess that satisfy Zermelo's theorem. With reference to the two game values, Zermelo's theorem can be formulated thus: under the hypotheses of the theorem, the maximin value is equal to the minimax value. This common value is called the *game value* (see Figure 18.4). Zermelo's theorem is thus frequently called the *minimax theorem*. Chess therefore differs fundamentally from a game like rock–paper–scissors, since in chess it is of no help to know the strategy of an optimal player. Thus chess can

Maximin Value = Minimax Value

Figure 18.4. Zermelo's theorem is valid for two-person zero-sum games with perfect information.

be played completely without reference to the opponent, all moves being planned according to objective criteria. Such a game can even be played by a computer that has been accordingly programmed.

For games that satisfy the conditions of Zermelo's theorem, the minimax strategies create a sort of balance that neither player can shift to his or her advantage. One can interpret the situation geometrically by placing the game values resulting from the two strategies in a two-dimensional coordinate system (an example can be seen later in the chapter in Figure 18.7). Maximin and minimax strategies thus form together with their game values a *saddle point*. One therefore speaks of saddle-point strategies, though also of minimax, maximin, and optimal strategies. A player who employs an *optimal* strategy can reveal that strategy without suffering any disadvantage.

It is significant that the value of a game with perfect information can be determined recursively. One does this by minimizing and maximizing move by move: if it is white's turn, the game value is equal to the maximum value of the positions that follow, while on black's turn, the subsequent positions are to be minimized. As we shall see, this process is one of the foundations of chess programs. The recursive maximin process is made clearer by representing it in the form of a *game tree*. Positions are indicated by nodes at the branch points in the tree, while moves are represented by the edges. A game is represented as running downward, and so the root of the tree, symbolizing the start position of the game, appears at the top. The end positions, or end nodes, correspond to the various possible endings of the game, and are labeled with the associated outcomes of the game. Figure 18.5 presents an example of a three-move game.

The value of a game can now be calculated from bottom to top using alternating maximizations and minimizations. Starting with the end positions and the winning values for white, one moves upward step by step.

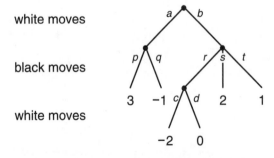

Figure 18.5. A three-move game represented as a game tree.

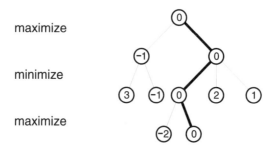

maximize

minimize

maximize

Figure 18.6. Mutually optimal play.

When white moves, the maximum value is taken for the next-lowest level, and when black moves, the minimum value is used. Thus, for example, we see in Figure 18.6, emphasized with a heavy line, the path through the tree that is optimal for both sides.

This example is also well suited for a more precise discussion of the notion of strategy. If white decides to open with move a in Figure 18.5, then she has no need to make any further plans. The situation is different for move b, for which white must have a plan, in the event of black's choosing move r, whether to make move c or move d. Therefore, white has a choice of three strategies: a, b-c, and b-d. Black must also plan for two positions with which he may be confronted. This represents six strategies, namely, p-r, p-s, p-t, q-r, q-s, and q-t. If we record the results of all the combinations of white's and black's moves in a table, we achieve the *normal form* of the game. This has immediate practical value only for games that are not much more complex than the one presented here. For chess, we would have an enormous table of astronomical complexity. The importance of the normal form is that a game without perfect information like rock–paper–scissors can be easily represented in normal form, but as a tree only with considerable difficulty.

In Table 18.1, the optimal strategies with the associated saddle point are enclosed in double lines. Neither player can improve his or her prospects

		Black					
		p-r	p-s	p-t	q-r	q-s	q-t
	a	3	3	3	-1	-1	-1
White	b-c	-2	2	1	-2	2	1
	b-d	0	2	1	**0**	2	1

Table 18.1. The normal form of the three-move game, showing the saddle point.

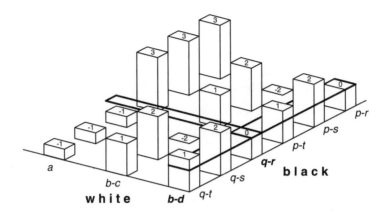

Figure 18.7. Normal form as a bar graph.

by altering this optimal srategy. If a normal form is displayed graphically, as in Figure 18.7, one can even, with a little imagination, see the "saddle": in one direction, there is no higher winning value, and in the cross direction, there is none smaller.

A Proof of Zermelo's Theorem

Using the principle of strong induction, we assume that the theorem is true for all games that take at most some number n of moves to play. Such an assumption must hold for $n = 0$, since the rules for such a game consist solely of instructions as to who wins how much from whom. In the induction step, one considers a game lasting at most $n+1$ moves, where on account of symmetry we may assume that white goes first. Every continuation of the game obtainable from the first move, being a game of at most n moves, is strictly determined. That is, the maximin value is always equal to the minimax value. We call the largest of these values v. As we shall now show, this value is equal to the total game value (see Note 1 at the end of the chapter):

- white moves first, resulting in a continuation with game value v. Then white can be assured of a win with value v within this continuation.

- black knows the move with which white has opened, based on perfect information. Therefore, it is possible for black to defend himself in the resulting continuation as he would do in such a game. Thus white wins at most the value of the achieved endgame, namely, at most the value v.

Chapter Notes

1. For games with alternating moves that can only be either won or lost, there is a simple proof going back to Hugo Steinhaus (1887–1972); see Marc Kac, Hugo Steinhaus, A reminiscence and a tribute, *American Mathematical Monthly* **81**, 1974, pp. 572–581, which can be reduced with the help of logical quantifiers to two lines: first, each game is increased to a uniform length, by adding neutral moves as necessary. If $w_1, w_2, \ldots, w_n$ are white's moves, and $b_1, b_2, \ldots, b_n$ those of black, then either the condition

$$\forall w_1 \; \exists b_1 \quad \forall w_2 \; \exists b_2 \quad \ldots \quad \forall w_n \; \exists b_n, \quad \text{white wins}$$

holds, or else the opposite condition, namely,

$$\exists w_1 \; \forall b_1 \quad \exists w_2 \; \forall b_2 \quad \ldots \quad \exists w_n \; \forall b_n, \quad \text{white does not win.}$$

The first statement means that white possesses a winning strategy; the second, that black does.

19

Chances of Winning and Symmetry

In order not to give any player an advantage, the rules of most board games are essentially symmetric. Is the goal of fairness ever achieved in a concrete case?

The games chess, backgammon, checkers, halma, reversi, go, and nine men's morris are—aside from the privilege of the first move—completely symmetric for both players. More the exception are games such as wolf and sheep[1] and Scotland Yard.[2]

In games like chess, the player to move first is considered to have a slight advantage. However, as Zermelo's theorem tells us, such a board game offers either completely even chances to both players or a winning strategy for one of the players. For intellectual competitions only fair games are considered. If a two-person zero-sum game with perfect information is not

---

[1]The game is played on a checkerboard. White has four checkers, the sheep, and black, the wolf, has a single checker. Sheep can move only forward, while the wolf can move both forward and backward. The game is also known under the name fox and geese. A description can be found in Claus D. Group, *Brettspiel/Denkspiele*, Munich, 1976, pp. 90–92. The sheep have a winning stragegy, a proof of which can be found in Elwyn Berlekamp, John H. Conway, Richard K. Guy, *Winning Ways*, New York, Academic Press, 1982.

[2]Scotland Yard is a well-designed game of pursuit, in which one player must be caught by a group of cooperating opponents. The game uses a map of London, in which one moves via public transportation. See Erwin Glonnegger, *Das Spiele-Buch*, Munich 1988, pp. 124–125; Jury "Spiel des Jahres," *Spiel des Jahres*, Munich 1988, pp. 56–58; Jury "Spiel des Jahres," *Die ausgezeichneten Spiele*, Hamburg 1991, pp. 55–60.

fair, or—as is more often the case in practice—its game value is unknown, then one may attempt to even out the odds. There are several ways in which this can be accomplished:

- the player to move first is decided by lot. Then the chances are equal, but at the cost of introducing an element of randomness. Even though randomness has a small place in the overall game, its influence can be large. In games that are not a priori fair, it could theoretically be the decisive factor.

- two games are played, the players alternating first move. Any advantage or disadvantage accruing to the first player is compensated in the second game. It is of no importance who goes first in which game.

- in the board game Twixt,[3] by Alex Randolph, the advantage to the first player is supposed to be compensated by the second player's right to decide after the first move whether to swich sides. The idea of this rule follows the principle of division, according to which if two children share a piece of cake, a fair division is for one child to cut the cake and the other to choose the piece that he or she prefers. In so-called Texas roulette, the principle is used in an altered form in business dealings: if two equal partners in an enterprise wish to grant each other the right of first refusal, then the price for one partner to buy out the other can be fairly set by the rule that the partner offering to buy out his partner must be prepared to have the partner instead buy him out at the same price.

The methods just described appear reasonable, and indeed, they are used in practice. However, there are significant differences among them, which can be clarified with the help of Zermelo's theorem. To do this, we can determine the associated game values, which is possible to accomplish without knowing any equalization strategies.

Determining the opening player by lot means that each player has a strategy that ensures an *expectation* of zero. What this strategy actually looks like must often remain unresolved, since it is composed of equalization

---

[3]Twixt belongs to the class of "border-to-border" games, in which the player to go first can always achieve at least a draw with optimal play. Two other games of this type, namely Hex and Bridge-it, will be discussed later in this chapter. For more on Twixt, see David Pritchard (ed.), *Modern Board Games*, London 1975, pp. 92–101 (author: David Wells); Andreas Kleinhans, *Twixt: ein kleines Expertenheft*, Stuttgart 1990 (mimeographed brochure); Erwin Glonnegger, *Das Spiele-Buch*, Munich, 1988, pp. 142–143; Werner Fuchs, *Spieleführer 1*, Herford 1980, pp. 106–108.

strategies of the original game, and even a perfect player can lose because of bad luck in the choice of first player.

In playing two games, the players' chances are equalized without the introduction of any random elements. That is, the value of the combined games is 0. In particular, there is no inherent advantage or disadvantage to going first in the first game. To realize the game value 0, a player can use the equalization strategies of a single game. Since it is usually not known what these are, a true intellectual battle will ensue. Moreover, a tiny change in the rules suffices to allow a player to specify an equalization strategy, whereby the two parties play simultaneously according to a specific order of playing:

- player A makes a move as white in the first game.

- player B makes a move as white in the second game.

- player A makes a move as black in the second game.

- player B makes a move as black in the first game, and so on.

With this order of play, player B can simply copy the moves of player A to ensure a value of 0. This trick was described in Sidney Sheldon's novel *If Tomorrow Comes* (in Chapter 20), in which a pair of crooks play against two chess masters, who are a distance apart, betting a naive public that they can achieve at least two draws or one win. It would have been even more clever to have set bets individually for the two games.

Using the rule of switching sides, equal chances are guaranteed only if there is an opening move that leads to a position with value 0. With such an opening move, white can introduce an equalization strategy. All other moves are disadvantageous for white. If the value is positive, then black exchanges sides, and if it is negative, then black has a direct winning strategy. The value of the game with the exchange rule is therefore at most 0.

What we have said about all three variants depends on a comparison of the actions available to the two players. With some games, this can be done directly, and not just in the symmetric variant. Thus with Sid Sackson's board game Focus,[4] which is played with checkers on a checkerboard with the corners removed, the second player can mimic the moves of the first player. Unlike chess, focus permits such play, since the checkers can move

---

[4]Focus was introduced by Martin Gardner in *Scientific American* 1963/10, pp. 124–130, and it appears in his *Sixth Book of Mathematical Games from "Scientific American,"* San Francisco 1971, Chapter 5. The game is also described in Sid Sackson, *A Gamut of Games*, New York 1969; Erwin Glonngger, *Das Spiele-Buch*, Munich 1988, p. 161; Jury, "Spiel des Jahres," *Spiel des Jahres*, Munich 1988, pp. 44–46.

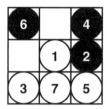

Figure 19.1. Tic-tac-toe: white wins on the seventh move. Black's move, number 2, was bad.

only horizontally and vertically, and the opening position is symmetric in that a reflection in the center of the board is equivalent to an exchange of colors. With a slight extension of the rules, uncertainty of play can be restored.

There is a large class of games that begin with an empty board, and players alternate by placing pieces of different colors, say, white and black, on an empty square. Once a piece is placed, it is neither moved nor captured, which makes such games suitable for playing with paper and pencil. The first player to achieve a particular configuration is the winner. Examples of this type of game are tic-tac-toe, go-moku, and Qubic, in which each player strives to achieve an unbroken line of pieces of a certain length, either vertical, horizontal, or diagonal:

- in tic-tac-toe, the object is to form a string of length 3 on a 3 × 3 board (see Figure 19.1).

- go-moku is played on a much larger board, and the object is to create a string of length 5.

- Qubic is won by linking four pieces within a 4 × 4 × 4 cube.

What all these games have in common is that an extra piece on the board is never a disadvantage. Since both sides have identical winning configurations, the result of this is that the player who goes second never has a winning strategy. If that were so, the player who moves first could place a piece on the board arbitrarily and then adopt the winning strategy of his opponent. If that strategy involves placing a piece on the square containing his first piece, the player can simply again place a piece at random. The player who moves first thereby possesses a winning strategy, which thus refutes the assumption of a winning strategy for the second player. This technique of stealing the opponent's strategy was first applied in 1949 by the American John Nash (1928–), who was awarded the Nobel Prize in economics in 1994.[5]

We note that the game values of these three games can be determined explicitly. In the case of tic-tac-toe, it is easy to analyze all possible moves. We see that the second player loses only if he makes an error, so that the value of tic-tac-toe is zero. If go-moku is played on a board of size at least 15×15, then the first player has a winning strategy available. This result was proved in 1993 by three Dutch computer scientists using extensive computer analysis that analyzed all possible moves of the second player. They did not need a separate analysis for moves that were symmetric to other moves, but even so, they analyzed about 15 million positions to come up with a winning strategy in the form of a library of 150 000 moves.[6] A similar extensive computer analysis for Qubic was done by Oren Patashnik in 1980, which shows that the first player has a winning strategy.[7] Both results depend on the principle, which is not undisputed in the serious realms of mathematical research, of proving theorems by means of extensive computer analysis. The first such theorem to be proved, and the most renowned, was the four color theorem (see Note 1 at the end of the chapter).

More Results on Games of Tic-Tac-Toe Type

Games related to tic-tac-toe were analyzed on a very abstract level in 1963 by Hales and Jewett.[8] An example of their research is the game in which one has to line up k game pieces in an n-dimensional playing field with k^n squares. Their results show that such games have two crucial properties in common:

- if the dimension of the playing field is large enough in relation to its length, then there is no position that is a draw. That is, no matter how the pieces are arranged, there is

[5] Martin Gardner, *Mathematical Puzzles and Diversions from "Scientific American,"* New York 1959; John Milnor, A Nobel Prize for John Nash, *The Mathematical Intelligencer* **17/3**, 1995, pp. 11–17; Sylvia Nasar, *A Beautiful Mind*, Touchstone, 1999.

[6] Victor Allis, Jaap van den Herik, Matty Huntjens, Eine Gewinnstrategie für Go-Moku, *Spektrum der Wissenschaft* **4** 1993, pp. 25–28; Victor Allis, *Searching for Solutions in Games and Artificial Intelligence*, Maastricht 1994.

[7] Oren Patashnik, Qubic: $4 \times 4 \times 4$ Tic-Tac-Toe, *Mathematics Magazine* **53**, 1980, pp. 202–216.

[8] A. W. Hales, R. J. Jewett, Regularity and positional games, *Transactions of the American Mathematical Society* **106**, 1963, pp. 222–229; reprinted in Ira Gessel, Gian-Carlo Rota, *Classic Papers in Combinatorics*, Boston 1987, pp. 320–327.

always a winning position for at least one of the players. It
follows that the first player always has a winning strategy.

- if the length of the playing field is large enough in relation to the dimension, then the second player can force a
  draw. To show this, Hales and Jewett use the "marriage
  theorem" (see Note 2 at the end of the chapter) to construct a pairing between the squares of the playing field.
  Whenever the first player places a piece on one square of
  a pair, the second player places a piece on the associated
  square. Since the pairing is constructed in such a way that
  every winning row contains a matched pair, and no square
  belongs to more than one pair, a winning row can never
  be achieved (see Note 3 at the end of the chapter). For a
  5×5 field, a pairing is shown in the following figure.

1	5	8	5	2
3	9	9	10	4
7	12		10	7
3	12	11	11	4
2	6	8	6	1

Quite a different approach was used by Erdős and Selfridge
in 1973 to improve quantitatively the second result of Hales
and Jewett. Their defensive strategy guarantees a draw to the
second player in cases in which the number of winning rows is
not too large in comparison to their length. The strategy is
based on a formula according to which a value is attached to
the winning rows still achievable by white. White must always
choose a move that makes that value as small as possible (see
Note 4 at the end of the chapter).

Additional representatives of games in which pieces once played are not
moved again and in which certain configurations are to be achieved include
such "border-to-border" games as Hex, Bridge-it, and Twixt. In all these
games, the object is to build a chain from one edge of the game board to the

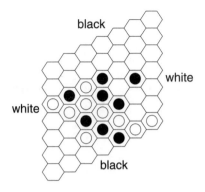

Figure 19.2. The game Hex.

opposite edge. Of particular mathematical interest are Hex and Bridge-it. In Hex,[9] the players lay their pieces on a hexagonal grid whose four sides are of equal length. Figure 19.2 shows a situation in which white, who went first, has won.

According to Martin Gardner,[10] Hex was invented by the Dane Piet Hein and then reinvented independently six years later by Nash, who was then a student at Princeton. A game of Hex cannot end in a draw, from which it follows that regardless of the size of the game board, white must possess a winning strategy. However, for large boards, no one knows what that strategy is.

And why is it that Hex cannot end in a draw? Since a strict mathematical proof, though not difficult (see Note 5 at the end of the chapter), is not particularly illuminating, we shall content ourselves with a plausibility argument: we begin with a game board completely filled with white and black game pieces. We assume that black does not have a winning position. We wish to prove that white has won. To do this, we begin to modify the position in such a way that does not improve white's position. More precisely, we exchange, piece by piece, as long as we can, individual white pieces for black pieces, provided that a winning position is not thereby achieved for black. Black's pieces then form two regions, neither of which contains "islands" and which are separated by a one-square-wide path of white pieces. This path, which is part of the original configuration, is in fact a path across the board. White has won.

---

[9]The most extensive reference on Hex is Cameron Browne, *Hex Strategy: Making the Right Connections*, Natick 2000.

[10]Martin Gardner, *Mathematical Puzzles and Diversions from "Scientific American,"* New York, Simon and Schuster, 1959, Chapter 8.

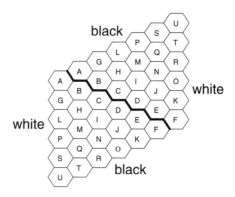

Figure 19.3. Asymmetric Hex with pairing strategy.

If one attempts in Hex to neutralize white's advantage of first move by shortening white's edge of the board by one square, so that white must build a path one square longer than the path that black has to build, then black is assured of a win by a pairing strategy. That is, as with some of the tic-tac-toe variants (see "More Results on Games of Tic-Tac-Toe Types"), the squares of the game board can be paired off in such a way that every white winning path contains such a pair. If black counters every move of white by playing on the paired field, then white cannot win. The division into pairs, as can be seen in the example of Figure 19.3, is the result of a reflection in the dark line running across the board.

Some similarities to Hex can be seen in the border-to-border game Bridge-it, invented by David Gale (1921–), like Nash a pioneer of game theory. This game was introduced in Martin Gardner's *Scientific American* column in October 1958.[11] The game is played with oblong pieces that are used to connect points of the player's color if the path is not crossed by an opponent's bridge. Figure 19.4 shows a game in which white, who played first, has won.

As with Hex, it can be seen that a draw in Bridge-it is impossible. Therefore, there must be a winning strategy for white. In contrast to Hex, it is possible to determine a winning strategy rather easily. Thus Oliver Gross has presented a pairing strategy in which white begins with the move depicted in Figure 19.5 and then counters black's moves pairwise as indicated by the thin lines: each line crosses a move that black can make as well as a move that white can make; these are basically the moves that

---

[11]See also Martin Gardner, *Second Book of Mathematical Puzzles and Diversions*, New York 1961, Chapter 7.

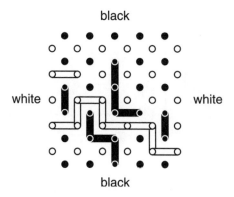

Figure 19.4. The game Bridge-it.

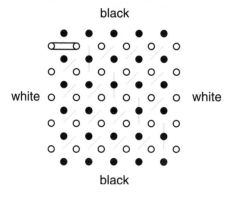

Figure 19.5. Pairing strategy for white after the first move.

are paired.[12] This strategy makes it impossible for black to connect both sides of his territory.

A game that is considered to have equal chances on the basis of a long tradition is nine men's morris.[13] A good player will lose neither as white nor as black. To be sure, white has the advantage of first move, which permits some aggressive threats. However, the tide frequently turns when black places the last stone and thereby puts his stamp on the subsequent play. How the various countervailing influences are balanced has been known in detail since the early 1990s. Using extensive computer analysis, the Zurich-

---

[12] *Martin Gardner's New Mathematical Diversions from "Scientific American,"* New York 1966, Chapter 18.

[13] For the rules and advice on play for nine men's morris, see *So gewinnt man Mühle,* Ravensburg 1980.

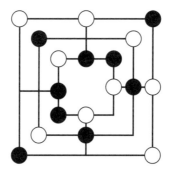

Figure 19.6. Whoever moves first, loses.

based computer scientist Jürg Nievergelt and his doctoral student Ralph Gasser proved that both white and black can force a draw.[14] However, it is easier for black to play optimally than it is for white. Up to symmetry, in the first eight (half) moves, there are 4393 positions that black has to take into account, while for white's strategy, that number is 15 513. Nievergelt and Gasser also carried out extensive statistical investigations into how the material strength of each side influences, on average, the chances of winning.

Figure 19.6 shows an example of a position in nine men's morris that demonstrates the power of computer analysis: whoever moves first, loses. If white moves first, she can defend her position for 37 double moves, while if black moves first, he can last "only" 30 more double moves.

Chapter Notes

1. The four color theorem asserts that any planar map consisting of adjacent countries can be colored with four different colors in such a way that no two countries sharing a common border have the same color. See Kenneth Appel, Wolfgang Haken, *Every Planar Map is Four-Colorable*, Providence, RI, American Mathematical Society, 1989; Keith Devlin, *Mathematics, the New Golden Age*, New York 1999; Ian Stewart, *The Problems of Mathematics*, Oxford 1987.

2. The marriage theorem deals with a situation in which m sets are given from each of which a single element is to be chosen. The theorem states that a selection of m *distinct* elements can be made from the m sets if and only

---

[14]R. Gasser, J. Nievergelt, Es ist entschieden: Das Mühlespiel ist unentschieden, *Informatik Spektrum* **17**, 1994, pp. 314–317; Ralph Gasser: Solving Nine Men's Morris, in: R. J. Nowakowski (ed.), *Games of No Chance*, Cambridge 1996, pp. 101–113.

if every union of k of the m sets contains at least k elements. The name "marriage theorem" comes from an interpretation of the problem in which each of the sets contains those women who would be willing to marry a particular man, and the problem is to choose one woman from each set, so that each woman is provided with a man whom she would consent to marry. Of course, the roles of men and women can be exchanged or modified to suit, and marriage can be replaced by any other activity in which two persons might wish to engage. The question is whether it is possible for each of the men to find a woman who would consent to marry him. The criteria of the marriage theorem are here equivalent to the scenario that in every possible constellation of a party in which any number of men and all women acceptable to them participate, there are at least as many women as men.

The marriage theorem was rediscovered in a variety of forms between 1910 and 1935. An overview can be found in Konrad Jakobs, *Selecta Mathematica I*, Heidelberg 1969, pp. 103–137.

3. For this to work, the number of winning rows to which a square belongs cannot be too large, since otherwise, a move on that square would result in a decisive multipronged threat. A measure of this is the largest number w_{max} of winning rows that run through a single square. Because of the diagonals, this maximum occurs in or around the center. For even squares $(n = 2)$, one has, for example, $w_{max} = 4$ for odd k, and $w_{max} = 3$ for even k.

The marriage theorem is now applied to the sets that contain all winning rows, each in two distinguishable instances. This duplication can be done by associating a marker with each winning row, such as a direction. If one has a selection of a such directed winning rows, then these encompass at least

$$\frac{a}{2} \times \frac{k}{g_{max}}$$

distinct squares. Even with this crude estimate, the marriage theorem can be applied for $k \geq 2g_{max}$, since then a directed winning rows always encompass a or more squares. One can thus find for each directed winning row a square, and consequently, for an even winning row a pair of squares, such that no square must be taken more than once. For even squares, then, the pairing strategy has been proved in the case $k \geq 6$.

4. Geometric considerations show that in the case of k^n tic-tac-toe, there are

$$\frac{1}{2}\left((k+2)^n - k^n\right)$$

winning rows. Namely, to each square of a one-square-side border there corresponds exactly one winning row, which in turn runs through two such border squares. If this number is less than 2^{k-1}, then black, the second player, can prevent a white victory.

This fact is a special case of a more general criterion that makes an assertion about how many squares the various winning rows contain: let $|A|$ denote the number of squares that the winning row A contains. Then if the inequality

$$\sum_{A \text{ is a winning row}} \frac{1}{2^{|A|}} < \frac{1}{2}$$

is satisfied, then black can prevent a white win.

This criterion can be explained as follows: black can play in such a way that one can easily project how this sum will change in the course of the game. In this, the winning rows are considered only from white's viewpoint. That is, in updating the sum, any winning rows blocked by black are excluded, while the winning rows partially completed by white are correspondingly reduced; concretely, if W and B denote respectively the sets of squares occupied by white and black, this corresponds to the sum

$$\sum_{\substack{A \text{ is a winning row} \\ \text{with } A \cap B = \varnothing}} \frac{1}{2^{|A-W|}}.$$

On her first move, white can at most double the value of the sum, so that all achievable summation values at this point in time are less than 1. In further play, based on the current sum, black can come up with a set of moves that keep him from losing. Concretely, for his move, black is searching for a square that limits this sum as much as possible. In a currently achievable sum of

$$\sum_{C} \frac{1}{2^{|C|}},$$

he should choose a square b in such a way that the resulting summands become as great as possible. If white then moves on square w, then the summands double on those winning rows that are still achievable by white and contain the square w. The total change resulting from these two moves is

$$-\sum_{C: b \in C} \frac{1}{2^{|C|}} + \sum_{\substack{C: w \in C \\ b \notin C}} \frac{1}{2^{|C|}} \leq -\sum_{C: b \in C} \frac{1}{2^{|C|}} + \sum_{C: w \in C} \frac{1}{2^{|C|}} \leq 0,$$

where the second inequality holds on account of the choice of move for black to the square b. It follows that the sum remains smaller than 1 at the times at which white moves, with the result that white cannot achieve a victory, since that would require that the sum be at least 1.

For the case of k^n tic-tac-toe, the assumption can be weakened, since on white's first move, instead of the assumed doubling, the most that is possible for an increase in the sum is

$$g_{\max} \times \frac{1}{2^k},$$

where as earlier, g_{max} denotes the maximal number of winning rows that run through a single square. In consequence, the condition that the total number of winning rows be less than $2^k - g_{max}$ already suffices to allow black to prevent a white victory.

For more information, see P. Erdős und J. L. Selfridge, On a combinatorial game, *Journal of Combinatorial Theory* **B 14**, 1973, pp. 298–301; J. Beck, Achievement games and the probabilistic method, in: Miklós, D., Sós, V. (eds.), *Combinatorics, Paul Erdős Is Eighty*, volume 1, Keszthely (Hungary) 1993, pp. 51–78.

5. Two different proofs that Hex cannot end in a draw can be found in David Gale, The game Hex and the Brouwer fixed-point theorem, *American Mathematical Monthly* **86**, 1979, pp. 818–827. In addition to a relatively elementary proof, it is shown that the impossibility of a draw is equivalent to Brouwer's (1881–1966) fixed-point theorem, whose planar version makes the following assertion: a continuous function that maps a 1×1 square, including the boundary, to itself has at least one fixed point, that is, a point that is mapped to itself. This doubtless quite abstract result has some down-to-earth consequences. Suppose one has two identical sheets of paper lying on top of each other. If you take the top piece, crumple it up into a ball without tearing it, and then place the ball on top of the second piece without its going over the edge, then there is at least one point on the crumpled sheet that is lying over its original position.

For the general version of Brouwer's fixed-point theorem, that is, no longer restricted to the plane, Gale generalized the game Hex to a higher-dimensional variant.

Brouwer's theorem has often proved useful in mathematics. It has been used, for example, to prove equilibrium theorems such as von Neumann's minimax theorem, first proved in 1928, and the multiperson variant proved in 1951 by Nash, the coinventor of Hex. We will have more to say about this in Chapters 34 and 43. A popular yet multifaceted representation of Brouwer's fixed-point theorem can be found in John Casti, *Five Golden Rules: Great Theories of 20th-Century Mathematics and Why They Matter*, Hoboken, NJ 1996.

20

A Game for Three

Three players alternate taking turns removing stones from a pile that initially contains a certain number of stones. Each player can take at most five stones. The player to remove the last stone wins one unit from the player who drew previously. The third player wins nothing and loses nothing. What should the players' strategies be?

The game that we have just described was first investigated by the chess master and mathematician Emanuel Lasker[1] (1868–1941) in his 1931 book *Brettspiele der Völker.*

In a chapter on mathematical "contact sports," Lasker first investigates two-person games and then attempts to generalize his results. Can it be determined for three players which among them can win with error-free play?

Following Lasker, let us analyze the game in reverse. Suppose the three players Xavier, Yves, and Zelda draw in alphabetical order. If there is only

---

[1]Lasker is known as a mathematician for his 1905 theorem in ideal theory, a branch of abstract algebra. Lasker's theorem is useful in the analysis of solution sets of systems of polynomial equations (E. Lasker, Zur Theorie der Moduln und Ideale, *Mathematische Annalen* **60**, 1905, pp. 20–116). A comprehensible explanation of this subject is given by Bartel L. van der Waerden, "Meine Göttinger Lehrjahre," *Mitteilungen der Deutschen Mathematiker Vereinigung* **2**, 1997, pp. 20–27; see also Markus Lang, Laskers "Ideale" und die Fundierung der modernen Algebra, in Michael Dreyer, Ulrich Sieg (eds.), *Emanuel Lasker: Schach, Philosophie, Wissenschaft*, Berlin, 2001, pp. 93–111.

Lasker was the world chess champion in the years 1894 and 1921. It is more difficult to comprehend the role of Lasker as the driving force behind game theory, as presented in Georg Klaus, Emanuel Lasker, Ein philosophischer Vorläufer der Spieltheorie, *Deutsche Zeitschrift für Philosophie* **13**, 1965, pp. I/976–988.

Size of Pile	Win for Player Drawing			
	First	Second	Third	Move
0	0	−1	1	
1	1	0	−1	take 1
2	1	0	−1	take 2
3	1	0	−1	take 3
4	1	0	−1	take 4
5	1	0	−1	take 5
6	−1	1	0	arbitrary
7	0	−1	1	take 1
8	1	0	−1	take 1
9	1	0	−1	take 2
10	1	0	−1	take 3
11	1	0	−1	take 4
12	1	0	−1	take 5
13	−1	1	0	arbitrary
14	0	−1	1	take 1
15	1	0	−1	take 2

Table 20.1. Equilibrium in a three-person game.

one stone remaining, the player whose move it is wins at once with the only possible move. Victory is also obtainable if two through five stones remain, all of which can be removed in one fell swoop. The other possible moves are certainly less attractive. If Xavier, say, is confronted with six stones, he has nothing nice to look forward to. Whatever he does, Yves can win and inflict a defeat on Xavier. A pile of seven stones offers a better scenario. Namely, if Xavier, say, removes a single stone, then Yves will lose to Zelda, and Xavier will break even. A pile of 8, 9, 10, or 11 stones is yet more favorable. If it is Xavier's turn, he will take the appropriate number of stones to reduce the pile to seven stones, granting a draw to Yves and victory to himself.

These results are collected in Table 20.1. For each size of pile, the game values are given for the players in the order in which they draw. In the right-hand column are listed the optimal moves. The resulting strategies of the players create an equilibrium (see "Multiperson Games with Perfect Information" at the end of the chapter).

Each row of the table is derived from the row above it. The player who is to draw, say it is Xavier, chooses the move that promises the greatest return. To decide how many stones to draw, Xavier looks in the five previous columns and chooses the move that will produce the row with the greatest win amount in the third column, since following his turn, he will be the third to draw from the new position. The result is a table that repeats itself periodically every seven rows.

Looked at from this point of view, it would appear that the three-person game that we have been considering is determined analogously to such two-person games as chess, go, and so on. But is that really so? In particular, we might ask what would happen if one of the players does not act according to the prescription that we have set down. In other words, what are the consequences of one of the players not playing optimally? Let us consider the following example: suppose that after Xavier's first move, in which three out of an original ten stones were removed, Yves thoughtlessly takes five stones? This gives Zelda, who originally was cast in the role of loser, an opportunity to win the game. Zelda's profit is paid for by both of the other two players. Yves's stupidity worsened his position from 0 to −1, but poor Xavier, who was uninvolved in the wrong move, who should have emerged the victor from a pile of ten stones, had his position reduced from 1 to 0.

You may be sure that such a possibility did not escape Lasker's attention. He remarks:

> player A wins, provided that B does not act against his own interest, and he does not lose if C does not make the same error.

In relation to larger piles, Lasker continues thus:

> of course, if B and C both make such errors, and then play without error, then A loses.

And in fact, if one starts with 15 stones, it is possible for errors of the second and third players not only to rob the first player of a victory, but to give him a defeat. The "errors" of these two players have the effect of increasing their mutual winnings from −1 to 1. Thus we may contemplate that what is going on is not erroneous play, but a successful cooperation between the second and third players. Such possibilities are revealed in Table 20.2, which again shows the prospects for three players, only this time tabulated with the winnings that a single player can get for himself on his own. Assumed is a *coalition* of the opponents, who attempt to optimize their total winnings. In other words, the variants "one against two" are investigated as a two-person game based on the minimax principle.

This table, like the previous one, can be calculated recursively, where again, one considers the five preceding rows. For the first win column, which contains the winning amount for the first player, the values of the

Size of Pile	Win for Player Drawing			Sum
	First	Second	Third	
0	0	−1	1	0
1	1	0	−1	0
2	1	0	−1	0
3	1	0	−1	0
4	1	0	−1	0
5	1	0	−1	0
6	−1	1	0	0
7	0	−1	0	−1
8	0	−1	−1	−2
9	0	−1	−1	−2
10	0	−1	−1	−2
11	0	−1	−1	−2
12	0	0	−1	−1
13	−1	0	−1	−2
≥ 14	−1	−1	−1	−3

Table 20.2. A three-person game: winning amounts achievable by one player against two.

last win column for the chosen move must be maximized. The winning amounts of the two players whose turn it is not are to be correspondingly minimized, and so for the second win column, the five preceding values of the first win column must be minimized. Analogously, the third win column is derived from the second. One sees at once that in piles of at least 14 stones, no player has a chance when the other two players are united against him.

It is clear, then, that three-person games can have quite a different character from that of the two-person zero-sum games that we have been considering. Although chance does not play a role, the results for the three players have nowhere near the stability that we found with Zermelo's theorem. While no individual player can be certain of achieving the result that we found in Table 20.1, such a result may be possible for two players cooperating, in two moves that taken individually appear unfavorable. Such games are seldom considered as fair intellectual contests. This also shows why there are few intellectual games for three or more players. There have at times appeared three-person variants of board games like chess, but none has achieved significant popularity.[2] An exception is board games for

---

[2]See D. B. Pritchard, *The Encyclopedia of Chess Variants*, Surrey 1994, in particular, the entries on three-person and four-person variants (pp. 310–313, and the cross

four players divided into two teams. Since the members of a team either win together or lose together, such games have more the character of a two-person game.

However, the idea of intellectually demanding multiperson games is not unthinkable. One must assume that the sum of the values that each player can achieve on his own is equal to zero. Then, no coalition can achieve more than can its individual members acting alone. In the theory of cooperative games, of particular interest in applications to economics involving coalitions, such games are called *inessential*. In this sense, an example of a game that is both inessential and at the same time balanced in its prospects for winning is a chess tournament in which each player plays every other participant twice, alternating the right of first move. On the one hand, this tournament game is symmetric, so that no player can be assured of a winning value greater than 0. On the other hand, each combination of two games has the value zero, due to the symmetry of the two games. Thus it is theoretically possible for each player to assure himself of at least a win value of zero over the entire tournament. In the end, this is due to the fact that in such tournaments, coalitions can operate only on the level of individual games, which offers no real benefit due to the zero-sum nature of the games.[3] However, caution is advised: if a victor is to be determined, then it must not be based on the player with the highest number of points, since then a coalition could assist a selected member. However, such support within a coalition is fruitless if each player is declared a winner of a match who has a total result of at least zero.

Multiperson Games with Perfect Information

One way of generalizing Zermelo's theorem to multiperson games is to investigate a game's equilibrium. An equilibrium involves a strategy for each player, and these strategies taken together

references to pp. 113–119); Siegmund Wellisch, Das Dreierschach, *Wiener Schachzeitung* **15**, 1912, pp. 322–330.
    [3]A nice example of how tournaments designed on other principles can have considerable significance from a game-theoretic point of view is the first round of the 1982 World Cup soccer tournament. Germany and Austria found themselves in the last game of their set at a score that would allow both of them to advance to the next round. They "agreed" to leave things as they were, provoking a loud protest in the media that the ball was simply being kicked leisurely about the field, behavior that is incomprehensible from a game-theoretic perspective.

should satisfy the following condition: if all the strategies except for that of one player use such an equilibrium, then this one player cannot do better than to do likewise. That is, for no individual player is it worthwhile to deviate from an equilibrium. Furthermore, strategic combinations that do not form an equilibrium are recognizable in that after the fact, at least one player should be unhappy with his strategy, since a different strategy would have brought a higher win value.

In principle, every finite multiperson game with perfect information, even one without a zero-sum character, can be analyzed like the three-person game considered in this chapter. In the reverse chronological order of the game, one obtains an equilibrium step by step, which shows that at least one such equilibrium exists. An existence theorem was first formulated in this generality by Kuhn (1925–).[4] We shall consider in Chapter 43 what more can be said about equilibria in games without perfect information.

In comparison to the special case of two-person zero-sum games, the equilibria of general multiperson games with perfect information have much less stability. Thus the expectation of victory, being linked to an equilibrium, can be destroyed both by poor play of a single opponent or the deliberate cooperation of several opponents; thus in our example, the player who was to draw from a pile of ten stones was no longer assured of victory. On the other hand, it is possible for there to be equilibria of different kinds. For example, how does a player act when he can choose between two moves that have the same outcome for him, but differing outcomes for his various opponents? In our example, such situations cannot exist, due to the distribution of winnings in the form $(1, 0, -1)$, though they would exist if the distribution were $(2, -1, -1)$ with the other rules unchanged. And even in games for two persons, though without the zero-sum property, there can exist several equilibria of varying winning distributions.

To form such coalitions is considered taboo in many games, even if the rules do not explicitly forbid it. In any case, it is of interest to measure the "power" of such coalitions. In 1928, John von Neumann came up with an approach to multiperson

---

[4]H. Kuhn, Extensive games, *Proceedings of the National Academy of Sciences of the USA* **36**, 1950, pp. 570–576. The theorem can also be found in most books on game theory.

zero-sum games that used coalitions to reduce the situation to a two-person game.[5] In the case of three-person games, it suffices to calculate the winnings that each player can ensure for himself. Because of the zero-sum property, this sum can never be positive. In our example, this winning sum, as we saw from Table 20.2, can take on the values $0, -1, -2, -3$, depending on the number of stones. If we change the signs of these numbers, the number provides information about the portion of the total winnings that no player can be sure of obtaining. This is the amount of "booty" that a two-person coalition can obtain above the individually obtainable winnings. Thus if the game starts with ten stones, the winning distribution $(0, -1, -1)$ shows that a bonus of two units of winnings accrues to the two-person coalition.

[5] John von Neumann, On the Theory of Games of Strategy, in: Contributions to the Theory of Games IV, *Annals of Mathematics Studies* **40**, Princeton 1959. In Chapter 43, we will discuss von Neumann's ideas about coalitions in somewhat more detail.

21

Nim: The Easy Winner!

Jack and Jill alternate removing stones from three piles, which initially contain six, seven, and eight stones. A turn consists in choosing one of the piles and removing one or more stones from it, up to the total number of stones in the pile. The winner is the player who removes the last stone. Jill goes first. How should she play?

Nim, as this game is called, is perhaps the best-known game for which a complete mathematical theory exists. Thus the reader may already know the answer to the question posed: Jill should remove seven stones from the largest pile.

Behind this solution stands a formula that was discovered in 1902 by Charles Bouton.[1] Bouton showed that the player to move second has a winning strategy available precisely when a certain "nim sum"[2] is equal to zero. This sum is the result of a base-2 addition without carry. An example of a position for which the second player has a winning strategy is three piles of sizes 6, 7, and 1. The binary representation of these numbers, namely, 110, 111, and 1, yields the nim sum 0. See "Nim Addition" for more on nim addition.

---

[1] Charles L. Bouton, Nim, a game with a complete mathematical theory, *Annals of Mathematics*, **Series II, 3**, 1901/02, pp. 35–39. Richard Guy writes in *Mathematical Reviews* 1982 f: 90101 that nim is probably not much older than Bouton's article. The oft-repeated "fact" in books on games that nim is an ancient game is not supported by any evidence.

[2] Not to be confused with dim sum, the Chinese dumplings.

Nim Addition

The reader is probably aware that the decimal system that we use in our everyday lives is not the only way in which numbers can be represented, and that there is also a binary system, which is used by computers for the internal representation of numbers. In binary representation, the number 2 is used in place of the decimal system's 10, so that one needs only two symbols, 0 and 1. Instead of representing numbers as powers of 10, as in the decimal representation

$$209 = 2 \times 100 + 0 \times 10 + 9 \times 1$$
$$= 2 \times 10^2 + 0 \times 10^1 + 9 \times 10^0,$$

one can use powers of 2, in which case we may write

$$209 = 1 \times 128 + 1 \times 64 + 0 \times 32 + 1 \times 16$$
$$+ 0 \times 8 + 0 \times 4 + 0 \times 2 + 1 \times 1$$
$$= 1 \times 2^7 + 1 \times 2^6 + 0 \times 2^5 + 1 \times 2^4$$
$$+ 0 \times 2^3 + 0 \times 2^2 + 0 \times 2^1 + 1 \times 2^0.$$

Thus we may write the number represented by 209 in base 10 as 11010001 in base 2. The arithmetic operations are carried out just as in the decimal system, except that the difference in the number of digits available means a difference in carrying: in base 10 a carry takes place if the sum of digits is greater than 9, while in binary, there is a carry if the sum of digits exceeds 1. Here is an example of binary addition of the numbers 5 and 7:

$$
\begin{array}{r}
1\,0\,1 \\
1\,1\,1 \\
+\,1\,1\,1 \\
\hline
1\,1\,0\,0
\end{array}
$$

In this particular example, there are three carries of 1. If we were simply to ignore all the carries, then we would no longer have a correct sum. But we may simply declare such carryless addition to be a new arithmetic operation, called *nim addition* and indicated with the symbol "$+_2$." We would then have $5 +_2 7 = 2$.

Nim addition satisfies most of the properties of usual addition. It is commutative, that is, $a +_2 b = b +_2 a$, and associative, meaning that $(a +_2 b) +_2 c = a +_2 (b +_2 c)$. Zero remains a neutral element in nim addition, since one always has $a +_2 0 = 0 +_2 a = a$. We have no need of negative numbers, since for every number we have $a +_2 a = 0$. When we compare nim addition to regular addition, we see that a sum in the former never exceeds the analogous sum in the latter; that is, we always have $a +_2 b \leq a + b$.

We note the following interesting, though rather complex, formula:

$$a +_2 b =$$
$$\min \left(\mathbb{N} - \{ a' +_2 b \mid 0 \leq a' < a \} - \{ a +_2 b' \mid 0 \leq b' < b \} \right).$$

This means that one can calculate a nim sum from two other numbers in the following way: for each summand, one forms all nim sums in which that summand is replaced by a smaller value. The nim sum is then the smallest natural number that is not one of the nim sums thus calculated.

One can proceed as follows to determine whether a winning move is available: first, given the sizes of the piles a, b, c, calculate the nim sum $s = a +_2 b +_2 c +_2 \cdots$. There is a winning move only if this sum is not equal to zero. One can recognize a pile from which stones can be removed by the fact that adding its size to the nim sum s results in a number smaller than that size. For example, if $a +_2 s < a$, then a win can be achieved by removing stones from the first pile. One should remove a quantity of stones resulting in a remainder of $a +_2 s$ stones. Then, just as desired, the resulting position will contain the nim sum $(a +_2 s) +_2 b +_2 c +_2 \cdots = s +_2 s = 0$.

In our example, with piles of sizes 6, 7, and 8, the nim sum is $s = 6 +_2 7 +_2 8 = 9$. If one now "nim adds" this value to each of the three piles in turn, one sees that in the case of the third pile, the result is a reduction, namely, $8 +_2 9 = 1 < 8$. The reduction of this pile to one stone, as depicted in Figure 21.1, is, in fact, the only winning move.

The fact that Bouton's rule actually works in all cases is based on two significant properties of nim sums of nim configurations:

- for every configuration with positive nim sum, there exists at least one move that reduces the nim sum to zero.

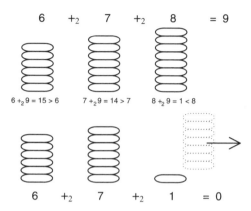

Figure 21.1. The only winning move: take seven stones from the right-hand pile.

- one can never remove stones from a configuration with nim sum zero and achieve another configuration with nim sum zero (see Note 1 at the end of the chapter).

Starting from a configuration with positive nim sum, a player can use these two properties to guarantee that there is always a move that will yield a nim sum of zero. After the opponent's move, the player is guaranteed a configuration with positive nim sum. This continues until the last stone is removed. Since the configuration of no stones has nim sum zero, such a configuration cannot be achieved by the opponent.

The special form of Bouton's formula makes nim a game that is particularly well suited for programming on a computer. Most programming languages allow bitwise operations on positive numbers. The nim sum of two integers a and b can then be calculated by means of

$$a \text{ XOR } b = (a \text{ OR } b) \text{ AND } (\text{NOT } (a \text{ AND } b)).$$

It is also easy to realize nim sums in switching circuits. It is thus not surprising that nim was the first game to be played by machine. In 1940, Westinghouse Corporation presented its "Nimatron" at the New York World's Fair.[3] This machine, which with its countless relays weighed in at over a ton, could play with a configuration of at most four piles each with at most seven stones. Another apparatus, this one called "Nimrod," appeared in

---

[3]See E. U. Condon, The Nimatron, *American Mathematical Monthly* **49**, 1942, pp. 330–332, US-Patent-Nr. 2 215 544. A different machine for playing nim was described in Raymond Redheffer, A machine for playing the game Nim, *American Mathematical Monthly* **55**, 1948, pp. 343–349.

1951, and attracted attention by defeating the German economics minister Ludwig Erhard.[4]

Chapter Notes

1. The second property is relatively clear. Namely, if one removes stones from a configuration with pile sizes $a, b, c, \ldots$, for which $a +_2 b +_2 c +_2 \cdots = 0$, resulting in the configuration $a', b, c, \ldots$, then one has the relation

$$a' +_2 b +_2 c +_2 \cdots = a' +_2 a +_2 a +_2 b +_2 c +_2 \cdots = a' +_2 a.$$

For $a' < a$ this number is never equal to zero.

To prove the first property, we must prove that in the case

$$s = a +_2 b +_2 c +_2 \cdots > 0,$$

at least one of the numbers $a +_2 s$, $b +_2 s$, $c +_2 s$, $\ldots$ is smaller than the original pile size $a, b, c, \ldots$. If z denotes the largest power of two less than or equal to s, then the binary representation of at least one of the pile sizes has a 1 in the place where z has a 1. If, for example, that is the case for the first pile, with a stones, then one has

$$a +_2 s = (a +_2 z) +_2 (s +_2 z) = (a - z) +_2 (s - z) \leq a - z + s - z < a.$$

---

[4]Digital computers applied to games, in B. V. Bowden, *Faster Than Thought*, London 1953, pp. 287, 394 ff.

22

Lasker Nim:
Winning Along a Secret Path

There is a variant of nim in which a move consists in either removing stones from a pile or in dividing a pile of at least two stones into two piles, not necessarily of equal size. The player who removes the last stone wins. Is there a simple analysis for this game as there was for nim?

The nim variant that we have just described was invented by Emanuel Lasker, and it is described in his book on games.[1] In honor of its inventor, the game has been given the name Lasker nim.

Lasker also investigated other variants of nim. In doing so, he attempted to extend to his variants the division of configurations into winning and losing positions. To do this, a position is evaluated from the perspective of the player whose turn it is. That is, a configuration is a winning position if it offers the next player to move a winning strategy. Lasker puts it thus:

> If a configuration that we are examining can be brought by a permissible move into a losing configuration, then the configuration under examination is a winning position. If we cannot do so, then it is a losing configuration. There is no third choice.

In contrast to chess, it is unnecessary to distinguish the players as white and black, since their choices of move are the same for a given configuration.

---

[1] Emanuel Lasker, *Brettspiele der Völker*, Berlin 1931. The nim variant is described on pp. 183 ff.; the first of the following quotations appears on pp. 177 f.

Lasker's idea for analyzing the game consists in creating a new configuration from two that have already been classified by simply laying the two sets of piles next to each other. He then seeks criteria for the winning character of a position thus achieved. He makes the following observation:

> Again, a losing configuration appended to a winning configuration results in a winning configuration, while a move that transforms the original winning configuration into a losing configuration also changes the extended configuration into a losing configuration.

What this means is that one can append a losing configuration to an arbitrary configuration without changing that configuration's character. We are left to consider the situation of two winning positions merged into a new position. In contrast to the previous situation, here we cannot state a result unambiguously: for example, the configuration $\{1, 1\}$ composed of the two winning configurations $\{1\}$ and $\{1\}$ is a losing configuration, while the combined configuration $\{1, 2\}$ is a winning configuration. Lasker comes to the following conclusion:

> We see first that two groups of piles can be "equivalent" in that in every losing configuration in which one group appears, it can be replaced by the other without altering the win character of that configuration. Merging two equivalent configurations results in a losing configuration.

And in fact, the converse of this statement holds as well: two configurations are *equivalent* precisely when merging them results in a losing configuration (see Note 1 at the end of the chapter). In particular, equivalent configurations always have the same win character. Moreover, all losing configurations are equivalent to one another, while the totality of winning configurations can be divided into equivalence classes. Using this concept of equivalent configurations, Lasker was able to classify a configuration by showing its equivalence to a configuration with a particularly simple structure. Thus for his nim variant, he found the following "prime," as he called them, "piles that are not equivalent groups of smaller piles": $\{1\}$, $\{2\}$, $\{3\}$, $\{7\}$, $\{15\}$, $\{31\}$, and so on. Except for $\{2\}$, these are all numbers that are 1 less than a power of 2. Lasker remarks:

> An arbitrary pile is either equivalent to one of these piles or to a group of these piles. For example, $\{4\}$ is equivalent to $\{1, 2\}$; $\{5\}$ is equivalent to $\{1, 3\}$; $\{6\}$ is equivalent to $\{2, 3\}$; $\{8\}$ is equivalent to $\{1, 2, 3\}$.

More extensive configurations such as $\{1, 3, 5, 8\}$ can be easily simplified to $\{1, 3, 1, 3, 1, 2, 3\}$ and finally to $\{1, 2, 3\}$. Since from this configuration one can obtain the losing position $\{1, 2, 1, 2\}$ by dividing the last pile, the original configuration is a winning configuration. Every configuration is equivalent to exactly one combination of prime configurations of different sizes. There are only a few ways to obtain a losing configuration, which is equivalent, of course, to the empty configuration: if a losing configuration was to result from a nonempty collection of prime piles of different sizes, then there would be an equivalence among these piles, and the largest pile would then not be prime. That is, in grouping prime piles of different sizes into a configuration, only the empty grouping yields a losing configuration! For this reason alone, the configuration $\{1, 2, 3\}$ investigated above cannot be a losing configuration. Therefore, once one has determined all the prime piles of a nim variant, that game can be analyzed like standard nim:

- every pile in a configuration is first replaced by equivalent prime piles.

- once this has been done, a losing position is present only if every prime pile occurs an even number of times.

The method thus described is not made explicit in Lasker's book. Although one obtains usable results, it is much more complex than Bouton's criterion for standard nim, whose prime pile sizes are powers of two. An improvement was obtained in 1935 by Roland Sprague (1894–1967), and independently in 1939 by Patrick Michael Grundy.[2] Their principal contribution beyond what Lasker did was to find a connection between generalized and standard versions of nim. It turns out that the variations of nim affect more the appearance of the game to the players than the underlying mathematical substance. In the theory developed over the years, the games investigated by Sprague and Grundy have become known as *impartial* games, in which the last player to move wins. The following properties are assumed for these games:

- the game is a two-person game with perfect information with no elements of randomness.

- the two players alternate first move with a fixed initial configuration in each game.

- the available moves for a particular configuration are independent of which player's turn it is; thus the name *impartial*. The configurations

[2]R. Sprague, Über mathematische Kampfspiele, *Tôhuko Mathematical Journal* **41**, 1935/6, pp. 438–444; P. M. Grundy, Mathematics and games, *Eureka* **2**, 1939, pp. 6–8, reprinted in *Eureka* **27**, 1964, pp. 9–11.

derived from a given configuration are called that configuration's *successors*.

- the player who makes the last move is the winner.

- the game is finite, in that it always ends after a finite number of moves. It is also frequently required that the number of possible moves from every configuration be finite.

- if several positions $G, H, L, \ldots$ are assembled into a new position, then this occurs in the form of a *disjunctive sum* $G + H + L + \cdots$. That is, confronted with such a configuration, a player moves by selecting one of the components $G, H, L, \ldots$ and moving under the rules applicable to that component.

Disjunctive sums can be formed from configurations in any number of nim variants. Thus Lasker's notion of equivalence can be extended to arbitrary configurations in the games under consideration here. This has the advantage that the basic assertions of the theory of Sprague and Grundy can be formulated rather simply:

In an impartial game in which the last player to move wins, every configuration is equivalent to a pile of standard nim.

The size of this pile is called the *Grundy value*. It serves to characterize the underlying position. The Grundy value has two characteristics that simplify its calculation:

- the Grundy value of a position is equal to the smallest natural number that does not occur among the Grundy values of its successors (see Note 2 at the end of the chapter).

- the Grundy value of a disjunctive sum of configurations is equal to the nim sum of the Grundy values of the configuration's components (see Note 2 at the end of the chapter).

To win, a player must always attempt to achieve a configuration with Grundy value 0, for such a position is equivalent to an empty nim pile, so that the opponent will be faced with a losing configuration. Here, one may use the two properties that we have formulated for determining the Grundy value in two steps: on the one hand, the sequence of Grundy values $g(0)$, $g(1)$, $g(2)$, ... is calculated for the configurations that consist of a single pile. On the other hand, using nim addition, this result can be used to determine the Grundy value of configurations with more than one pile. In our example of Lasker nim, the process looks like this:

The empty pile has no successors, and so its Grundy value is the smallest natural number, namely, $g(0) = 0$. For the Grundy value of a pile containing a single stone, the value 0, as the Grundy value of the empty position, must be excluded, so that one has the Grundy value $g(1) = 1$. If the pile encompasses two stones, then in a single turn, either one or two stones can be taken, or the pile can be split in two. The successors have the Grundy values 1, 0, and $1 +_2 1 = 0$, which yields the Grundy value 2 for the two-stone pile. A pile of three stones offers four possible moves: to remove one, two, or three stones, or to divide the pile. This leads to successors with Grundy values 2, 1, 0, and $1 +_2 2 = 3$, so that the three stones have a Grundy value of $g(3) = 4$. In general, the Grundy values for Lasker nim satisfy the recursion relation

$$g(n) = \min \left(\begin{array}{c} \mathbb{N} - \{ g(0), g1), \ldots, g(n-1) \} \\ - \{ g(1) +_2 g(n-1), g(2) +_2 g(n-2), \ldots \} \end{array} \right).$$

We collect some of the results of this formula in Table 22.1.

The Grundy values in Table 22.1 are periodic: the Grundy value of a pile is larger by four than the Grundy value of a pile that is smaller by four stones; that is, $g(n) = g(n-4) + 4$. We demonstrate how to find a winning position using these Grundy values using the example configuration $\{ 1, 3, 5, 8 \}$. Its Grundy value is the nim sum of the Grundy values of the individual piles, namely,

$$g(1) +_2 g(3) +_2 g(5) +_2 g(8) = 1 +_2 4 +_2 5 +_2 7 = 7.$$

A winning move for the given position in Lasker nim is seen to be analogous to such a move for the position $\{ 1, 4, 5, 7 \}$ in standard nim. There are three such moves:

- based on the winning move in standard nim of $\{ 1, 4, 5, 7 \}$ to $\{ 1, 3, 5, 7 \}$, the three stones constituting the second pile are split into two piles of sizes 1 and 2, to realize the Grundy value 3 (see Figure 22.1).

- based on the winning move in standard nim of $\{ 1, 4, 5, 7 \}$ to $\{ 1, 4, 2, 7 \}$, three stones are removed from the third pile, of size 5, so that with the remainder of 2 stones, the Grundy value 2 is achieved.

- based on the winning move in standard nim of $\{ 1, 4, 5, 7 \}$ to $\{ 1, 4, 5, 0 \}$, the fourth pile is removed completely.

This example demonstrates that disjunctive variants of an impartial game can be won almost as easily as in standard nim. However, someone uninitiated into the winning principle would be unlikely to figure it out.

Size of Pile	Grundy Value
n	$g(n)$
0	0
1	1
2	2
3	4
4	3
5	5
6	6
7	8
8	7
9	9
10	10
11	12
12	11

Table 22.1. Grundy values for Lasker nim.

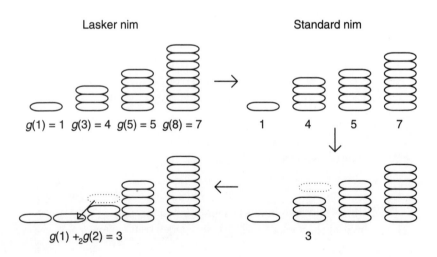

Figure 22.1. Lasker nim: first, the configuration is transformed into an equivalent configuration in standard nim; the winning move found in the standard nim configuration is then transformed back to Lasker nim.

Nim Variants en Masse

Of course, in addition to Lasker nim, there are many other variants of nim that one can imagine. We mention two of them here:

- in bowling nim,[3] either one or two stones may be removed from a pile. Thereafter, the pile can be divided.

- in subtraction games, certain positive integers $s_1, s_2, \ldots, s_n$ are specified, indicating how many stones may be removed from a pile on a given turn. Piles with fewer than the smallest of these numbers may be removed completely, since there would otherwise be no legal move for that pile.

These and other nim variants, including the original nim and Lasker nim, can be categorized in a very large class of nim games whose rules can be encoded in the form of a sequence of integers. In such a sequence $A_0 A_1 A_2 A_3 \ldots$, the number A_i gives information about the circumstances in which, for the nim variant in question, i stones may be removed from a pile. This is allowed precisely when the reduced pile is then divided into t piles, where the number $t \geq 0$ must satisfy the condition $2^t +_2 A_i > 0$. If the required number of piles is not achievable by division, it is not permitted to remove i stones:

- $0.333\ldots$ is the code for standard nim, where corresponding to the numbers $3 = 2^1 + 2^0$, an arbitrary number of stones may be removed, so that $t = 0$ or $t = 1$ pile remains.

- 0.77, short for $0.77000\ldots$, encodes bowling nim. The first two digits $7 = 2^2 + 2^1 + 2^0$ indicate that only one or two stones may be removed, while the remaining pile may then be divided so that $t = 0$, 1, or 2 piles result.

- $4.333\ldots$ is the code for Lasker nim. Moves are as in standard nim, except that a move of removing no stones is permissible if in the process, $t = 2$ piles result.

---

[3]The name is derived from the fact that this game can be seen as a variant of bowling: one lines up the pins in a row, so that on a single throw, one or two pins can be hit. Pins standing next to one another without a gap constitute a "pile."

- 0.03003 stands for a special subtraction game in which either two or five stones may be removed, and as a result, either $t = 0$ or $t = 1$ pile results. In particular, piles with only a single stone can no longer be drawn from.

Nim codes of this type were discovered in 1955 by Richard Guy and Cedric Smith.[4] The nim variants thus created, sometimes called *octal games*, were used by Guy and Smith in their investigations of Grundy values. Using the EDSAC computer, a significant computer at the time, they calculated the sequence of Grundy values $g(0), g(1), g(2), \ldots$ and looked for patterns. For example, they found for the game 0.137, known as Dawson's chess,[5] a period of length 34, namely, the Grundy values

$$8\ 1\ 1\ 2\ 0\ 3\ 1\ 1\ 0\ 3\ 3\ 2\ 2\ 4\ 4\ 5\ 5\ 9\ 3\ 3\ 0\ 1\ 1\ 3\ 0\ 2\ 1\ 1\ 0\ 4\ 5\ 3\ 7\ 4,$$

where at the beginning, the values $g(0) = g(14) = g(34) = 0$ and $g(16) = g(17) = g(51) = 2$ deviate from the pattern of periodicity. Much more complex is the game 0.16, whose Grundy values, starting from 105 350, exhibit a period of 149 450.[6] In addition to nim games with periodic Grundy values, among which is bowling nim, with a period length 12 beginning with $g(72)$,[7] there are also variants in which the growth of the Grundy values is periodic, as we saw with the example of Lasker nim. If the prescribed set of pile reductions $\{s_1, s_2, \ldots\}$ for a subtraction game is finite, then its sequence of Grundy values is periodic. Moreover, they possess some rather interesting additional properities, even in the infintite case, as discovered by Ferguson in 1974:[8]

- $g(n) = 1$ precisely when $g(n - s_1) = 0$ for the smallest number s_1 in the permissible set $\{s_1, s_2, \ldots\}$.
- in the case $g(n) = 0$ with $n \geq s_1$, there is a permissible reduction s_k with $g(n - s_k) = 1$.

[4]Richard K. Guy, Cedric A. B. Smith, The *g*-values of various games, *Proceedings of the Cambridge Philosophical Society* **52**, 1956, pp. 514–526. See also Richard K. Guy, *Fair Game*, Arlington 1989; John H. Conway, *On Numbers and Games*, Natick, MA 2002, Sections 11, 12; E. Berlekamp, J. Conway, R. Guy, *Winning Ways*, second edition, Natick, MA 2001, volume 1, Chapters 2–4.

[5]The game is discussed in volume 1 of *Winning Ways*.

[6]A. Gangolli, T. Plambeck, A note on periodicity on some octal games, *International Journal of Game Theory* **18**, 1989, pp. 311–320.

[7]A table of all Grundy values for bowling nim will be given in Chapter 26.

[8]T. S. Ferguson, On sums of graph games with last player losing, *International Journal of Game Theory* **3**, 1974, pp. 159–167.

The second property, which can be derived from the first (see
Note 4 at the end of the chapter), was used by Ferguson to
investigate the *misère* version, in which the last player to move
loses. We shall return to this issue in Chapter 26.

Chapter Notes

1. A proof of this statement is not particularly difficult: let us begin with the
   two equivalent configurations $\{G\}$ and $\{H\}$. Without changing the win
   character of the combined configuration $\{G, H\}$, we may replace H by its
   equivalent G. The resulting configuration $\{G, G\}$ is a losing configuration,
   since the second player will be able to imitate the moves of the first player.
   Conversely, if the combined configuration $\{G, H\}$ is a losing configuration,
   then we need to show that for every configuration $\{L\}$, the two combined
   configurations $\{G, L\}$ and $\{H, L\}$ have the same win character. That this
   is indeed so can be seen by considering the configuration $\{G, G, H, L\}$: this
   arises from $\{G, L\}$ by adjoining the losing configuration $\{G, H\}$ and from
   $\{H, L\}$ by adjoining the losing configuration $\{G, G\}$.

2. A proof can be derived from the following considerations: if the successors $G_1, G_2, \ldots$ of a configuration G are equivalent to nim piles of
   sizes $m_1, m_2, \ldots$, then we must show that the disjunctive sum $G + \star m$
   is a losing configuration, where $\star m$ represents the nim pile with $m =$
   $\min(\mathbb{N} - \{m_1, m_2, \ldots\})$ stones. This is so, because every possible successor is seen to be a winning configuration: if one draws from the left
   component of $G + \star m$, then the result is a configuration that is equivalent
   to the position $\star m_i + \star m$ for some index i. Since $m_i \neq m$, this is a winning
   position. On the other hand, if one draws from the right component of
   $G + \star m$, then this leads to a configuration $G + \star n$ for some integer $n < m$.
   Since m is the smallest natural number not among $m_1, m_2, \ldots$, it follows
   that $G + \star n$ is one of the configurations $G + \star m_i$. This is a winning configuration, since one can attain a losing configuration after the move to
   $G_i + \star m_i$.

3. If m and n are the Grundy values of two configurations G and H, then
   $G + \star m$ and $H + \star n$ are losing configurations. Moreover, Bouton's theory
   tells us that the standard nim configuration comprising three piles $\star m +$
   $\star n + \star(m +_2 n)$ is a losing configuration. It therefore follows that $G + H +$
   $\star m + \star n$ and hence $G + H + \star(m +_2 n)$ are losing configurations.

4. The first statement can be proved in both directions by complete induction,
   where the beginning of the induction is clear due to $g(0) = g(1) = \cdots =$
   $g(s_1 - 1) = 0$ and $g(s_1) = 1$.

In the case $g(n) = 1$, we have for all permissible reductions the inequality $g(n - s_k) \neq 1$ (provided that $n - s_k \not< 0$). By the induction hypothesis, we thus always have $g(n - s_k - s_1) \neq 0$ (provided that $n - s_k - s_1 \not< 0$), from which it follows that $g(n - s_1) = 0$.

Converserly, if $g(n - s_1) = 0$, then for all permissible reductions s_k, we have the inequality $g(n - s_1 - s_k) \neq 0$ (provided that $n - s_1 - s_k \not< 0$). From the induction hypothesis, we obtain $g(n - s_k \neq 1$ (provided that $n - s_k \not< 0$; the cases $0 \leq n - s_k < s_1$ are clear) for all reductions s_k. Together with the initial condition $g(n - s_1) = 0$, it finally follows that $g(n) = 1$.

We now come to the second assertion, in which we begin with a number $n \geq s$ with $g(n) = 0$. In particular, in such a case, we have $g(n - s_1) \neq 0$, which again has as a consequence the existence of a reduction s_k with $g(n - s_1 - s_k) = 0$. From what we have already proved, we may conclude that $g(n - s_k) = 1$.

23

Black-and-White Nim: To Each His (or Her) Own

Black-and-white nim is played with towers of black and white stones. A turn consists in a player choosing a stone of his own color and removing it together with all the stones lying above it from the pile. The player who makes the last move wins. How can one find a winning move, for example, from the configuration shown in Figure 23.1?

Black-and-white nim, which is a simplified version of the game hackenbush that was studied by Berlekamp, Conway, and Guy,[1] has a significant characteristic that differentiates it from the nim variants that we have been investigating thus far: like most board games, it is not impartial; that is, the available moves depend on whose turn it is. Thus, for example, white's moves from the configuration of Figure 23.1, shown on the left-hand side of Figure 23.2, differ from those of black, four of whose seven possible moves appear on the right-hand side (the remaining three are at best as favorable as the second move from the top).

Figure 23.1. A configuration in black-and-white nim.

---

[1]See the bibliography at the end of the chapter.

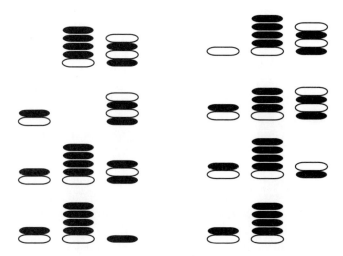

Figure 23.2. Black and white do not have the same moves in the configuration of Figure 23.1. White's moves are shown on the left, four of black's seven moves on the right.

Aside from the fact that the moves in a given configuration may be different for the two players, the assumptions of the previous chapter are in force. Therefore, for the general part of our investigations, we assume the following:

- the game is a two-person game with perfect information without random effects.

- the players alternate first move from a fixed opening configuration.

- the player to make the last move wins.

- the game is finite; that is, it always ends in a finite number of moves. It is generally also required that the number of possible moves on each turn be finite.

- if several configurations $G, H, L, \ldots$ are assembled into a larger configuration, then this occurs in the form of a disjunctive sum $G + H + L + \cdots$. A player moves in such a sum configuration by choosing a component $G, H, L, \ldots$ and making a permissible move from that component.

Such games were first studied systematically in the 1970s by John Horton Conway, who extended the theory of nim to nonimpartial games, obtaining a large number of results that are interesting from both the gaming and mathematical points of view. It later became clear that some aspects of Conway's results had been discovered earlier, in 1953 by John Milnor (1931–) and in 1959 by Olof Hanner (1922–),[2] whose findings were scarcely noticed at the time. As with nim, it is the configurations that are the actual objects of study. In the case of nonimpartial games, this has the result that there are always *two* games to be studied that have a given initial configuration: the version in which white plays first, and the version in which black plays first. One seeks winning strategies for *both* games: *who* can win, and *how*?

As with the impartial nim variants, the generalized theory, often called *combinatorial game theory*, deals principally with finding winning strategies for configurations given by disjunctive sums by analyzing the sums' components. This is accomplished in two steps: first, the components of a disjunctive sum are replaced by equivalent but less complicated configurations. Then, one attempts to determine for a disjunctive sum the options for winning, calculating them with a mathematical formula if possible.

If losing configurations can be determined for a disjunctive sum in an impartial game, then for a nonimpartial game, the winning moves for both variants—with white and black each having the first move—must be considered. This leads to the notion of a *null configuration*, a configuration in which the player to go second has a winning strategy:

> Given a null configuration H, a player who as first or second to move has a winning strategy for a configuration G can also win if the game starts in configuration $G + H$ (provided that the player to move first remains the same). That is, with optimal play, the addition of a null configuration does not alter the outcome of the game.

We can see this by quoting Lasker's work on nim almost word for word: if the second player to move has a winning strategy from configuration G, then he can also win from initial configuration $G + H$ by countering every move of the first player in one of the two components as if the other one was not present. On the other hand, if the first player has a winning strategy for configuration G, then he will also win if the game begins in configuration $G + H$ by first choosing the winning move from configuration G, thereby arriving at the situation that we have just analyzed.

A trivial example of a null configuration is the one denoted by 0, in which neither player can move. Additional examples of null configurations

---

[2]John Milnor, Sums of positional games: in H. W. Kuhn, A. W. Tucker (eds.), Contributions to the Theory of Games II, *Annals of Mathematics Studies* **28**, 1953, pp. 291–301; Olof Hanner, Mean play of sums of positional games, *Pacific Journal of Mathematics* **9**, 1959, pp. 81–89.

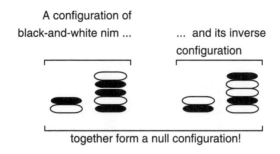

Figure 23.3. A null configuration formed from a given configuration and its inverse.

can be obtained by taking the sum of an arbitrary configuration G and its *inverse configuration* $-G$. This is the configuration in which the two players, beginning in configuration G, switch roles; that is, the possible moves remain unchanged, but white makes the moves that black would make and vice versa. In the case of black-and-white nim, this can be achieved simply by exchanging the colors of the stones. If one forms the sum $G + (-G)$, then this is a null configuration, since the second player can mimic the moves of the first player in the other component. This situation is depicted in Figure 23.3.

A more interesting example of a null configuration is the following three towers:

Whoever is to move is in trouble. If white goes first, then she has only one move, up to symmetry, namely, to the configuration:

Black can now win by removing the top stone from the left-hand tower. On the other hand, if black goes first, his best move is to create the configuration:

But if white now removes the middle tower, black loses.

Null configurations like that in this example, consisting of a sum of configurations, can often be used to simplify a given configuration. Thus if $H + L$ is a null configuration, then in every sum of configurations, H and $-L$ are interchangeable, in that the winning possibilities of the players are not altered. That is, if G is an arbitrary configuration, then $G + (-L)$ has the same winning possibilities as $G + (-L) + (H + L)$, and thus the same as $G + H$. Thus the configurations H and $-L$ are equally favorable. As in the case of impartial games, they are called *equivalent*, and the notation $H = -L$ is used.

We shall now examine the configuration

to see how equivalent configurations can be used to simplify a given configuration. Without changing the outcomes for the two players, the towers of a single color can be replaced by a single stone. Moreover, pairs of one white and one black stone can be removed, since together they form a null configuration. Finally, two of the right-hand towers can be replaced by a single stone. We can do all of this because of the following equivalences:

Once all of the simplifications have been carried out, we are left with the single tower:

Thus white wins whether she moves first or second.

If one is prepared to invest a certain amount of effort, almost every configuration of black-and-white nim can be analyzed in this way. Indeed, we have the following theorem:

> Every configuration of black-and-white nim is equivalent to a
> sum of towers of the following form:

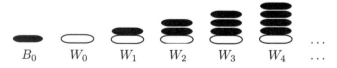

where we note that $W_i = W_{i+1} + W_{i+1}$.

If we express the individual towers of a given configuration as a sum of towers from the collection just presented, then the sum can be easily analyzed. Thus for the configuration of Figure 23.1, considered at the start of this chapter, we arrive at

$$\text{⬭} = W_1 \qquad \text{⬭} = W_4 \qquad \text{⬭} = B_0 + W_2 + W_3$$

and so finally,

$$W_1 + W_4 + B_0 + W_2 + W_3 = -W_4.$$

Therefore, black can win whether he goes first or second.

Although we have solved our initial problem, instead of resting on our laurels, let us look more closely into two points that arose in our considerations. First, we should look more closely at the equivalence $W_i = W_{i+1} + W_{i+1}$. And then we should close an even bigger gap in our explanation, occurring in the first step, in which each individual tower was replaced by a sum of towers. In principle, it is possible to establish a presumed equivalence of two configurations G and H by proving the difference $G - H$ to be equivalent to the null configuration by an analysis of moves. Since this is not very efficient, we shall develop a simpler procedure in the remainder of the chapter: this will allow us to reduce an arbitrary configuration of black-and-white nim to a simpler configuration involving sums of configurations $B_0, W_0, W_1, W_2, W_3, \ldots$, given that such representations already exist for all configurations reachable in a single move.

Since in an impartial game it suffices simply to distinguish winning configurations from losing ones, in the case of nonimpartial games, we may define four cases, depending on the winning chances of the players, since each configuration corresponds to two possible games. One of the four classes is the previously analyzed null configuration:

- a configuration is called *positive* if white has a winning strategy regardless of who goes first.

- a configuration is called *negative* if black has a winning strategy regardless of who moves first.

- a configuration is called a *null configuration* if the second player has a winning strategy.

- a configuration is called *fuzzy* if the first player has a winning strategy.

That every configuration falls in one of the four classes follows directly from Zermelo's theorem. Figure 23.4 shows simple examples of configurations for each of the four classes. Since there are no fuzzy configurations in black-and-white nim, we offer an example from standard nim, whose stone we have drawn half black, half white, to indicate that either player may take the stone.

Configuration	 ⬭	 ⬭	 ⬭ ⬭	 ⬭
Winning strategy for...	white	black	the second player	the first player
Configuration type	positive $G > 0$	negative $G < 0$	null configuration $G = 0$	fuzzy $G \parallel 0$

Figure 23.4. The four win classes, with an example configuration for each.

The notation that we have used in Figure 23.4 can be interpreted along the lines of the way we have interpreted the "=" sign, namely, as a way of comparing a configuration G with the end configuration 0. If we extend this idea to the comparison of two arbitrary configurations G and H, we can make an even finer distinction among configurations. We let $G > H$, $G = H$, $G < H$, and $G \parallel H$ denote the corresponding statement with respect to the configuration $G + (-H)$ (which we shall write as $G - H$) in relation to the end configuration 0. How are these relationships to be interpreted in practice? Here is a simple example: for the configurations

$$G = \text{⬭} \text{ and } H = \text{⬭} \text{ we have } G > H,$$

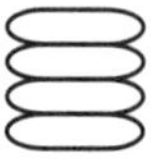

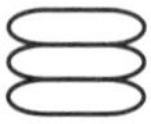

$$\text{since } G - H = \text{⬭} > 0.$$

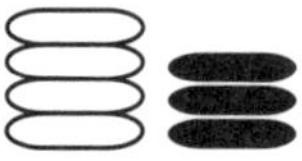

The configurations G and H are both positive, and so they offer white a winning strategy whether she moves first or second. Nonetheless, in certain special situations, configuration G can be more favorable for white than H, namely, when they appear as components of a disjunctive sum. Thus, for

$$L = \text{⬭} \text{ we have } G + L > 0, \text{ but } H + L = 0.$$

That is, white can win from configuration $G + L$, but that is not the case for configuration $H + L$.

Moreover, we may interpret combined comparisons such as $G \geq 0$, that is, $G > 0$ or $G = 0$, strategically as follows:

- $G \geq 0$: white to move second has a winning strategy.

- $G \leq 0$: black to move second has a winning strategy.

- $G \parallel > 0$: white to move first has a winning strategy.

- $G < \parallel 0$: black to move first has a winning strategy.

If we leave aside for a moment the somewhat unusual-looking relation $G \parallel H$ as the fourth alternative after $G > H$, $G < H$, and $G = H$, we can manipulate these symbols more or less the way we do in other mathematical contexts. Thus, for example, the following laws are easy to prove:

- a relation such as $G > H$ retains its validity by the addition to one side of a null configuration. That is, $G + L > H$ follows from $G > H$.

- the sum of two positive configurations is again positive. That is, if $G > 0$ and $H > 0$, then $G + H > 0$.

- the "greater than" relation is transitive: for three configurations satisfying $G > H$ and $H > L$, it follows that $G > L$.

With the help of the comparison relations and its properties that we have just described, a configuration that occurs as a disjunctive sum can often be analyzed "locally," that is, component by component. If we wish to compare the components G and H of a configuration, then we have merely to investigate the difference configuration $G - H$. The following table shows how the relations $G = H$, $G \geq H$, and $G \leq H$ operate in disjunctive sums. The table compares the configurations $G + L$ and $H + L$ for an arbitrary configuration L:

	Whoever Moves First, for Every Configuration L ...
$G = H$	...white finds $G + L$ just as favorable as $H + L$.
$G \geq H$	...white finds $G + L$ at least as favorable as $H + L$.
$G \leq H$	...white finds $H + L$ at least as favorable as $G + L$.

The converses of the three statements hold as well, which one can see at once if one substitutes the configuration $-H$ for L. In the remaining cases, that is, when the relation $G \parallel H$ holds, one can make no blanket statement as to whether $G + L$ or $H + L$ is more favorable for white. That depends on the structure of L and on who moves first.

There is good reason why the symbols "$=$," "$<$," "$>$," "$\leq$," "$\geq$," "$+$," "$-$," and "0" can be used in the familiar way: many configurations, and in particular, all configurations of black-and-white nim, can be represented as numbers. Thus, the configurations B_0, W_0, W_1, W_2, W_3, ... correspond to the numbers -1, 1, $1/2$, $1/4$, $1/8$, etc. The fact that the operations on configurations agree with those on the associated numbers and are therefore "compatible"[3] with one another is a result of the equalities $W_i = W_{i+1} + W_{i+1}$ and $B_0 = -W_0$. One may now interpret the value of

---

[3]The reader who knows what a homomorphism is may have noticed that we have here a homomorphism between ordered groups.

a configuration as the degree of advantage that white has over black. An advantage of 2 in the configuration

means that white can move two turns longer than black can. Of course, an advantage of $-1/4$ of a move as in the configuration

defies such a simple interpretation. Nevertheless, even a value like $-1/4$ can be interpreted indirectly as a degree of advantage. To do this, simply add four such towers together to achieve a configuration equivalent to a single black stone.

Even configurations that cannot be represented by a number can satisfy size relations. For example, the relation

holds for the nim pile in standard nim, which, like all winning configurations in impartial nim games, is not equivalent to any number. In particular, the Grundy values of nim can never be interchanged with the numbers that we have just described.

To make possible the investigation of configurations on a more abstract level, one frequently writes configurations in the form

$$(\{\, G, H, \ldots \,\}, \{\, P, Q, \ldots \,\}),$$

or simply

$$\{\, G, H, \ldots \mid P, Q, \ldots \,\}$$

for short.

Here, $G, H, \ldots$ and $P, Q, \ldots$ are configurations, in particular, those from which white, respectively black, can move. See "Conway's Universe of Games" for a brief look at how abstractly one can "play" using these ideas.

Whenever possible, in what follows, we will represent configurations as simply as we can. Since for both players, equivalent configurations are in general equally favorable, that is, as components of a disjunctive sum, we shall usually not distinguish between equivalent configurations. Configurations that correspond to a number—and this is something that we have already done with the end configuration 0—will simply be denoted by that number. Thus instead of

where the horizontal line indicates the empty end configuration, we simply write

$$-\frac{1}{4} = \left\{ -\frac{1}{2}, -1 \middle| 0 \right\}.$$

Although the advantages of this notation over the tried and true

$$\{\,|\,\} = 0, \quad \{0\,|\,\} = 1, \quad \{1\,|\,\} = 2, \quad \{\,|\,0\} = -1, \quad \{0\,|\,1\} = \frac{1}{2},$$

may not be apparent, the situation changes dramatically when the relationships become more complicated. This is particularly true for the two following "calculation rules." The first of these concerns *dominated move options*:

> A move that is less advantageous in comparison to other moves can be removed from a configuration. For example, if white can choose between moves to configuration G or configuration H, for which the relation $G \leq H$ holds, then the option to move to G can be dispensed with:
>
> $$\{\,G, H, \ldots \mid P, \ldots \} = \{\,H, \ldots \mid P, \ldots \}.$$

It is almost self-evident that disadvantageous moves can be omitted (see Note 1 at the end of the chapter). As an example, we will simplify the configuration that consists of the tower W_{i+1} in black-and-white nim. We may delete all of black's moves except for the most favorable one:

$$\frac{1}{2^{i+1}} = \begin{matrix} \bullet \\ \vdots \\ \bullet \\ \bullet \end{matrix} = \left\{ 0 \middle| \frac{1}{2^i}, \frac{1}{2^{i-1}}, \ldots, 1 \right\} = \left\{ 0 \middle| \frac{1}{2^{i-1}} \right\}.$$

One can also find similar simple representations for other fractions, such as 3/4. Beginning with the disjunctive sum

$$\frac{3}{4} = \frac{1}{2} + \frac{1}{4} = \{0 \mid 1\} + \left\{ 0 \middle| \frac{1}{2} \right\},$$

two options present themselves for each player, namely,

$$\frac{3}{4} = \left\{ 0 + \frac{1}{4}, \frac{1}{2} + 0 \middle| 1 + \frac{1}{4}, \frac{1}{2} + \frac{1}{2} \right\},$$

which, after the dominated move options are deleted, yields the representation

$$\frac{3}{4} = \left\{ \frac{1}{2} \,\middle|\, 1 \right\}.$$

Similarly, every other fraction whose denominator is a power of 2 can be
represented by a pair of move options. For integers n and k with $n \geq 0$,
we have

$$\frac{2k+1}{2^{n+1}} = \left\{ \frac{k}{2^n} \,\middle|\, \frac{k+1}{2^n} \right\}.$$

Furthermore, this equation, together with the equations

$$0 = \{\,|\,\}, \quad n+1 = \{\,n\,|\,\}, \quad -(n+1) = \{\,|\,-n\,\},$$

forms a complete set of "standard representations" of the numbers that
appear in black-and-white nim. In transforming a given configuration into
such a form, we may make use of the following *simplicity theorem*:

> If a number is representable as a configuration in the form
> $\{\,G\,|\,H\,\}$ with numbers $G < H$, then this number is also repre-
> sentable by all configurations $\{\,P\,|\,Q\,\}$ whose move options P
> and Q satisfy the conditions $G \leq P < \|\,\{\,G\,|\,H\,\} < \|\,Q \leq H$.

More important than this abstract formulation and the proof of the
simplicity theorem (see Note 2 at the end of the chapter) is its most frequent
application, in which a configuration $\{\,P\,|\,Q\,\}$ with $P < Q$ is equivalent
to the "simplest" number s that lies between P and Q, that is, for which
$P < s < Q$ holds. Here a number is by definition simpler, the earlier it
comes in the following sequence:

$$0, \ 1, \ -1, \ 2, \ -2, \ \ldots, \ \frac{1}{2}, \ -\frac{1}{2}, \ \frac{3}{2}, \ -\frac{3}{2}, \ \cdots \ \frac{1}{4}, \ -\frac{1}{4}, \ \frac{3}{4}, \ -\frac{3}{4}, \ \ldots$$

To make clear the connection to the abstract version of the simplicity
theorem, we shall compare a given configuration $\{\,P\,|\,Q\,\}$ with the standard
representation $s = \{\,G\,|\,H\,\}$. For example,

$$\left\{ -\frac{3}{2} \,\middle|\, -\frac{3}{4} \right\} = \{\,-2\,|\,0\,\} = -1, \quad \text{since} \quad -2 \leq -\frac{3}{2} < -1 < -\frac{3}{4} \leq 0;$$

$$\left\{ \frac{9}{16} \,\middle|\, \frac{29}{32} \right\} = \left\{ \frac{1}{2} \,\middle|\, 1 \right\} = \frac{3}{4}, \quad \text{since} \quad \frac{1}{2} \leq \frac{9}{16} < \frac{3}{4} < \frac{29}{32} \leq 1;$$

$$\left\{ \{\,0\,|\,0\,\} \,\middle|\, \frac{1}{2} \right\} = \{\,-1\,|\,1\,\} = 0, \quad \text{since} \quad -1 \leq \{\,0\,|\,0\,\} \,\|\, 0 < \frac{1}{2} \leq 1.$$

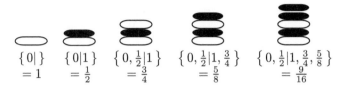

$\{\,0\,|\,\}$ $\{\,0\,|\,1\,\}$ $\{\,0,\frac{1}{2}\,|\,1\,\}$ $\{\,0,\frac{1}{2}\,|\,1,\frac{3}{4}\,\}$ $\{\,0,\frac{1}{2}\,|\,1,\frac{3}{4},\frac{5}{8}\,\}$

$=1$ $=\frac{1}{2}$ $=\frac{3}{4}$ $=\frac{5}{8}$ $=\frac{9}{16}$

Figure 23.5. Stepwise calculation of a black-and-white nim tower.

Once again, this time in other words: if a number appears in the form of a configuration, then this number also appears even when the move options of both players or of one player improve only slightly. In addition, for white in particular, such moves are allowed that are smaller than the given number, and for black, such moves that are greater than this number.

With these two instruments at our disposal, we can now investigate arbitrary towers in black-and-white nim. For example, to calculate the number associated with the tower on the right-hand side of Figure 23.5, we go through all towers that can result from the tower under consideration.

There is thus an efficient procedure for carrying out an analysis of black-and-white nim:

- if the numerical values that can arise from all possible moves are known, then the numerical value of the configuration is known as a result of the simplicity theorem.

- if a given configuration is represented as a disjunctive sum, the procedure needs to be applied only to the individual components, and then the values found for the individual components are added.

Conway's Universe of Games

In the sense of the definition established by Conway, for each configuration,[4] we are dealing with a pair of elements from two sets, each containing exclusively configurations. The reflexivity inherent in the definition is by no means as nonsensical as one might think at first: as the first configuration, the end configuration, denoted by 0, arises directly from "nothing," that

---

[4]For Conway, the notions of *configuration* and *game* are equivalent. Here, however, the word *game* is used in the game-theoretic sense, so that a *configuration* corresponds to two games, one with white moving first, and the other with black moving first.

is, solely in reference to the empty set $\varnothing$, via the construction $(\varnothing, \varnothing)$. In the next step, one obtains three additional configurations, namely,

$$1 = (\{0\}, \varnothing), \quad -1 = (\varnothing, \{0\}), \quad * = (\{0\}, \{0\}).$$

The last of these, the configuration denoted by $*$, corresponds to a nim pile with a single stone. The nim pile of size 2, denoted by $*2$, arises in the next step, in which, among others, the following configurations arise. The usual notation as well as the actual definitions are used, which are abbreviated as $\{\ldots \mid \ldots\}$:[5]

$$2 = \{1 \mid\}, -2 = \{\mid -1\}, \frac{1}{2} = \{0 \mid 1\}, -\frac{1}{2} = \{-1 \mid 0\},$$
$$\uparrow = \{0 \mid *\}, \downarrow = \{* \mid 0\}, \pm 1 = \{1 \mid -1\}, *2 = \{0, * \mid 0, *\}.$$

For a configuration $G = (G_1, G_2)$, we define the inverse configuration $-G$ formally by

$$-G = (\{-G'' \mid G'' \in G_2\}, \{-G' \mid G' \in G_1\}),$$

and together with an additional configuration $H = (H_1, H_2)$, the sum $G + H$ is taken to mean

$$(\{G' + H \mid G' \in G_1\} \cup \{G + H' \mid H' \in H_1\},$$
$$\{G'' + H \mid G'' \in G_2\} \cup \{G + H'' \mid H'' \in H_2\}).$$

The order relations can also be explained without reference to the existence of a game or even the notion of winning. By definition, $G \geq 0$ holds precisely when there is no configuration $G'' \in G_2$ with $G'' \leq 0$. Analogously, $G \leq 0$ is defined by there being no configuration $G' \in G_1$ with $G' \geq 0$. By combining these two relations, it is possible to define the relations "$>$," "$<$," "$=$," and "$\parallel$." For example, by definition, $G \parallel H$ is satisfied when neither $G + (-H) \geq 0$ nor $G + (-H) \leq 0$ holds.

Not all configurations are numbers. For example, the configurations $*$, $\uparrow$, $\downarrow$, ± 1, and $*2$ appearing above are not numbers. However, for every number there is a configuration. More on this can be found in "Conway's Universe of Numbers."

---

[5] A complete investigation of the configurations generated in this and the following step can be found in David Moews, Sums of games born on days 2 and 3, *Theoretical Computer Science* **91**, 1991, pp. 119–128.

Conway's Universe of Numbers

Conway's definition of configuration can be restricted in a way that all configurations generated can be interpreted as numbers. In particular, the definition encompasses all real numbers. This definition is that a number is a configuration $G = (G_1, G_2)$ in which the sets G_1 and G_2 contain numbers exclusively. Moreover, for none of these numbers $G' \in G_1$ and $G'' \in G_2$ is the relation $G'' \leq G'$ allowed to hold. With addition, equality, and order relations as already defined for general configurations and an analogously defined multiplication, a class[6] of numbers arises that forms a totally ordered field. In particular, two numbers G and H, in contrast to two configurations, may always be compared as to size; that is, one of the following relations always holds: $G > H$, $G < H$, $G = H$.

It is clear that all numbers are configurations. Conversely, however, such configurations as $*$, $\uparrow$, $\downarrow$, and $* + 1$ are not encompassed by the numerical definition, and are therefore not numbers in the formal sense. Included, however, are all real numbers, including some that we have not met in black-and-white nim, such as $2/3$, $\sqrt{2}$, and π. They are represented by infinite sets of "move options." For example,

$$\frac{2}{3} = \left\{ \frac{1}{2}, \frac{1}{2} + \frac{1}{8}, \frac{1}{2} + \frac{1}{8} + \frac{1}{32}, \dots \;\middle|\; \frac{2}{2}, \frac{2}{2} + \frac{1}{8}, \frac{2}{2} + \frac{1}{8} + \frac{2}{32}, \dots \right\},$$

for which there is a tower in a version of black-and-white nim that allows infinite towers:

In addition, some rather bizarre numbers can be constructed, such as infinitely large numbers like

$$\omega = \{ 0, 1, 2, 3, \dots \mid \}$$

---

[6]In the sense of naive set theory, one may consider a class to be simply a set. However, such a point of view leads to logical contradictions, such as in the construction of the set of all sets, since in Conway's sense there are simply "too many" numbers.

and even
$$\omega + 1 = \{\, 0, 1, 2, 3, \ldots, \omega \mid \,\},$$

as well as infinitely small numbers like

$$\frac{1}{\omega} = \left\{\, 0 \,\middle|\, 1, \frac{1}{2}, \frac{1}{4}, \frac{1}{8}, \ldots \,\right\}.$$

We may agree with Leopold Kronecker (1823–1891), who re-marked in 1886, "The integers were created by God; all else is the work of Man." Or in the words of Donald Knuth, of course not to be taken seriously:[7]

> In the beginning, everything was void, and J. H. W. H. Conway began to create numbers. Conway said, "Let there be two rules which bring forth all numbers large and small. This shall be the first rule: Every number corresponds to two sets of previously created num-bers, such that no member of the left set is greater than or equal to any member of the right set. And the second rule shall be this: One number is less than or equal to another number if and only if no member of the first number's left set is greater than or equal to the second number, and no member of the second number's right set is less than or equal to the first number."
>
> And Conway examined these two rules he had made, and behold! they were very good.

From a mathematical point of view, Conway's approach is note-worthy in that he contrives to accomplish in a single step what is usually a multistep process of constructing sets of numbers. At the same time, his definition is extremely multifaceted. Thus, on the one hand, one can find the analogues to the construction of the natural numbers

$$0 = \varnothing, \; 1 = \{\, \varnothing \,\}, \; 2 = \{\, \varnothing, \{\, \varnothing \,\} \,\}, \; 3 = \{\, \varnothing, \{\, \varnothing \,\}, \{\, \varnothing, \{\, \varnothing \,\} \,\} \,\},$$

and so on, as proposed in 1923 by the then 19-year-old John von Neumann. On the other hand, the definition also contains a generalization of the *Dedekind cuts*, special pairs taken from two

---

[7]Donald E. Knuth, *Surreal Numbers*, Reading, MA 1974. In this book, Conway's ideas were popularized, while being made mathematically precise.

sets of fractions. Their discoverer, Richard Dedekind (1831–1916), used such cuts, for example,

$$\{\, x \in \mathbb{Q} \mid x^2 < 2 \,\}, \quad \{\, x \in \mathbb{Q} \mid x^2 > 2 \,\},$$

to define the real numbers formally for the first time, using the rational numbers as a basis. Finally, we note that Conway also defines the calculational and comparison operators for all types of numbers, including the infinitely large and infinitely small. For more details on this extensive theory, the reader is directed to the wider literature.[8]

Chapter Notes

1. Nonetheless, it behooves us, at least once, to convince ourselves formally of the rightness of this assertion. We must show, then, using the notation described in "Conway's Universe of Games," that

$$\{\, G, H, \ldots \mid P, \ldots \,\} + \{\, -P, \ldots \mid -H, \ldots \,\} = 0,$$

that is, that the player to move second can always win. If white moves first and moves to configuration G in the first component, black can reply with a move to configuration $G - H \leq 0$, which leads to victory for black. Every other opening move, whether by white or by black, can be countered by the complementary move in the other component. Thereby, a null configuration can always be achieved that ensures victory to the second player.

2. To prove the simplicity theorem, we must show that $\{\, G \mid H \,\} + \{\, -Q \mid -P \,\}$ is a null configuration. By symmetry, it suffices to find a winning strategy for the second player white in this configuration: if black plays to the configuration $H + \{\, -Q \mid -P \,\}$, then white counters with a move to $H - Q \geq 0$, which ensures a victory just as if black opened to $\{\, G \mid H \,\} - P \parallel > 0$.

---

[8]The construction of sets of numbers is the topic of the book H. D. Ebbinghaus, H. Hermes, F. Hirzebruch, et al., *Zahlen*, Berlin 1983. (In English translation as *Numbers*, New York 1991.) In Chapter 12 ("Zahlen und Spiele" (numbers and games)), Hans Hermes provides an overview of Conway's theory.

Further Literature on Conway's Games

[1] John H. Conway, *On Numbers and Games*, second edition, Natick, MA 2001.

[2] E. Berlekamp, J. Conway, R. Guy, *Winning Ways for Your Mathematical Plays*, second edition, Natick, MA 2001–2004. Printed in four volumes.

[3] J. H. Conway, All games bright and beautiful, *American Mathematical Monthly* **84**, 1977, pp. 417–434.

[4] John Horton Conway, A gamut of game theories, *Mathematics Magazine* **51**, 1978, pp. 5–12.

[5] Richard K. Guy, Graphs and games, in L. W. Beincke, R. J. Wilson (eds.), *Selected Topics in Graph Theory*, vol. 2, London 1983, pp. 269–295.

[6] Elwyn Berlekamp, Two-person, perfect-information games, in *The Legacy of John von Neumann, Proceedings of Symposia in Pure Mathematics* **50**, 1990, pp. 275–286.

[7] Richard K. Guy, Combinatorial games, *Handbook of Combinatorics* (R. L. Graham, M. Grötschel, L. Lovász, eds.), Amsterdam 1995, vol. 2, pp. 2117–2162.

[8] Richard K. Guy (ed.), Combinatorial games, *Proceedings of Symposia in Applied Mathematics* (AMS Short Course Lecture Notes) **43** 1991.

The first two of these books are references on combinatorial game theory. The articles cited next also give an overview of the field. We mention in particular the last-mentioned anthology. A readable, historically oriented book is the following:

[9] R. K. Guy, Mathematics from fun & fun from mathematics: An informal autobiographical history of combinatorial games, in J. H. Ewing, F. W. Gehring (eds.), *Paul Halmos: Celebrating 50 Years of Mathematics*, pp. 287–295.

In addition to these original articles and compilations, we recommend the following:

[10] John D. Beasley, *The Mathematics of Games*, Oxford 1990, pp. 120–136.

24

A Game with Dominoes: Have We Run Out of Space Yet?

Two players alternate placing dominoes on a playing field divided into squares like a chessboard. Each domino takes up two squares of the board. White has to place her dominoes vertically on two unoccupied squares, while black has to place his on two unoccupied horizontal squares. Whoever can move last in placing a domino on the board is the winner. Who can win, starting from the position depicted in Figure 24.1?

The game that we have just described originated with Göran Andersson, who communicated it to Martin Gardner for his mathematical games column in *Scientific American*.[1] The impartial variant, in which each player may place his or her domino either vertically or horizontally, was christened "cram" by Gardner, and Andersson's nonimpartial variant became known as "crosscram." In the investigations of Conway, Berlekamp, and Guy mentioned earlier,[2] the game was called "domineering."

A particularly interesting aspect of crosscram is that in the course of a game, disjunctive sums of configurations appear naturally. By this, we mean that many configurations can be easily understood as sums of simpler configurations. This is so because it is always only the spaces between the

---

[1] *Scientific American* 1974/2, p. 106, and 1976/9, p. 206. See also Martin Gardner, *Knotted Doughnuts and Other Mathematical Entertainments*, New York 1986, Chapter 19.

[2] E. Berlekamp, J. Conway, R. Guy, *Winning Ways for Your Mathematical Plays*, Natick, A K Peters, 2002.

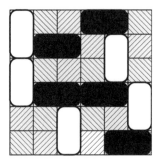

Figure 24.1. Who wins the domino game?

dominoes already placed that are significant in determining the possibilities for a win. Therefore, if the totality of unoccupied squares breaks into several regions that touch at most at the corners, then their disjunctive sum is equal to the total configuration. That is, a player can play a domino in only one of the regions. In the problem posed at the beginning of the chapter, for example, we have the following sum:

$$\square\square \ + \ \square \ + \ \square\square\square \ + \ \square\square \ + \ \square\square$$

Some of the components are known to us already in equivalent form from black-and-white nim, while the others were referred to at least in "Conway's Universe of Games" and "Conway's Universe of Numbers" at the end of the previous chapter:

$$\square\square = \{\,|\,0\,\} = -1$$

$$\square = \{\,\square\,|\,\} = \{\,0\,|\,\} = 1$$

$$\square\square = \{\,\square\,|\,\square\,\} = \{\,0\,|\,0\,\} = *$$

$$\square\square = \left\{\,\square\,|\,\square\square\,\right\} = \{\,1\,|\,-1\,\} = \pm 1$$

$$\square\square\square = \left\{\,\square\square\,|\,\square + \square, \square\,\right\} = \{\,-1\,|\,0,1\,\} = -\tfrac{1}{2}.$$

For the configurations to be investigated, we have altogether

$$-1 + 1 - \frac{1}{2} + (\pm 1) + * = -\frac{1}{2} + * + (\pm 1),$$

where the components $* = \{\, 0 \mid 0 \,\}$ and $\pm 1 = \{\, 1 \mid -1 \,\}$ exhibit a character that is completely unknown in black-and-white nim. Namely, whereas in black-and-white nim it is always more favorable to let the opponent move first, the right of first move in the configuration ± 1 is very lucrative, which has great consequences for the total configuration under investigation: if it is white's turn, she can move to configuration $-1/2 + * + 1 = 1/2 + *$ and be assured of a win on account of $* > -1/2$. On the other hand, if black moves first, then he can achieve the configuration $-1/2 + * - 1 = -3/2 + *$ and thereby win.

We can even "calculate" with configurations that are not numbers. For example, $* + * = 0$ and $(\pm 1) + (\pm 1) = 0$. Not so obvious is the equality

$$\boxed{} + \boxed{} = \{\, 1 \mid 0 \,\} + 1 = \{\, 2 \mid 1 \,\} = \boxed{}$$

The last of the above equalities results from the fact that the addition of a number to a configuration that is not itself a number shifts the move options of the position by the amount of the number:

> For every number x and every configuration $G = \{\, H, \ldots \mid P, \ldots \,\}$
> that is not itself a number, the following equality holds:
>
> $$\{\, H, \ldots \mid P \ldots \,\} + x = \{\, H + x, \ldots \mid P + x, \ldots \,\}.$$

Therefore, in the sum of a number and a nonnumber, the move options of the numerical component can be omitted without changing the chances of winning for both players. It is thus possible for each player to find an optimal move within the nonnumerical component, which in our example is the left component $\{\, 1 \mid 0 \,\}$. Depending on its interpretation, this statement is called either the *translation principle* or the *number avoidance theorem*. The theorem is based on the fact that a move in a numerical configuration does not improve one's winning position, though it does so in a nonnumerical configuration (see Note 1 at the end of the chapter).

In cases in which sums cannot be directly calculated, one may appraise the effect of the individual components on the winning prospects by comparing the components with numbers. For example, for every positive number ε, no matter how small, one has the inequalities

$$-\varepsilon < * < \varepsilon,$$

which can be clarified via the nim positions

Other examples are

$$-1 + \varepsilon < \pm1 < 1 + \varepsilon,$$
$$-\varepsilon < \{1 \mid 0\} < 1 + \varepsilon,$$

which are the narrowest limits for such comparisons with numbers. The configuration $*$ is also tightly linked to the number 0, while an analogous result holds for the configurations ±1 and $\{1 \mid 0\}$ for the closed intervals $[-1, 1]$ and $[0, 1]$, respectively (see also Figure 24.2). The larger the interval "interwoven" with a configuration, the more indeterminate the configuration is. By this is meant that the influence of such configurations on the win profile of a disjunctive sum also depends on the other components. In such cases, therefore, it no longer suffices to investigate the individual components in isolation.

How can such comparison intervals be found in general? To answer this question, let us start with an arbitrary configuration G that is finite, in the sense that from it, by arbitrary, not necessarily alternating, sequences of moves by the two players, at most a finite number of configurations can arise. How does the configuration G behave with respect to the real numbers?

We assume that configuration G is not itself a number, which we may do, since numbers are easily compared with every other number. In comparison

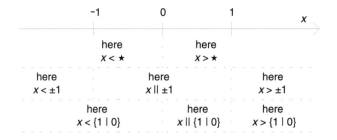

Figure 24.2. Three configurations compared by size in the set of real numbers.

with an arbitrary configuration x that is equivalent to a number, we have, since $G = x$ is impossible, either $G > x$, $G < x$, or $G \parallel x$. Which of these three relations holds for a given number x can be determined by analyzing the win profile of the configuration $G + (-x)$, which can be interpreted as a configuration extended by a handicap. Even if the win profile is not yet known, the number avoidance theorem gives information on how a player should move from the handicapped configuration $G + (-x)$. Since the configuration G is not a number, the player should restrict his play, regardless of the value of x, to the G component, that is, to play first in the configuration G, and to continue to do so in the further course of the game, in the configurations that follow, until finally, a number arises. Who wins, then, is clear if the sum of the resulting numerical configuration and the number x is formed. White is then interested in moving the G component to as large a number as possible. Conversely, black seeks to minimize the number that finally arises from the G component. There thus arises a minimax situation, typical for two-person zero-sum games, relating to a variable win level, where the variable win levels are surprising in that the games under investigation are either won or lost.

Therefore, depending on who moves first, every configuration can be associated with two game values, which can be calculated, due to the alternating right of first move, in a common minimax process. One should note that this is the perspective from which Milnor investigated disjunctive sums in 1953.[3] To conform to Conway's designation of "left" for white and "right" for black, these values will be called *left stop* and *right stop* and denoted respectively by $L_0(G)$ and $R_0(G)$.[4] Apart from configuration G, for which we are dealing with numbers and for which we have $L_0(G) = R_0(G) = G$, the stop values for games $G = \{\, G', \ldots \mid G'', \ldots \,\}$ can be computed recursively with the help of the two minimax equalities

$$L_0(G) = \max(R_0(G'), \ldots),$$
$$R_0(G) = \min(L_0(G''), \ldots).$$

For example, we have

$$L_0(\{\, 1 \mid 0 \,\}) = R_0(1) = 1 \qquad \text{and} \qquad R_0(\{\, 1 \mid 0 \,\}) = L_0(0) = 0,$$
$$L_0(*) = R_0(0) = 0 \qquad \text{and} \qquad R_0(*) = L_0(0) = 0,$$

---

[3] John Milnor, Sums of positional games, in: H. W. Kuhn, A. W. Tucker (eds.), Contributions to the Theory of Games II, Annals of Mathematics Studies **28**, 1953, pp. 291–301.

[4] In the books of Conway and Guy cited above, symbols $L(G)$ and $R(G)$ are defined that in comparison to $L_0(G)$ and $R_0(G)$ provide more information, since they also tell which player reaches the number in question.

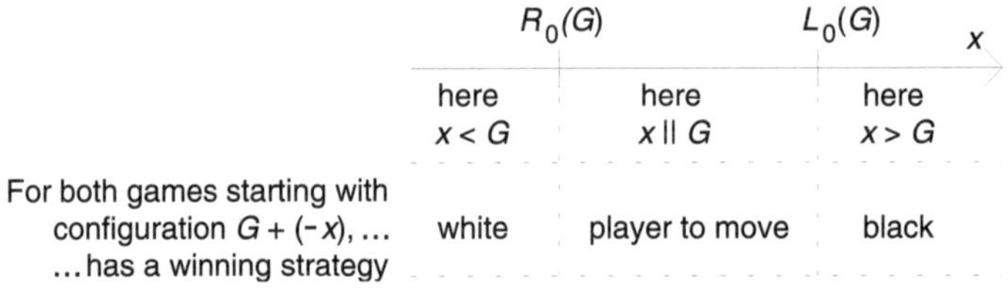

	$R_0(G)$		$L_0(G)$	x
	here	here	here	
	$x < G$	$x \parallel G$	$x > G$	
For both games starting with configuration $G + (-x)$, ... has a winning strategy	white	player to move	black	

Figure 24.3. Configurations in comparison to real numbers

and for $G = \{\,\{\,5\mid 3\,\}, 4\mid\{\,4\mid 1\,\}, \{\,3\mid 2\,\}\,\}$, we have

$$L_0(G) = \max(R_0(\{\,5\mid 3\,\}), 4) = \max(3, 4) = 4,$$
$$R_0(G) = \min(L_0(\{\,4\mid 1\,\}), \quad L_0(\{\,3\mid 2\,\}) = \min(4, 3) = 3.$$

For the case that white goes first, the left stop $L_0(G)$ forms the boundary between the numbers x for which the advantage in the handicap configuration $G + (-x)$ switches between white and black: namely, if $x > L_0(G)$, then black, going second, has a winning strategy for $G + (-x)$; that is, we have $x \geq G$, and therefore $x > G$. For numbers x with $x < L_0(G)$, on the other hand, white, moving first, must have a winning strategy for $G+(-x)$. Therefore, either $x < G$ or $x \parallel G$.

If we do the same for the right stop $R_0(G)$, then we obtain qualitatively the situation depicted in Figure 24.3. In particular, we have $L_0(G) \geq R_0(G)$, where the difference $L_0(G) - R_0(G)$ is a measure of how lucrative it is for a player to make the first move in configuration G.

Aside from the stop values $R_0(G)$ and $L_0(G)$ themselves, every number x can immediately be compared to the configuration G. Thus, as shown in Figure 24.3, for every positive number ε, no matter how small, we have

$$R_0(G) - \varepsilon < G < L_0(G) + \varepsilon, \quad R_0(G) + \varepsilon \parallel > G, \quad L_0(G) - \varepsilon < \parallel G,$$

where we may now include the situation in which the configuration G is a number.

What statements can we now make based on stop values that offer concrete advice in practice? The whole matter would be relatively easy to explain if the stop values of a disjunctive sum could be calculated from the stop values of the individual components. Alas, that is not the case. However, as Milnor has already remarked, one can make approximate statements. Thus for two configurations G and H and every positive number ε, we always have $G + H < L_0(G) + L_0(H) + 2\varepsilon$, from which follows the inequality

$$L_0(G + H) \leq L_0(G) + L_0(H).$$

Analogously, for the right stop values, we have

$$R_0(G + H) \geq R_0(G) + R_0(H).$$

Of much greater interest is the fact that for disjunctive sums, the difference of the two stop values, that is, $L_0(G + H) - R_0(G + H)$, remains relatively small. Namely, for each configuration G, one can construct a value $d_L(G)$ such that it is always true that (see Note 2 at the end of the chapter)

$$L_0(G + H) - R_0(G + H) \leq \max\big(d_L(G), d_L(H)\big).$$

In particular, for n-fold disjunctive sums $nG = G + \cdots + G$ (n times), the difference $L_0(nG) - R_0(nG)$ always remains bounded by $d_L(G)$. The two sequences

$$\frac{L_0(nG)}{n} \quad \text{and} \quad \frac{R_0(nG)}{n}$$

approach closer and closer to each other, and in fact, converge to a common limit (see Note 3 at the end of the chapter), the *mean value* $m(G)$ of configuration G. Together with another parameter, called the *temperature* and denoted by $t(G)$ (see "The Temperature"), the mean value makes possible an approximate characterization of the configuration G in disjunctive sums, in a way similar to that of random variables in terms of the expectation and variance: for arbitrary configurations G and H and a positive number ε, we have the relations

$$m(G) - t(G) - \varepsilon < G < m(g) + t(G) + \varepsilon,$$
$$m(G + H) = m(G) + m(H),$$
$$t(G + H) \leq \max\big(t(G), t(H)\big).$$

That is, within sums, every configuration behaves approximately like a numerical configuration whose value is equal to the mean value.[5] The precision of this approximation depends on the temperature.

Thus the mean value and temperature allow for the approximate ordering of a configuration among the real numbers in the form of possible handicaps, not exactly, to be sure, as is possible with stop values, but with properties that are relatively easy to deal with. The moderate growth of temperatures in disjunctive sums is due to the fact the players move

---

[5] The equality $m(G+H) = m(G)+m(H)$ follows without reference to the temperature from the representation of the mean value as a limit, as well as the chain of inequalities

$$R_0(nG) + R_0(nH) \leq R_0(nG + nH) \leq L_0(nG + nH) \leq L_0(nG) + L_0(nH).$$

The other two equalities follow from an analysis of the thermograph schema used in the definition of the temperature.

alternately, and so the advantage offered by its being one's move in "hot" components passes back and forth between the players. Therefore, a ramping up of the "heat" in the sense of an accumulation of advantages accruing to the player with the first move does not occur.

As an example of the application of the not-so-simple temperature theory, we investigate the configuration

$$G = \{\, \{\, 3 \mid 2 \,\} \mid 1 \,\} + \{\, -2 \mid -3 \,\} + \{\, 0 \mid -2 \,\} + 3.$$

It is not clear how to analyze this configuration directly, due to its computational complexity. However, if we investigate with methods like the mean values and the temperatures of the four individual configurations, as described in "The Temperature," then we obtain the following:

$$m(G) = \frac{7}{4} - \frac{5}{2} - 1 + 3 = \frac{5}{4}, \quad t(G) \leq \max\left(\frac{3}{4}, \frac{1}{2}, 1, 0\right) = 1.$$

It follows that $G > m(G) - t(G) - \varepsilon = 1/4 - \varepsilon$ for every positive number ε, and finally, that $G > 0$. That is, for both games with initial configuration G, left, a.k.a. white, possesses winning strategies. In many other cases, a similar analysis is possible, for example, when in the domino game that we investigated, played on a large playing field, only small gaps—though perhaps a large number—remain open. However, this works only if one player possesses a large advantage, namely, one that exceeds the temperature, in comparison to the imprecision of the approximation due to the temperature.

How one can actually find a good move using the temperature analysis is another matter entirely, one that can be solved according to the procedure presented in "The Thermostrat." However, it often suffices to compare the effects of the various move options. Thus white, starting from configuration $G + H$, can move only to configurations of the form $G' + H$ or $G + H'$. With regard to the configuration $G + H$ that obtains before the move, this corresponds to the addition of a configuration of the form

$$G' - G \quad \text{or} \quad H' - H.$$

Such a configuration of the form $G' - G$ is called a (left) *incentive* of configuration G. The incentives of a configuration together provide information on how well white can improve her position with one move (see Note 4 at the end of the chapter). If these configurations depending on only a single component are comparable among one another, then the best move can be recognized. As an example, we return to the configuration of the problem stated at the beginning of this chapter: for the components of this position,

$-1/2 + * + (\pm 1)$, there are three incentives for white for the representations $-1/2 = \{ -1 \mid \}$, $* = \{ 0 \mid 0 \}$, and $(\pm 1) = \{ 1 \mid -1 \}$, namely,

$$-\frac{1}{2}, \quad *, \quad 1 + (\pm 1).$$

On account of $1 + (\pm 1) \geq -1/2$ and $1 + (\pm 1) \geq *$, white can ensure her prospects of winning by moving according to the third incentive to $-1/2 + * + 1$. It is of interest that this statement can be made without the winning prospects being made explicit.

The Temperature

The *temperature* is a measure of how advantageous it is for a player to move first from a given position. An approach by which this advantage can be formally and precisely measured was proposed by Hanner in 1959.[6] The incentive is reduced to the right of first move by setting a "tax" on each move in the form of transfer payments between the players. The taxation is handled on the basis of the stop values, which are interpreted as the score of the two games starting at the configuration in question. With increasing taxation, the stop values change and approach each other. The taxation works as follows:

- for each move, the player whose move it is pays a tax to his opponent.

- for the first move, the taxation is set explicitly. For later moves, the amount of tax demanded is the amount that was actually paid by the opponent on the previous move.

- if the amount of tax seems too high to the player whose move it is, he can seek tax relief; he offers to pay his opponent a smaller amount, in exchange for which the opponent takes over the right to move by paying a higher tax.

- the player who did not make the last move will have to make the next move together with a tax payment.

[6] Olof Hanner, Mean play of sums of positional games, *Pacific Journal of Mathematics* **9**, 1959, pp. 81–89.

With the method that we have just described, a player is never worse off by having to move; thus it being one's turn is never a disadvantage. On the other hand, a player can reduce his tax at will. For a position $G = \{\, G', \ldots \mid G'', \ldots \}$ and a starting tax set at the level $t \geq 0$, one thereby obtains the "cooled," that is, taxed, stop values of

$$L_t(G) = \max(R_t(G') - t, \ldots) \quad \text{and}$$
$$R_t(G) = \min(R_t(G'') + t, \ldots).$$

However, one must exclude the cases in which there is already an equilibrium of the cooled stop values for a smaller value u with $0 \leq u < t$, that is, $L_u(G) = R_u(G)$. Since the tax is too high, it is no longer worthwhile to make the first move, and therefore, we have $L_t(G) = R_t(G) = L_u(G) = R_u(G)$. The temperature $t(G)$ is by definition the beginning required tax from which the stop values cooled via taxes agree and no longer change, but remain at a fixed value, namely, the mean value $m(G)$.

One obtains the best and fastest overview of the entirety of all cooled stop values $L_t(G)$ and $R_t(G)$ by means of a graphical representation in a coordinate system. The following figure shows for the example of the configuration $G = \{\, \{\, 3 \mid 2 \,\} \mid 1 \,\}$ how its so-called *thermograph* arises from the already represented thermographs of the two possible moves, namely, to $\{\, 3 \mid 2 \,\}$ and 1.

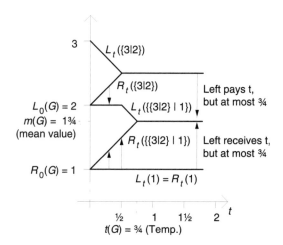

As can be readily seen in the figure for this example, the cooled stop values $L_t(G)$ and $R_t(G)$ are continuous as functions of the parameter t and are either piecewise constant or linear with slope -1 in the case of $L_t(G)$, or 1 in the case of $R_t(G)$. As a result of these possible slopes, which are inherited recursively move for move, we end up with the inequalities

$$m(G) - t(G) \le R_0(G) \le m(G) \le L_0(G) \le m(G) + t(G),$$

and thus $m(G) - t(G) - \varepsilon < G < m(G) + t(G) + \varepsilon$ for every positive number ε, as can be seen in the following figure:

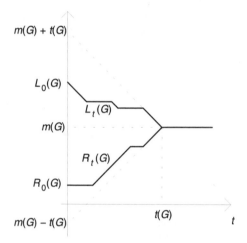

Instead of taxing the stop values, it is also possible to *cool* the game itself. This leads to the definition of a game

$$G_t = \{ G'_t - t, \dots \mid G''_t + t, \dots \},$$

cooled by the value $t \ge 0$, where again one must exclude the cases of excessive tax that arise when in the process a number results for a lesser cooling u. The stop values of the cooled configurations $L_0(G_t)$ and $R_0(G_t)$ then agree, with the cooled stop values $L_t(G)$ and $R_t(G)$. For example, for the cooled configuration $\{ 3 \mid 2 \}_t$ with $t \le 1/2$, we are dealing with the configuration $\{ 3 - t \mid 2 + t \}$, while stronger cooling results in $\{ 3 \mid 2 \}_t$; for the boundary case $t = 1/2$, we have $\{ 3 \mid 2 \}_t = \{ 5/2 \mid 5/2 \} = 5/2 + *$. Of particular interest is the fact that with cooling we are dealing with a homomorphism of configurations; namely, for two configurations G and H we always have $(G+H)_t = G_t+H_t$, and in the case of $G \geq H$, this relation carries over to the cooled configurations; that is, we have $G_t \geq H_t$. Of further interest is the behavior in the case of multiple cooling, which is also attainable in a single step: $G_{t+u} = (G_t)_u$, whereby on account of $L_t(G_u) = L_0((G_u)_t) = L_0(G_{u+t}) = L_{u+t}(G)$, the thermograph of cooled configurations results from a translation to the right of the vertical coordinate. The significant features of configurations and relations between configurations remain fixed under cooling, despite the fact that cooling generally results in simplified configurations. In particular, some less-hot configurations "freeze" into numbers, while numbers themselves remain unchanged under cooling. Altogether, in cooling, little significant data of a configuration are lost, while the most important properties are made more easily recognizable.

The Thermostrat

In a disjunctive sum of configurations $G = G_1 + \cdots + G_n$, the mean value $m(G) = m(G_1)+\cdots+m(G_n)$ represents a good and relatively easily calculated approximation to the stop values. Based on the bound

$$t(G) \leq \max_{i=1,\ldots,n} (t(G_i))$$

for the temperature $t(G)$, the data for the individual configurations yield the following:

$$m(G)- \max_{i=1,\ldots,n} (t(G_i)) \leq R_0(G) \leq L_0(G) \leq m(G)+ \max_{i=1,\ldots,n} (t(G_i)).$$

Left, playing second, can win every game that starts in a configuration of the form

$$G - m(G) + \varepsilon.$$

How, then, should left play to win? An answer was given by Hanner, and again by Conway, who constructed strategies for such situations by which left can force a win. Here left chooses each of her moves based on the thermographs corresponding to the currently reached components within the sum. More precisely, first, in a manner to be described, a suitable component offering victory is selected in which the move is to be made. In this component, the best move is determined based solely on local considerations. That is, it is chosen as though the other components were not present.

We still must describe how left can find the components G_i for each of her moves within the disjunctive sum $G_1 + \cdots + G_n$ in which under the given assumptions there is a winning move: starting with the individual stop values $L_t(G_1)$, ..., $L_t(G_n)$, $R_t(G_1)$,..., $R_t(G_n)$ recorded in the thermograph, as well as the derived values

$$W_t(G_1, \ldots, G_n) = \max(L_t(G_1) - R_t(G_1)), \ldots, L_t(G_n) - R_t(G_n)),$$

first the least value $t' > 0$ is determined for which the sum

$$\mathbf{L}_t = R_t(G_1) + \cdots + r_t(G_n) + W_t(G_1, \ldots, G_n)$$

attains its maximum on the interval $t \geq 0$. Now left chooses for her turn the component G_i whose thermograph exhibits the greatest width at the t' that was found:

$$W_{t'}(G_1, \ldots, G_n) = L_{t'}(G_i) - R_{t'}(G_i).$$

In this component, left moves as though the other components were not present. Left continues to follow this "recipe" until the sum of the configurations is equal to a number; then left obtains, as the player to move from configuration $G = G_1 + \cdots + G_n$, a number that is at least as large as the mean value $m(G)$ (see Note 5 at the end of the chapter).

Since with the thermostrat the components in which to move are selected whose thermographs show the greatest width at the previously determined t' level, this frequently, but by no means always, involves the hottest components.

Chapter Notes

1. Since we have to find good moves for the component G, we may first assume for the numerical position x that it is written in the standard representation discussed in Chapter 23, for example, as $0 = \{\ |\ \}$, $1 = \{0\ |\ \}$, $-1 = \{\ |\ 1\}$, and $1/2 = \{0\ |\ 1\}$. For such representations, one now shows inductively— in order of "simplicity"—that for an arbitrary nonnumerical configuration $G = \{H, \ldots\ |\ P, \ldots\}$, the number avoidance theorem holds:

> If white to move from configuration $G+x$ has a winning strategy, then she can find a winning move within the G component.

For numerical configurations x that in the standard representation offer white no opportunity to move, the assertion is clear, since in that case, when faced with configuration $G + x$, white has moves only in the G component.

On the other hand, if white can move in configuration x, then there is only one possible move, and that leads to a simpler numerical configuration x' with $x' < x$, for which the assertion has already been proved by the induction hypothesis. For the corresponding move that white can execute from configuration $G+x$ to $G+x'$, there are two possibilities: if it is not a winning move, then there must, as asserted, be one in the other component, G. In the other case, a winning configuration arises for white, who is now to play second; that is, we have $G + x' \geq 0$, and therefore, since G is not a number, actually $G + x' > 0$. Thus white moving into the position $G + x'$ certainly can win. Therefore, from the induction hypothesis, there must be a move from configuration G to a configuration H such that $H + x' \geq 0$. Since $x > x'$, it follows that $H + x > H + x' \geq 0$; that is, the move from G to H is also a winning move from the original configuration $G + x$.

2. The upper bound $D_L(G)$ can be defined for every configuration G that is not a number by

$$d_L(G) = \max\left(\left\{ L_0(G' - G)\ \middle|\ \begin{matrix} G' \text{ is reachable from } G \text{ by} \\ \text{left in a single move} \end{matrix} \right\}\right).$$

On account of

$$G = G' + (G - G'),$$

the handicap improvement that left can obtain with a move to configuration G' is bounded:

$$R_0(G) \geq R_0(G') + R_0(G - G') = R_0(G') - L_0(G' - G) \geq R_0(G') - d_L(G).$$

If left moves optimally, then we have

$$R_0(G) \geq L_0(G) - d_L(G).$$

For application to disjunctive sums, we note the obvious inequality

$$d_L(G + H) \leq \max\left(d_L(G), d_L(H)\right).$$

An extension of the last two inequalities to arbitrary configurations is achieved when for numerical configurations G, we have $d_L(G) = 0$.

3. First, we clearly have the chain of inequalities

$$R_0(G) \leq \frac{R_0(2G)}{2} \leq \frac{R_0(4G)}{4} \leq \cdots \leq \frac{L_0(4G)}{4} \leq \frac{L_0(2G)}{2} \leq L_0(G).$$

If we combine this with the already noted inequality

$$0 \leq \frac{L_0(nG)}{n} - \frac{R_0(nG)}{n} \leq \frac{d_L(G)}{n},$$

we obtain that the two partial sequences for the indices $n = 1, 2, 4, 8, 16,$ ... converge, and indeed, to the same limit. For natural numbers q, s, and $r = 0, 1, \ldots, 2^s - 1$, we also have

$$L_0\left((q2^s + r)\,G\right) \leq qL_0\left(2^sG\right) + rL_0(G),$$

and thus

$$\frac{L_0\left((q2^s + r)\,G\right)}{q2^s + r} \leq \frac{L_0\left(2^sG\right)}{2^s} + \frac{L_0(G)}{q}.$$

If the number $n = q2^s + r$ can be chosen arbitrarily large, then for the values of q and s, every predetermined size can be achieved. In passing to the limit $n \to \infty$, we achieve the desired result, together with the analogous inequality for the right stop values.

4. The analogous construction for player black, a.k.a. right, is $G - G''$, where G'' is an arbitrary configuration reachable by black from configuration G in a single move. The sign difference here indicates that it is in black's interest to choose a move with the greatest possible incentive.

Incentives, which we have gotten to know implicitly in our investigation of stop values, depend on the concrete form of a configuration. That is, equivalent configurations can have quite different sets of incentives.

5. To prove this, as taken from volume 1 of *Gewinnen*, pp. 160 f., 179–181, for every value $t \geq 0$, a strategy is constructed with which left, according to whether or not she moves in the disjunctive sum $G_1 + \cdots + G_n$, brings the game to a number that is at least as great as

$$\mathbf{L}_t = R_t(G_1) + \cdots + R_t(G_n) + W_t(G_1, \ldots, G_n),$$

respectively

$$\mathbf{R}_t = R_t(G_1) + \cdots + R_t(G_n) - t.$$

The desired conclusion is obtained for the particular value $t = \max(t(G_i))$ for which $\mathbf{L}_t = m(G)$ and $\mathbf{R}_t = m(G) - \max(t(G_i))$. One can make an even stronger statement if one chooses the t value in such a way that the expression for $\mathbf{L}_t$ attains its maximum.

The proof of both statements is obtained by complete induction. One begins with the case that the sum $G_1 + \cdots + G_n$ is equal to a number x, that is, that the play to numbers is over: one may assume, then, that the underlying maximum of the value $W_t(G_1, \ldots, G_n)$ is attained at the first component G_1. Since

$$R_t(G_2 + \cdots + G_n) = R_t(x - G_1) = x - L_t(G_1),$$

it follows that for the value x, we have

$$x = L_t(G_1) + R_t(G_2 + \cdots + G_n) \geq L_t(G_1) + R_t(G_2) + \cdots + R_t(G_n)$$
$$= R_t(G_1) + \cdots + R_t(G_n) + W_t(G_1, \ldots, G_n) = \mathbf{L}_t \geq \mathbf{R}_t.$$

If it is right's move at a configuration $G_1 + \cdots + G_n$ differing by a number and he moves to $G_1'' + G_2 + \cdots + G_n$, then by the induction hypothesis, left can then play in such a way that the game ends in a number that is at least as great as

$$R_t(G_1'') + R_t(G_2) + \cdots + R_t(G_n) + W_t(G'', G_2, \ldots, G_n).$$

Since

$$R_t(G_1'') + W_t(G_1'', G_2, \ldots, G_n) \geq L_t(G_1'')$$

and

$$R_t(G_1) = \min(L_t(G_1''), \ldots) + \min(t, t(G_1)) \leq L_t(G_1'') + t,$$

the quoted mean value is greater than or equal to $R_t(G_1) + \cdots + R_t(G_n) - t$. If it is left's turn at a configuration $G_1 + \cdots + G_n$ that differs by a number, then in the case $t \leq \max(t(G_1), \ldots, t(G_n))$ she searches for her move in the component G_i for which the difference $L_t(G_i) - R_t(G_i)$ is maximal; in the special case $t = \max(t(G_1), \ldots, t(G_n))$, she chooses a component with maximal temperature. In general, one then has $t \leq t(G_i)$. If we again assume, without loss of generality, that $i = 1$ and left then chooses in the selected component G_1 the optimal move to a configuration G_1', then in the further course of the game, left can always move, according to the induction hypothesis, so that the game ends with a number that is at least as large as

$$R_t(G_1') + R_t(G_2) + \cdots + R_t(G_n) - t.$$

Since

$$L_t(G_1) = \max(R_t(G_1'), \ldots) - \min(t, t(G_1)) = R_t(G_1') - t,$$

there follows, as desired, the attainability of a number of at least as big as

$$L_t(G_1) + R_t(G_2) + \cdots + R_t(G_n) =$$
$$R_t(G_1) + R_t(G_2) + \cdots + R_t(G_n) + W_t(G_1), \ldots, G_n).$$

There remains the case in which it is left's turn in a configuration $G_1 + \cdots + G_n$ that differs by a number and also with $t > u = \max(t(G_1), \ldots, t(G_n))$.

In these circumstances, left can move, based on the case just considered, so that a number is attained that is at least as large as

$$R_u(G_1) + \cdots + R_u(G_n) + W_u(G_1, \ldots, G_n).$$

Since $R_t(G_i) = R_u(G_i) = m(G_i)$ and $W_t(G_1, \ldots, G_n) = W_u(G_1, \ldots, G_n) = 0$, the assertion follows in this case as well.

25

Go: A Classical Game
with a Modern Theory

What is white's or black's best move from the go position shown in Figure 25.1? How many points can each achieve?

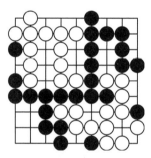

Figure 25.1.

Go is one of the oldest games. It is known to have been played as long ago as 300 B.C.E. in China, and its origins may well be one or two thousand years earlier. Fifteen hundred years ago, go fever spread to other lands in Asia, including Korea, and, above all, Japan.[1] Go arrived relatively late

---

[1]Information on the history of go and how to play can be found in the following sources: Siegmar Steffens, *Go spielend lernen*, Berlin 1990; Michael Koulen, *Go: Die Mitte des Himmels*, Cologne 1986; Jörg Digulla, Alfred Ebert, Horst Timm, *Go: Anfängerbuch*, Kassel 1994; Gilbert Obermair, *Klassische Spiele aus dem Fernen Os-*

in Europe, at the end of the 19th century. One of the great advocates of go was Emanuel Lasker, whose book *Brettspiele der Völker* cited earlier devoted 80 pages to the game. Lasker had the following to say about go:[2]

> Go has a much more penetrating logical structure than chess. It surpasses chess in simplicity and is its equal in its demands on the players' imagination.

The Rules of Go

In comparison to chess, which models a battle between opposing armies, the course of a game of go is much more abstract. The standard game is played on a square 19 × 19 board, on which the players alternately place a stone, black for one player, white for the other. The stones, which except for their color are all identical, are placed on unoccupied intersections of horizontal and vertical grid lines. The goal of the game is to surround as large an area as possible with one's own stones, where the opponent's stones that are thus enclosed are captured and removed from the board. Except for such removal, once a stone has been played, it is not moved again for the duration of the game.

Neighboring stones, joined horizontally or vertically, directly or indirectly, form a chain. Thus in the left-hand diagram below, the white stones form two chains, and the black stones one. An unoccupied intersection point that is a horizontal or vertical neighbor of a stone in a chain is called a *liberty* of the chain. In the left figure, the liberties of the black chain are indicated with a cross. An intersection point can be a liberty of several chains.

If the placement of a stone results in the last liberty of one or more chains of the opposing color being occupied, then all the stones of the affected chain are captured. According to the rule against suicide, a stone cannot be placed so as to remove

---

ten, Munich 1986, pp. 35–56; Frederic V. Grunfeld, *Games of the World*, New York 1975; Erwin Glonnegger, *Das Spiele-Buch*, Munich 1988, pp. 132–139; Erhard Gorys, *Das grosse Buch der Spiele*, Hanau ca. 1987, pp. 218–225; Richard Bozulich, *The Go Player's Almanac*, Tokyo 1992.

[2]Pages 89–169. The quotation comes from page 89.

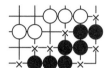 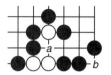

the last liberty of its own chain unless at least one opposing
stone is thereby captured. Thus in the right-hand figure, white
cannot play in the corner b. Occasionally, the possible moves
are limited in another manner by the rule of *ko*, according to
which a move cannot reproduce the exact situation from which
the opponent had just moved. Thus the repetition of moves by
capturing and recapturing of a single stone are prevented.

There is no zugzwang (the requirement to move) in the game
of go. If neither of the players wishes to move, the game ends
and is scored: first, the stones whose capture cannot be prevented are considered captured and removed from the board.
These stones, together with those already removed, are counted
as one point each for the opponent. Most of the points scored
come from captured territory, where each unoccupied intersection point that is surrounded by a player's stones counts as one
point. An intersection point is considered surrounded if every
path from the point following horizontal and vertical lines is
broken by the player's own stones. The effect of this rule will
become clear in the simple examples presented in the main text.

Since the number of points won decides the victor in go, and not who
makes the last move, it is clear that we are not dealing with a game in the
sense of Conway. On the other hand, it can happen that parts of the board
can become completely stable many moves before the end of the game, so
that a number of isolated battlefields emerge, each completely independent
of the others. Since each move is made in one battlefield or another, we are
dealing with a disjunctive sum of these subconfigurations. The analysis of
sums of positional games with point scoring was first undertaken in 1953, in
the work of Milnor previously cited.[3] Milnor, and later Hanner, analyzed

[3] John Milnor, Sums of positional games, in: H. W. Kuhn, A. W. Tucker (eds.),
Contributions to the Theory of Games II, *Annals of Mathematics Studies*, **28**, 1953,
pp. 291–301.

winning positions that each player can secure in a sum of positional games (see "Go as a Point-Scoring Game"). Although the article makes no direct reference to go, it is clear that application to that game was a prime motive for both researchers.[4]

We have seen that with the help of stop values, one can "translate" games in which the last player to move wins into winning levels of a zero-sum game. But the reverse process is also possible, as will be shown in "Blockbusting." However, one must assume that it is advisable to move. That is, the minimax value for the player whose turn it is can never be less than the minimax value that would obtain for the same player moving second.

Blockbusting

The game blockbusting is played on a board whose squares are arranged in a single row. Players black and white alternate placing stones on an unoccupied square. The game ends when the board is full. White wins one point from black for each boundary line between two squares both of which are occupied by white stones. Black wins nothing. He can only attempt to

---

[4]Thus on page 298, Milnor uses the go term "sente" to describe the situation in which a move must be countered in the same component. In the introduction to the collection in which Milnor's article appeared, the editors Kuhn and Tucker note that go endgames in particular frequently have the character of a sum of isolated games as investigated by Milnor (p. 191). Milnor, who later became well known for his work in topology, for which he was awarded the Fields Medal in 1962, was at the time a student at Princeton. As he wrote elsewhere, go was a game that he played frequently (A Nobel Price for John Nash, *The Mathematical Intelligencer* **17/3**, 1995, pp. 11–17). For a mathematical interpretation of the sente see Elwyn Berlekamp, John Conway, Richard K. Guy, *Winning Ways*, second edition, Natick, MA 2001, volume 1.

In Hanner's article there is also no direct reference to go. However, Olof Hanner cheerfully discussed the origin of his work: Hanner first encountered the game of go on a visit to the United States in 1949/50. But he became more deeply interested in the game on encountering Takagawa's book *How to Play Go*. The idea formed in Hanner's mind that end configurations in go could be assigned a single number. Hanner assembled multiple copies of a configuration; he was not yet acquainted with Milnor's work. When such multiple configurations are played out, it is often advisable to play in one or another different configuration. But how are such sente–gote questions to be resolved? Hanner assigned values to the various moves and determined which values led to a contradiction. He eventually arrived at a formal definition in which the right to move is "auctioned."

An explicit application of Milnor's results to go can be found in John Miller, The End Game of Go, in: *Proceedings of Northwest 76, ACM/CIPS Pacific Regional Symposium*, Seattle, 1976, pp. 228–233.

points scored ↓ ↓
by white: 1 1

minimize his loss. The following illustration shows a midgame configuration in which white has won two points.

For a complete analysis of this game, the minimax values in the sense of Zermelo's theorem must be calculated for each configuration. Here one must consider all possible configurations that can arise in the course of the game, and indeed, in the two variants in which black or white moves first. Clearly, all of the minimax values are integers. Moreover, the right of first move cannot be detrimental, since adding a stone can only improve one's situation. On the other hand, the right to move can never be worth more than two points (see Note 1 at the end of the chapter).

The crucial conditions for evaluating an intermediate configuration are the following:

- the points already scored by white;
- the remaining boundaries between squares and the bordering stones (one can imagine a black stone on the far left and far right of the board).

Thus a configuration can be thought of as a disjunctive sum of its subconfigurations. In our example, we have

$$2 + \boxed{} + \boxed{} = 2 + W3B + B1B,$$

where the notation should be self-explanatory. Apart from the empty spaces $W0W = 1$ and $W0B = B0W = B0B = 0$, every subconfiguration BnB, BnW, WnW, for $n = 1, 2, 3, \ldots$, is uniquely determined by the available moves on either side. For example,

$$B1B = \{ B0W + W0B \mid B0B + B0B \} = \{ 0 \mid 0 \} = *$$

and

$$W3B = \left\{ \begin{matrix} W2W, W1W + W1B, \\ 1 + W2B \end{matrix} \;\middle|\; \begin{matrix} W2B, W1B + B1B, \\ B2B \end{matrix} \right\}.$$

In comparison to go, blockbusting is a trivial pursuit. The variety of situations that arise in go is much, much greater, and in blockbusting, the endgame is uniquely determined. Blockbusting can serve as a good example of a translation from a game in which points are fought over to a game in the sense of Conway, in which the last player to move wins. The translation is made configuration by configuration, whereby each point won by white in the original game offers the possibility of an additional move. This has the effect that the resulting stop values agree with the two game values of the original configuration. The reason that this construction works is found in the number avoidance theorem. It ensures that the additional move options that arise in the numerical components are immediately unattractive and are therefore not used until the end of the actual game. For intermediate spaces of size at most four squares, the following table contains the configurations translated into the Conway system. The configurations have been simplified as much as possible, which is not always a simple task:

n	WnW	WnB	BnB
0	1	0	0
1	$\{2\,|\,0\}$	$\{1\,|\,0\}$	$*$
2	1	$\{\{2\,|\,1\}\,|\,*\}$	0
3	$\{2\,|\,1\}$	$\{\{\{3\,|\,2\}\,|\,1+*\}\,|\,0\}$	$\{1\,|\,0\}$
4	$\{\{3\,|\,2\}\,|\,1+*\}$	1	$\{1+*\,|\,*\}$

The winning possibilities of a blockbusting configuration can be quickly analyzed if the mean values and temperatures of all intermediate spaces are known. It turns out that the temperature of a configuration with one intermediate space is at most 1, a property that carries over at once to all configurations. The mean values of the configurations tabulated above are collected in the following table:

n	$m(WnW)$	$m(WnB)$	$m(BnB)$
0	1	0	0
1	1	$1/2$	0
2	1	$3/4$	0
3	$3/2$	$7/8$	$1/2$
4	$7/4$	1	$1/2$

In our example, we obtain the mean value

$$m(2 + W3B + B1B) = 2 + \frac{7}{8} + 0 = \frac{23}{8},$$

from which we obtain, since the temperature is at most 1, the two stop values $L_0(2 + W3B + B1B) = 3$ and $R_0(2 + W3B + B1B) = 2$. The minimax value of the original game, starting with the configuration $2 + W3B + B1B$, is therefore 3 if white begins, and 2 if black goes first. Although in this case the result can be derived much more quickly by an analysis of the possible moves, it is already clear how simple the procedure used here can be in complex cases.

One can use a technique of Berlekamp in which all configurations are cooled by 1 to simplify greatly the recursive analysis of configurations with one intermediate space.[5] Of course, some information is lost in cooling, for example, when the configurations $W1B = \{\, 1 \mid 0 \,\}$ and $B4B = \{\, 1 + * \mid * \,\}$ are both cooled by $1/2$. However, the most important characteristics of the configurations are preserved as before. Aside from $(W1W)_1 = 1 + *$, cooling by 1 already suffices to freeze all configurations of one intermediate space to their mean values.

In general, in the translation every end configuration of the given zero-sum game is replaced by a configuration in a game in the sense of Conway, and in fact, by a numerical configuration corresponding to the amount won by white. For example, if white wins two points, then this configuration is replaced by the sequence of moves $2 = \{\,\{\,\{\mid\}\mid\}\mid\,\}$; that is, white can move twice, while black can move neither immediately before nor after each of white's moves (see Note 2 at the end of the chapter). If this happens for all end configurations that can occur starting with a configuration of the given zero-sum game, and if all other possible moves remain unchanged, then this construction of Conway games has the following properties:

[5]Elwyn R. Berlekamp, Blockbusting and Domineering, *Journal of Combinatorial Theory* **A 49**, 1988, pp. 67–116. Of special significance in Berlekamp's procedure is that the cooling process by 1 can be essentially reversed for the blockbusting configurations. Thus it is possible to parameterize the configurations with one intermediate space in a simple manner using the mean value. One can proceed similarly in an analysis of go. It can be argued plausibly on the level of the original (point-counting) version that in cooling by 1, no significant information about winning possibilities in disjunctive sums is lost.

- the two minimax values of the zero-sum game depending on who moves first always agree with the two stop values of the configuration that arises.

- the construction is compatible with the creation of disjunctive sums. That is, disjunctive sums of zero-sum games can be translated individually or en masse into Conway games. In each case, the result is the same.

As we saw in our investigation of blockbusting, the configuration transformation of zero sums to Conway games works on account of the number avoidance theorem. It guarantees that the moves that arise within the Conway game are unattractive and therefore are used only immediately before the end of the game. Therefore, points won in the zero-sum game are translated into possible moves for the winning player.

In the special case of go, there are some particularities to be taken into account. There are several variants of the rules, differing in details, that occur in cyclic repetitions of moves. Thus two-move repetitions are prevented by the rule of ko, although at the price that the disposition of the playing stones does not contain all the information about the configuration. Repetitions of four or six moves are possible in principle, however, and they, too, require an extension of Conway's theory in which loops are allowed. In our brief overview, we will therefore generally exclude ko situations. Also problematic is the peculiar way in which the game ends: play continues until neither player wishes to move.

With regard to this last point, consider the following diagram, which shows a simple situation called *dame*. As in all the subconfigurations that follow, the stones on the boundary are considered to be "alive," that is, not subject to capture:

Both black and white can make a move, or not make one, without changing the result of the game. The translation of this configuration into a Conway game can take either the form $0 = \{\,|\,\}$ or the form $* = \{0\,|\,0\} = \{\{\,|\,\}\,|\,\{\,|\,\}\}$. Thus the convention according to which all configurations in which good players would no longer choose to move is usable in individual cases, but is too vague for general application. For in more complex situations, the pointlessness of further moves is not so apparent as in the pictured example, in particular when a player hopes that his opponent will make an error. On the other hand, the theoretically absolute criterion that the minimax values of both players' right to move first agree is in a practical sense unverifiable for complex configurations.

One possibility that yields the requisite uniqueness for mathematical investigations for relatively simple configurations in which no ko and seki[6] situations arise was found at the beginning of the 1990s by Berlekamp and Wolfe.[7] To this end, the rules of go are modified to a Conway game called *mathematical go*. That is, starting with the geometric construction of go configurations, the rules of stone placement are altered in such a way that a Conway game results. In the resulting stone-by-stone translation of a configuration into the Conway game, one can always obtain a configuration that is equivalent in the sense that the minimax values of the go configuration agree with the stop values in mathematical go. Here are the rules of mathematical go:

- each player must move. A player is not permitted to pass.

- as usual, chains of stones are captured when they lose their last liberty by an opponent's move.

- instead of placing a stone, a player may return an opponent's captured stone.

- moves are prohibited that reproduce a previous configuration.

- prohibited are suicide moves; a stone that is placed must either possess at least one liberty or capture at least one enemy stone.

- stones that surround one or more "eyes" of unoccupied squares remain free from capture for the remainder of the game.[8]

- the last player who is able to move is the winner.

---

[6] A seki in go is a configuration of stones in which neither player can place a stone without loss. The squares are evaluated as undecided.

[7] Elwyn Berlekamp, Introductory overview of mathematical go endgames, in: Richard K. Guy (ed.), *Combinatorial Games, Proceedings of Symposia in Applied Mathematics* (AMS Short Course Lecture Notes) **43**, 1991, pp. 73–100; Elwyn Berlekamp, David Wolfe, *Mathematical Go*, Wellesley 1994; Elwyn Berlekamp, The economist's view of combinatorial games, in: Richard J. Nowakowski (ed.), *Games of No Chance*, Cambridge 1996, pp. 365–405. An overview is provided by the go master Robert High, Mathematical Go, in: Richard Bozulich, *The Go Player's Almanac*, Tokyo 1992, pp. 218–224; David Gale, Go, *The Mathematical Intelligencer* **16/2**, 1992, pp. 25–29; J. Nievergelt, Das Go-Spiel, *Mathematik und Computer, Informatik Spektrum* **17**, 1994, pp. 106–110.

[8] This rule prevents a player from losing a position with two eyes by playing in one of the eyes. One would never play thus in normal go, but a player must do so in mathematical go if he wishes to transform the surrounded fields into moves.

Since in mathematical go, surrounded areas can have stones placed in them, these rules will seem highly irregular to an experienced go player. These rules can be clarified by means of a few simple examples, in which it will quickly become clear how elegantly the scoring of points, whether by surrounding area or capturing stones, can be transformed into additional moves:[9]

$$\text{(figure)} = \left\{ \text{(figure)} \;\middle|\; \text{(figure)} \right\} = \{\, 0 \mid 0 \,\} = *$$

$$\text{(figure)} = \left\{ \text{(figure)} \;\middle|\; \right\} = \{\, 0 \mid \,\} = 1.$$

In the next figure, white captures a black stone and thereby obtains an additional move for the following play:

$$\text{(figure)} = \left\{ \text{(figure)} + \bullet \;\middle|\; \text{(figure)} \right\} = \{\, 2 \mid 0 \,\},$$

since

$$\text{(figure)} + \bullet = \left\{ \text{(figure)} + \bullet, \; \text{(figure)} \;\middle|\; \right\} = \{\, 1 \mid \,\} = 2.$$

[9]In contrast to chess, in which white moves first, in go, black begins. Thus black would in fact have the left role and the positive range of numbers, and that is how Berlekamp and Wolfe handled it. Here, however, as we did in the previous chapters, we will stick with the convention that we used for chess, in which white corresponds to left. Readers who delve into the works cited should be aware of this difference.

A bit more multifaceted and therefore more interesting configuration is the following:

$$= \{\, 1, * \mid *, \{\, 2 \mid 0 \,\} \,\} = \{\, 1 \mid * \,\},$$

which involves dominated moves in the two omitted configurations. Thus on account of $1 \geq *$, white can avoid the move to $*$ without loss. For black, the relation $* \leq \{\, 2 \mid 0 \,\}$ is decisive, since white, moving second, possesses a winning strategy for the configuration $\{\, 2 \mid 0 \,\} - *$. From the resulting form $\{\, 1 \mid * \,\}$ one finally recognizes without difficulty the mean value $1/2$ and the temperature 1.

The simple technique of comparing two possible moves as we have just done is well known to us from Conway games. Using the variant rules for mathematical go, it can be used indirectly for normal go: a simple analysis of moves, yielding the result that white, going second in the difference configuration

$$\{\, 2 \mid 0 \,\} - * = \quad \text{}$$

can always force the last move for herself, suffices for us to be certain that the subconfiguration $*$ can never be worse for black than $\{\, 2 \mid 0 \,\}$, independent of the rest of the board, and indeed, in normal go as well. One may now object that a mathematical theory is not required for such obvious results, since no go player with a certain amount of experience would ever conjecture otherwise. However, the objection is justified only as long as the prospects for winning in the configuration being examined are as easy to compare with one another as in this example. We shall return to this topic later.

With increasing skill at recognizing such dominated moves, one can turn to configurations with a greater variety of available moves. Thus in the next configuration, for both white and black, only the move at the left intersection point comes into question:

$= \{\, 2 \mid \{\, 1 \mid * \,\} \,\}.$

The mean value of this configuration is 5/4, and the temperature is 3/4, which can be seen from the thermograph:

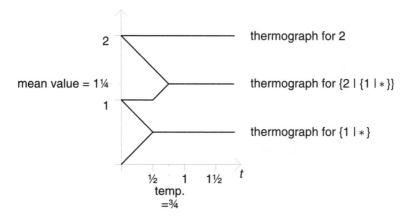

If the configuration appears five times in a row, then there is a net mean value of 25/4 and a temperature of at most 3/4. Since the stop values must be integers, one sees that 7 is the only possible value of the left stop; correspondingly, the right stop equals 6. In normal go, white can therefore achieve seven points with the first move, and six with the second.

In the next configuration as well, the move on the left intersection point dominates the other possibility, and this holds regardless of whether white or black moves first:

$$= \{ 3 \mid \{ 2 \mid 0 \} \}.$$

The mean value of the configuration is 2, and the temperature is equal to 1. Analogously, for the configuration

$$= \{ 4 \mid \{ 3 \mid \{ 2 \mid 0 \} \} \}$$

the mean value is 3 and the temperature 1.

We will need to simplify things significantly if we are to get a handle on the enormous multiplicity of subconfigurations in go endgames. Extrapolating from the most successful results in blockbusting, Berlekamp cooled the configurations of mathematical go that he was investigating by 1. The result is somewhat related to the function of a technical drawing, which clarifies the structure of an object by providing a unique perspective that highlights important features while disregarding what is of lesser importance. In cooling, what remains in particular are mean values and mutual associations based on the greater than or equal relation or on a disjunctive sum. Why the value 1 for cooling is appropriate for go will be discussed in "Go as a Point-Scoring Game."

Go as a Point-Scoring Game

Disjunctive sums, greater than or equal relations, and cooling can be used for the direct evaluation of go positions, without recourse to a transformation into mathematical go, and thus without the reverse translation into a game with point scoring implicitly required for the definition of the stop values. This direct method corresponds to the approach taken by Milnor and Hanner in the 1950s.

For each go configuration G—depending on whose move it is— there are two minimax values $L_0(G)$ and $R_0(G)$, each of which reflects the score that left obtains when both players play optimally. Since one may pass in go, left moving first can score at least as much as left moving second. That is, the relation $L_0(G) \geq R_0(G)$ always holds.

For the disjunctive sum $G + H$ of two subconfigurations G and H, we have Milnor's inequalities

$$L_0(G) + L_0(H) \geq L_0(G + H) \geq$$
$$R_0(G) + L_0(H) \geq R_0(G + H) \geq R_0(G) + R_0(H)$$

and

$$L_0(G) + L_0(H) \geq L_0(G + H) \geq$$
$$L_0(G) + R_0(H) \geq R_0(G + H) \geq R_0(G) + R_0(H).$$

Each individual inequality is derived from strategic considerations like those used by Lasker in his investigations of nim variants (see Chapter 22). A player counters his opponent's move in the same component in which that player has just moved, and indeed, with the move that is optimal in the component in question without regard to the other components.

In particular, Milnor's inequalities make it possible to estimate the influence of individual subconfigurations on the winning prospects of a global configuration. In this regard, we may consider how the winning prospects of a configuration G change when it is enlarged by a subconfiguration H to a global configuration $G + H$: on account of

$$L_0(G) + L_0(H) \geq L_0(G + H) \geq L_0(G) + R_0(H),$$
$$R_0(G) + L_0(H) \geq R_0(G + H) \geq R_0(G) + R_0(H),$$

the changes experienced by the two minimax values are limited by the minimax values of the additional configuration H.[10] For left, the configuration $G + H$, in comparison to configuration G, is:

- at least as favorable in the case of $R_0(H) \geq 0$;
- equally favorable in the case of $L_0(H) = R_0(H) = 0$;
- at most as favorable in the case of $L_0(H) \leq 0$;
- less or equally favorable in the case of $R_0(H) < 0 < L_0(H)$ according to whose move it is and the starting configuration G.

Configurations in point scoring games whose two minimax values are equal to zero are called null configurations. They do not alter the winning prospects as subconfigurations of a disjunctive sum. Examples of such null configurations are obtained when one adds a configuration H and its inverse configuration $-H$: the second player to move can then imitate the move of his opponent in the other component.

---

[10]On this basis, J. Mark Ettinger (A metric for positional games, *Theoretical Computer Science* **230**, 2000, pp. 207–219) defines a "distance," that is, a mathematical metric, for two arbitrary such point scoring configurations G and H of "Milnor type" (that is, with $L_0(J) \geq R_0(J)$ for all subsequent configurations J). This is done by forming the maximum

$$\rho'(G, H) = \max_X |L_0(G + X) - L_0(H + X)| = \max_X |R_0(G + X) - R_0(H + X)|$$

for arbitrary configurations X of Milnor type. Two configurations then are separated by a small distance precisely when an exchange of the two configurations within a sum changes the minimax values by at most a correspondingly small amount.

With the help of Milnor's inequalities, two configurations G and H can be compared by considering how as subconfigurations they influence the winning prospects of disjunctive sums. Using the inverse configuration $-H$, one forms the difference configuration $G + (-H)$, or $G - H$ for short. Since

$$G = H + (G - H),$$

one has the following for the subconfiguration G in comparison to the subconfiguration H within an arbitrary global configuration:

- G is at least as favorable as H in the case $R_0(G - H) \geq 0$;

- G is equally favorable to H in the case $L_0(G - H) = R_0(G - H) = 0$;

- G is at most as favorable as H in the case $L_0(G - H) \leq 0$;

- depending on who moves first and the remaining configuration, G can be less or equally favorable to H in the case $R_0(G - H) < 0 < L_0(G - H)$.

In a game, a player must compare the winning prospects of configurations before each move. If he has the choice within a configuration G to move to point a or to point b, say, and thereby achieve configuration G_a or G_b, then he should choose the move that would yield him as second player the greater minimax value. Since the two configurations G_a and G_b arise from the same configuration, they differ at only a few locations, so that the difference configuration $G_a - G_b$ is relatively simple in composite configurations. The two minimax values provide information as to whether one of the two moves is better, independent of the remainder of the configuration, and if so, which move it is. For example, if we ask whether the move to point a or point b is better for black in the configuration

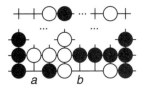

then we need to investigate the following configuration:

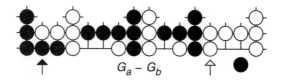

The parts of the board outside the depicted subconfiguration consist of two mutually inverse subconfigurations, and they therefore together form a null configuration. Since these regions therefore have no influence on the winning prospects of the total configuration, they have been omitted from the diagram. If one now analyzes the possible subsequent moves obtainable from the depicted difference configuration, one obtains

$$L_0(G_a - G - b) = 1 \quad \text{and} \quad R_0(G_a - G_b) = -1.$$

Consequently, it cannot be decided without knowledge of the rest of the board which move, a or b, is better for black. In fact, depending on the remaining configuration, either move could be the better one. Even if this result falls short of expectations, the example clearly shows qualitatively which situations are possible. In particular, moves are not always locally comparable!

How lucrative the right of first move can be arises from the difference $L_0(G) - R_0(G)$. The incentive to have the first move is lessened if one places a tax on moves, as described in the previous chapter. The more the initial tax requirement is increased, the more the taxed minimax values $L_t(G)$ and $R_t(G)$ approach each other, where the change from the initial value $L_0(G)$ or $R_0(G)$ is at most t. If the amount of cooling is sufficiently large, then the two cooled minimax values agree. This value fixes the mean value $m(G)$ of the configuration and with it the requisite cooling of the temperature $t(G)$. As described in the previous chapter, the winning prospects of a disjunctive sum can be approximated to the extent that the mean values and temperatures of the individual components are known. If one also knows the thermographs of the components, that is, the behavior of the minimax values under the influence of a continuously increasing cooling, then using the thermostrat, one can actually find approximately good moves.

A particular phenomenon occurs when the configurations are cooled by the amount $1 - \varepsilon$, a value just short of 1: if with this cooling one is able to determine the minimax values $L_{1-\varepsilon}(G)$ and $R_{1-\varepsilon}(G)$, then one can find from these the original, always integral, resulting minimax values $L_0(G)$ and $R_0(G)$, since an interval of length $1 - \varepsilon$ can contain at most one integer. Although with such cooling no information about the winning prospects is lost, nonetheless, within a compound configuration, drastic simplifications can result, so that in individual components some move variants cool early on to a fixed value. We shall see in what follows that on the level of mathematical go with cooling 1, a similar effect is obtained.

In order to be able to tell from a go diagram whether the Conway configuration itself or the configuration cooled by 1 is meant, a special notation is used, in which the stones at the edge of the diagram that are assumed to be alive are depicted only partially. For our first example, we return to a configuration that we have already studied in its uncooled state:

$$\text{◀⬡▶} = \{\, 3 - 1 \mid \{\, 2 \mid 0 \,\}_1 + 1 \,\} = \{\, 2 \mid 2 + * \,\} = 2 + \{\, 0 \mid * \,\}$$

Here the configuration called "up," denoted by $\uparrow = \{\, 0 \mid * \,\}$, is positive, but is nonetheless "almost" equal to zero: both the mean value and the temperature, and thereby both stop values, are equal to zero. For every small positive number ε, we therefore have $0 <\uparrow< \varepsilon$.

In other cases, a cooling by less than 1 leads to the "freezing" of a configuration to its mean value. Thus for the configuration of the following diagram a cooling in the amount t that is just a bit larger than $1/2$ suffices:

$$\text{◀⬡▶} = \{\, 1 \mid * \,\}_1 = \{\, 1 - t \mid *_t \,\} = \{\, 1 - t \mid t \,\} = \frac{1}{2}$$

The required taxation on the stop values, which are interpreted as a score, in the amount 1, is thereby reduced to $1/2$. However, in go configurations such tax reductions can be dispensed with. That is, aside from the case in which both stop values agree, one can always charge the full tax of 1. This is noteworthy in that under some circumstances the incentive to move first is not only completely removed, but is even reversed: to move is punished. However, the temperature does not fall much below the value 1, in any case not so far that the uncompromising tax of 1 changes the result with respect to normal cooling (see Note 3 at the end of the chapter). Thus for our example, we have

$$\{1 - 1 \mid *_1 + 1\} = \{0 \mid 1\} = \frac{1}{2} = \{1 \mid *\}_1.$$

As one can see, the tax raised by 1/2 is again compensated by a corresponding bonus, which furthermore would no longer function in the case of a tax demand of more than 1. Since with the simplified recursion formula only whole points are to be calculated, "cold go," the game of go cooled by 1, can be easily played: as long as it is advisable to play in normal go, for each move in cold go, one point in the form of a captured stone must be paid to the opponent:

$= \left\{ \text{} + \bigcirc \;\middle|\; \text{} + \bullet \right\} = \{0 \mid 1\} = \frac{1}{2}.$

In the next two configurations as well, for both white and black only the moves to the intersection point furthest to the left come into question, since that move dominates the other possible moves:

$= \{4 - 1 \mid \{3 \mid \{2 \mid 0\}\}_1 + 1\} = \{3 \mid 3 + \uparrow\} = 3 + \{0 \mid \uparrow\},$

where for the configuration $\{0 \mid \uparrow\}$, the equality $\{0 \mid \uparrow\} = \uparrow + \uparrow + *$ can be proved. Although the configuration $\uparrow$ is not comparable to the configuration $*$, it is comparable for the duplicated configuration $\uparrow + \uparrow$: altogether, we have $\uparrow \parallel *$ and $0 < \uparrow < \uparrow + \uparrow < \varepsilon$ for every small positive number ε. Much smaller yet than $\uparrow$ is the next configuration in go cooled by 1:

$= \{5 \mid \{4 \mid 0\}\}_1 = \{5 - 1 \mid \{4 \mid 0\}_1 + 1\}$

$$= \{4 \mid \{3 \mid 1\} + 1\} = \{4 \mid \{4 \mid 2\}\}$$

$$= 4 + \{0 \mid \{0 \mid -2\}\}$$

On the one hand, the subconfiguration $\{0 \mid \{0 \mid -2\}\}$ is positive, since white can win moving first or second. On the other hand, even the infinitesimally small configuration $\uparrow$ is many times larger than $\{0 \mid \{0 \mid -2\}\}$:

$$\{0 \mid \{0 \mid -2\}\} + \cdots + \{0 \mid \{0 \mid -2\}\} < \uparrow.$$

Moreover, we could have encountered the configurations $\uparrow$ and $\{0 \mid \{0 \mid -2\}\}$ in the domino game, since we have

$$\uparrow = \quad\text{and}\quad \{0 \mid \{0 \mid -2\}\} = \quad.$$

Between 0 and $\{0 \mid \{0 \mid -2\}\}$ there are additional configurations that can also be discerned in cold go:

$$= \{7 \mid \{6 \mid 0\}\}_1 = \{7 - 1 \mid \{6 \mid 0\}_1 + 1\}$$
$$= 6 + \{0 \mid \{0 \mid -4\}\}.$$

In general, the configurations $+_r = \{0 \mid \{0 \mid -r\}\}$, definable for every positive fraction r, satisfy the chain of inequalities

$$0 < \cdots < +_4 < +_3 < +_2 < +_1 < +_0 = \uparrow,$$

where the changes in size are marked. For two numbers $s > r$, one has namely

$$0 < +_s + \cdots + +_s < +_r.$$

As one can easily imagine, this catalog of configurations is nowhere near complete. Nonetheless, many endgame situations can be investigated with the types of configurations that we have examined, since configurations of stones can frequently be represented by equivalent Conway games. Thus the book by Berlekamp and Wolfe cited previously contains pages of overviews of stone configurations together with the Conway configurations that arise from them in cold go. However, there are also configurations that after their cooling still possess a positive temperature and for which therefore the first move is very lucrative. Examples are

$$= \left\{1\tfrac{1}{8} \mid 1\right\} \quad\text{and}\quad = \left\{3\tfrac{1}{4} \mid 1\tfrac{1}{2}\right\}.$$

Yet how are the statements that one obtains for a given go configuration with the help of subconfigurations cooled by 1 to be interpreted? The answer is surprisingly simple. One first forms the disjunctive sum of the subconfigurations cooled by 1, from which the winning prospects can be read off at once:

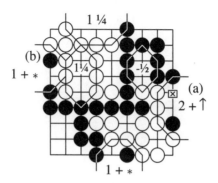

Figure 25.2.

- the left stop, that is, the minimax value for white moving first, is equal to the smallest integer that is greater than or equal to the disjunctive sum.

- the right stop, that is, the minimax value for white moving second, is equal to the largest integer that is less than or equal to the disjunctive sum.

As an example, let us consider the position that appeared at the beginning of the chapter. In addition to the certain points, namely, 3 for white and $5 + 7 = 12$ for black, we have the splitting into the disjunctive sum shown in Figure 25.2. Also shown are the Conway configurations cooled by 1, as we have already seen in the previous examples.

If the configuration shown is cooled by 1, then one obtains the configuration

$$-3 + \uparrow,$$

whose ordering by size in comparison to the integers is given by

$$-3 < -3 + \uparrow < -2.$$

Thus the minimax value for white is -2 points if white begins, and -3 points if black moves first. Moves with which these minimax values are realizable are recognizable analogously:

- if it is white's turn, she moves to the field marked ⊠ in Figure 25.2 and thereby attains the configuration 3 in this component. In cold go, the position -2 is therefore achieved, which is also the minimax value of the now second player white.

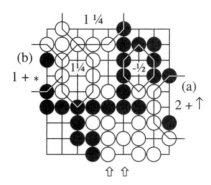

Figure 25.3.

- if black begins, he also plays on the marked point, attaining in cold go $1 + *$ in the relevant component and $-4 + *$ altogether, which guarantees him a win of at least three points; namely, as first to move, white has a minimax value of -3, the smallest integer that is greater than or equal to $-4 + *$.

Moreover, the move in one of the components cooling to $1 + *$ is bad for black: a move there would bring the configuration 2 to this component in the cooled version, and thus $-2 + * + \uparrow$ altogether. The greatest integer that is less than or equal to this configuration is -3, and thus the minimax value for the second player white. With a move in the component marked with (b), white thus gives away a point.

However, if, as one can see in Figure 25.3, one of the two $*$ components is no longer available, then the situation changes fundamentally. Although the configuration agrees with that under investigation up to the intersection points marked with $\Uparrow$, a completely different situation results, and indeed, in its long-term effect on the remaining components (a) and (b): with one move in the formerly optimal component (a), white can attain in cold go, starting from $-3 + * + \uparrow$, the configuration $-2 + * + \uparrow$, and with a move in (b), the configuration $-2 + \uparrow$. Since now black can move, these configurations correspond to minimax values of -3 and -2. In the latter case, white's loss is reduced by one point.

The incommensurability of Conway configurations such as $*\|0$ and $*\| \uparrow$ is therefore no mathematical fantasy, but a strategic reality: whether the move to (a) or (b) is better for white depends on the structure of the remaining go configuration. Isolated statements are impossible to make in such a situation.

Whether local assertions are possible and what they might be can be determined with the difference procedure mentioned above: a player forms the difference between two cooled configurations that can be reached through some moves. Equivalent to this is the pairwise comparison of incentives, as described at the close of the previous chapter. In the following table are collected the incentives of several configurations in cold go, in relation to the specified possible moves. The greater the incentives, the better the move. The moves within a numerical configuration are unattractive, as is already known from the number avoidance theorem.

Configuration $G = \{\, G', \ldots \mid G'', \ldots \}$	Incentives for ...	
	White: $G' - G$	Black: $G - G''$
$1 = \{\, 0 \mid 2 \,\}$	-1	-1
$1/2 = \{\, 0 \mid 1 \,\}$	$-1/2$	$-1/2$
$1/4 = \{\, 0 \mid 1/2 \,\}$	$-1/4$	$-1/4$
$* = \{\, 0 \mid 0 \,\}$	$*$	$*$
$\uparrow = \{\, 0 \mid * \,\}$	$-\uparrow$	$\uparrow + *$
$\uparrow + \uparrow + * = \{\, 0 \mid \uparrow \,\}$	$-\uparrow = \uparrow + *$	$\uparrow + *$

Let us summarize: endgame situations in go that remain ko and seki free can be transformed into an equivalent Conway game of mathematical go. If these configurations are cooled by 1, then in the components there frequently arise numbers or other familiar standard configurations known from other games, such as $*$, $\uparrow$, and $+_s$, which are easier to deal with than the uncooled configurations of mathematical go. Since furthermore, the properties of interest, namely, the original stop and minimax values, continue to be determinable, one can shift one's investigations completely over to the level of cold go. Table 25.1 shows how the various variants of go are related. The mathematical background, which is by no means obvious, is discussed further in "Why Cold Go Is So Informative."

To conclude, we should say that configurations in which ko situations can arise in further play can be analyzed on the basis of mathematical go. To this end, Martin Müller and Ralph Gasser[11] defined two rule variants in which a player is restricted in his choice of moves in all situations involving ko in such a way that the repetition of moves is excluded. In relation to winning prospects, these two variants form a limitation for the given configuration; that is, in comparison to the normal rules, one variant is at

---

[11] Martin Müller, Ralph Gasser, Experiments in computer Go endgames, in: Richard J. Nowakowski (ed.), *Games of No Chance*, Cambridge 1996, pp. 273–284; Martin Müller, Computer Go as a sum of local games: An application of combinatorial game theory, dissertation ETH Nr. 11006, Zurich 1995.

Game	Win Prospects of a Configuration in (Classical) Go	
Classical go; *game with point scoring*	Minimax value for white moving first	Minimax value for white moving second
Mathematical go; *last player to move wins*	Left stop	Right stop
Cold go (cooled by 1); *last player to move wins*	Smallest integer greater than or equal to the configuration	Largest integer less than or equal to the configuration

Table 25.1. Three go variants and their interrelations

least as favorable for black, the other at most as favorable. For example, for the configuration

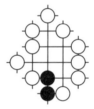

the result is in one case a mean value of $77/32$ and temperature $53/32$, and in the other, a mean value of $76/32$ and temperature $52/32$.

With its application to a classical and significant game like go, Conway's theory has doubtless attained an important summit, and that with a game that is not even a Conway game.[12] Aside from the mathematical theory, there is still much that remains open—from the opening, through the midgame, until deep in the endgame—in which a decomposition into independent subconfigurations leads only seldomly to such manageable small-scale configurations as in the examples studied here. In all seriousness, one

---

[12]In fact, go is a positional game with point scoring. As already sketched in "Go as a Point-Scoring Game," the investigations of Milnor and Hanner were based directly on minimax values. There the cooling by 1 of a configuration in mathematical go finds its correspondence when the minimax values are cooled by an amount $1 - \varepsilon$ close to 1. A disadvantage of this procedure is that no knowledge of "infinitesimal" Conway games such as $*$, $\uparrow$, $\{0 \,|\, \uparrow\}$, and $+_s$ can be used. However, in exchange, cooling on the level of point scoring games is more suggestive. Thus it is relatively plausible why optimal stratgies in point scoring go cooled by just under 1 are also optimal in normal go. It is also immediately clear how much a player can lose at most if he uses a strategy that has proven optimal in a cooling of $2 - \varepsilon$.

could not have expected more. However, it is significant that the value of a configuration depending on the right of first move can be expressed exactly through properties of independent subconfigurations. In particular, go concepts such as sente and gote have found a formal exact analogue. To that extent, we may hope that the approaches of mathematical go have also been able to make a small contribution to the creation of computer programs for go that in the not too distant future might be as strong as those of today for chess.[13]

Why Cold Go Is So Informative

In order to recognize how the stop values of mathematical go are reflected in cold go, one may translate strategies for a player of mathematical go into the cooled version (see Note 4 at the end of the chapter). To this end, in cold go as well, the player in question also places a stone where he would have placed it in uncooled go, and does this until a configuration is reached for which in uncooled go both stop values are the same.

In this way, we obtain the following assertions for configurations G of mathematical go, as we shall soon see:

- from $L_0 G)$ follows $G_1\| > 0$; that is, in cold go, white moving first wins.

- from $L_0(G) = 0$ follows $G_1 \leq 0$; that is, black moving second can win in cold go.

If one also considers possible translations by integers, the two assertions can be collected into the chain of inequalities $L_0(G) - 1 < \|G_1 \leq L_0(G)$, so that $L_0(G)$ is the smallest number greater than or equal to G_1.

Let us look at the first assertion. If white makes the first move in a configuration G of mathematical go, then, depending on

---

[13]Information on go programming can be found in Anders Kierulf, Smart game board: a workbench for game-playing programs, with Go and Othello as case studies, dissertation, ETH Nr. 9135, Zurich 1990; Christian M. Hamann, Chronologie der Programmierung des japanischen Brettspiels Go: eine Herausforderung an die Künstliche Intelligenz, *Angewandte Informatik* **12**, 1985, pp. 501–511; David Erbach, Computer and Go, in: Richard Bozulich, *The Go Player's Almanac*, Tokyo 1992, pp. 205–207; Martin Müller, Review: Computer Go 1984–2000, in: *Computers and Games, Lecture Notes in Computer Sciences 2063*, Berlin 2001, pp. 405–413.

the play of white and black in mathematical go, one obtains a game of type

$$G \xrightarrow{W} H \xrightarrow{B} L \xrightarrow{W} M \xrightarrow{B} \cdots \longrightarrow Z \longrightarrow \cdots,$$

where we assume that the configuration Z is the first in the game to arise with the property $L_0(Z) = R_0(Z)$. If both players translate their strategies into cold go, then the result is the game

$$G_1 \xrightarrow{W} H_1 - 1 \xrightarrow{B} L_1 \xrightarrow{W} M_1 - 1 \xrightarrow{B} \cdots \xrightarrow{B} Z_1$$

or

$$G_1 \xrightarrow{W} H_1 - 1 \xrightarrow{B} L_1 \xrightarrow{W} M_1 - 1 \xrightarrow{B} \cdots \xrightarrow{W} Z_1 - 1.$$

Here the configuration Z is cooled to the integer $Z_1 = L_0(Z)$.

Because of the assumption $L_0(G) = 1$, white in mathematical go can play so well that the configuration Z with $L_0(Z) \geq 1$ is attained. For the strategy translated to cold go, it follows that $Z_1 \geq 1$, which in both cases ensures white the last move and thereby proves $G_1 \| > 0$.

The second assertion is obtained analogously. To obtain a corresponding characterization of the right stop value, it suffices to replace the configuration G by the inverse configuration $-G$.

Environmental Go:
An Extended Theory of Temperature

In the previous chapter, we defined the cooling of configurations in a Conway game by the "taxation" of moves. However, this approach has two disadvantages in application to go: on the one hand, a generalization to configurations in which subsequent play can lead to ko situations is possible only by forbidding repetition of moves to one of the players, selected in advance. And on the other hand, it turned out that go players interested in the combinatorial theory have generally found the taxation approach not very suggestive, and not only because people do not like to pay taxes.

For both of these reasons, Elwyn Berlekamp came up with an alternative construction for cooling configurations.[14] He considered a configuration to be analyzed according to the temperature as a locally bounded component within larger, quasi "environmental" configurations. More precisely, he investigated sums of the given configurations using several standardized configurations. It suffices, then, for such sums to use exclusively so-called switching games, that is, games of the form $\{t, -1\}$. It is then possible using such sums to weigh the advantage of first move in the further course of the game. In this regard, one should recall that a weighing of winning prospects, though in another form, as it relates to who moves first in the entire game (for the purpose of handicapping), belongs to the tradition of go through the komi system of handicaps.

To make his *environmental go* construction, whose exact definition we defer for now, playable in practice, Berlekamp packaged his idea in the following form: in addition to the go board, on which the stones are placed according to the usual rules, a sorted deck of cards is used with values 10, 20, 19.5, 19, 18.5, 18,..., 1.5, 1, 0.5. A player whose turn it is may either place a stone on the board or take the top card from the deck, ensuring himself additional points at the value of the card. As compensation for not being able to make the first move, the second player—in go this is white, in contrast to usual practice in other games—receives the top card, with value 10.

To get an approximate idea of how the temperature develops in the course of a game of go, Berlekamp organized games of environmental go among professional go players. The first such game was played in 1998 by Rui Naiwei and Jiang Zhujiu, both players of the highest rank, 9-dan-pro (and now married to each other). The game ended very quickly, namely, depending on the country-specific rules, with an advantage of 2.5 for white or 0.5 for black. We can get an idea of the current temperature from the points at which cards were drawn by the players, at least if both go professionals were not making the same erroneous judgment as to the value of the current move.

Before the first move, the cards with values down to 14 were taken, and in a later game, down to 15. This allows us to con-

---

[14] Elwyn Berlekamp, The economist's view of combinatorial games, in: Richard J. Nowakowski (ed.), *Games of No Chance*, Cambridge 1996, pp. 365–405, in particular, pp. 394 ff.

jecture that the temperature of the empty board as a starting configuration was valued by the players at 14 or a bit higher. After 17 moves, the cards down to 10.5 had been taken. Thereafter, over 200 moves were made on the board.

For the theoretical investigation of the cooling of a configuration it would doubtless be helpful to refine the temperature grid of the "environmental" switching games. To analyze a cooling by the value t, a sum of switching games of the form

$$E_t = \{\, t \mid -t \,\} + \{\, t - \delta \mid -t + \delta \,\} + \{\, t - 2\delta \mid -t + 2\delta \,\} + \cdots$$

is used. Here the sum extends over all switching games of the given form with positive temperature. The grid size $\delta > 0$ is chosen to be sufficiently fine.

If the game E_t that serves as the environment is played by itself, it is advantageous for the player whose turn it is to select from among the remaining switching games the one with the highest temperature. Since the player to move thereby obtains, in each of his moves, a score that is higher by δ than that of his opponent, the result, if we ignore a small imprecision of at most δ arising for an odd number of summands, is the minimax values $L_0(E_t) = t/2$ and $R_0(E_t) = -t/2$.

The cooled minimax values $L_t(G)$ and $R_t(G)$ of a configuration G can now be approximated, as proven by Berlekamp (see Note 5 at the end of the chapter), in the following manner, where the error for a small enough grid size δ can be made arbitrarily small:

$$L_t(G) = L_0(G + E_t) - L_0(E_t) \quad \text{and} \quad R_t(G) = R_0(G + E_t) - R_0(E_t).$$

Applications of the generalized thermograph theory to go configurations with ko positions were given by Bill Spight, Martin Müller, and Elwyn Berlekamp.[15]

---

[15]Bill Spight, Extended thermography for multiple kos in Go, in: *Computer Games*, Lecture Notes in Computer Sciences **1558**, Berlin 1999, pp. 232–252; also in *Theoretical Computer Science* **252**, 2001, pp. 23–43; Martin Müller, Elwyn Berlekamp, Bill Spight, Generalized thermography: algorithms, implementation and applications to Go endgames, Technical Report 96-030, International Computer Science Institute, Berkeley 1996.

And Chess?

Configurations that can be represented as disjunctive sums of subconfigurations almost never occur in chess. Among the few exceptions are the zugzwang situations, as in the following position, which comes from a game played in 1929 between Schweda and Sika in Brno; it was analyzed in a book by Euwe and Hooper.[16]

The player whose king is forced to move loses. Therefore, each player attempts to make the last pawn move, and of course, the promotion of a pawn is also to be prevented. For this reason, the position can be described and investigated using Conway's approach. We begin with some simple configurations, and in order to simplify things, we turn the board by 90 degrees so that white moves from left to right:

$$\text{♙♟} = \text{♙♟} = \{\,|\,\} = 0$$
$$\text{♙\_\_♟} = \{\,\text{♙♟}\,|\,\text{♙♟}\,\} = \{\,0\,|\,0\,\} = *$$
$$\text{♙\_\_\_♟} = \{\,\text{♙\_\_♟}\,|\,\text{♙\_\_♟}\,\} = \{\,*\,|\,*\,\} = 0.$$

If the black pawn is still on its starting square on the seventh row, then we have

$$\text{♙\_\_\_♟}_{(7)} = \{\,\text{♙\_\_♟}\,|\,\text{♙\_\_\_♟}\,,\,\text{♙\_♟}\,\}$$
$$= \{\,*\,|\,*\,,\,0\,\} = \{\,*\,|\,0\,\} = -\!\uparrow,$$

---

[16]Noam D. Elkies, On numbers and endgames: Combinatorial game theory in chess endgames, in: Richard J. Nowakowski (ed.), *Games of No Chance*, Cambridge 1996, pp. 135–150.

where the second-to-last identity follows because the additional possible black move to $*$ is no improvement over $-\uparrow = \{\, * \mid 0 \,\}$, since it can be immediately "reversed" by white to the original sole option of 0. For the subconfiguration shown in the diagram in row h, we have, finally,

$$\text{♟}\,\text{▨}\,\text{▨}\,\text{♙}_{(7)} = \{\,\text{♟}\,\text{▨}\,\text{▨}\,\text{♙}_{(7)}, \mid \text{♟}\,\text{▨}\,\text{▨}\,\text{♙}, \text{♟}\,\text{▨}\,\text{♙}\,\}$$
$$= \{\, -\uparrow \mid *, 0 \,\} = \{\, -\uparrow \mid 0 \,\} = -\uparrow - \uparrow + *,$$

where the last identity has already been used in the analysis of go configurations, and the second-to-last identity is again explained by omitting a reversible move.

Significantly more complex and correspondingly difficult to analyze is the subconfiguration comprising the pawns on rows a and b. It is equal to $\uparrow$, so that for the depicted configuration, one has $- \uparrow + *$. On account of the relation $\uparrow \parallel *$, known already from our analysis of go, the player to move, whether white or black, can win the endgame. White moves h3-h4, to achieve $\uparrow - \uparrow = 0$. On the other hand, black cannot win by moving the h-pawn. However, the move a6-a5 does the trick.

Chapter Notes

1. If white is able to ensure a profit of at least v by moving in a particular configuration, then she can modify her underlying strategy in such a way that if her opponent moves first, she can be sure of at least $v - 2$ points. To do this, white proceeds as follows:

   - if black moves at any time during the game to the original "starting square," by which is meant the square on which white would have played had it been her turn, then white passes. The right to pass, which was not specifically provided for in the rules, can be granted to white, since passing does not improve her situation.

   - otherwise, white plays as though she had moved first, making her first move according to her original strategy as though she had already made the first move, independent of whether the starting square is empty or is occupied by a black stone.

With this strategy white attains a final configuration that is almost identical to what she could have obtained had she made the first move. The only difference is in the start square, which can reduce her winnings by at most two points.

2. All values that are integers or fractions with a power of two in the denominator can be translated directly without duplicating the winning values. Thus, for example, in the case that black wins a half a point, there arises the sequence of moves $-1/2 = \{-1 \mid 0\} = \{\{\mid\{\mid\}\} \mid \{\mid\}\}$ instead of the corresponding end configuration. However, in most games, including go and blockbusting, only integer winning values are possible.

3. That the simplified process of cooling a configuration by 1 in fact leads to the correct result is not at all obvious. The main reason for this special property of go configurations is the fact that the score is always a whole number. Since furthermore, the sides of thermographs are always arranged horizontally or at a $45°$ angle, they can run only along such lines as appear in the following diagram in the range from 0 to 7/8. Then the mean value of a go configuration must have a denominator of the form 2^n if the temperature has a value of at least $1 - 1/2^n$. Since for coolings below or up to a bit above the temperature one always has the equality $G_t = \{G'_t - t, \dots \mid G''_t + t, \dots\}$, in a cooling of 1, only the last phase of $1/2^n$ is critical. Its effect can be examined: no simpler number in the sense of the simplicity theorem can, as the mean value, lie between the configurations $G'_1 - 1, \dots$ on the one hand and $G''_1 + 1, \dots$ on the other. Concretely, from the assumed existence of a minimum counterexample, that is, a configuration with the shortest possible length of the game, one can draw conclusions that lead to a contradiction.

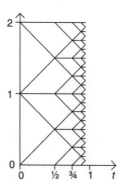

4. Less direct, but making use of more general concepts, is the approach of Berlekamp and Wolfe: essentially, the cooling homomorphism $G \to G_1$ for go configurations is reversed. This takes place by means of a universally recursively definable mapping u for an arbitrary Conway configuration $H =$

$\{\,H',\ldots\mid H'',\ldots\,\}$, which can be understood either as a special case of a so-called overheating operator or as a type of product with $(1 + *)$:

$$u(H) = \begin{cases} H & \text{when } H \text{ is even,} \\ H + * & \text{when } H \text{ is odd,} \\ \{\,-1 + u(H'),\ldots,\mid 1 + u(H''),\ldots\,\} & \text{otherwise.} \end{cases}$$

Like cooling, overheating is also a homomorphism; that is, it is compatible with disjunctive summation and the greater than or equal relationship:

- $u(H + L) = u(H) + u(L)$;
- the relation $H \geq L$ always implies $u(H) \geq u(L)$.

Moreover, for configurations of mathematical go in which the subsequent play has no ko or seki situations, one always has

$$G = u(G_1) \quad \text{or} \quad G = u(G_1) + *.$$

5. Elwyn Berlekamp, Sums of $N \times 2$ Amazons, in: *Game Theory, Optimal Stopping Probability Statistics, Papers in Honor of Thomas S. Ferguson*, Institute of Mathematical Statistics Lecture Notes Monograph Series **35**, Beechwood 2000, pp. 1–34, in particular, pp. 31 ff. A simpler proof follows. We will show inductively that the approximation errors

$$D_L(G, t) = L_0(G + E_t) - L_t(G) - L_0(E_t),$$
$$D_R(G, t) = R_0(G + E_t) - R_t(G) - R_0(E_t),$$

are bounded by $2N\delta$ for values of t involving an integral multiple of the grid size δ defined for the environment E_t. Here N denotes the maximal number of moves that can occur in a game beginning in configuration G. We begin with the cases $G = 0$ and $t = 0$, for which the assertion is obviously true. For the induction step we analyze the possible moves in configuration $G + E_t$ in order to be able to estimate the approximation errors recursively. For the configuration $G = \{\,G',\ldots\mid G'',\ldots\,\}$, we obtain

$$D_L(G, t) = \max_{G'} \left(R_0(G + E_{t-\delta}) + t, R_0(G' + E_t) \right) - L_0(E_t) - L_t(G)$$

$$= \max_{G'} \left(\begin{array}{l} R_0(G + E_{t-\delta}) + t - R_0(E_{t-\delta}) - t, \\ R_0(G' + E_t) + R_0(E_t) \end{array} \right) - L_t(G)$$

$$= \max_{G'} \left(\begin{array}{l} R_{t-\delta}(G) + D_R(G, t - \delta), \\ R_t(G') + D_R(G', t) + 2R_0(E_t) \end{array} \right) - L_t(G)$$

$$= \max_{G'} \left(\begin{array}{l} R_{t-\delta}(G) - L_t(G) + D_R(G, t - \delta), \\ R_t(G') - t - L_t(G) + t + 2R_0(E_t) + D_R(G', t) \end{array} \right).$$

Here $t + 2R_0(E_t)$ is equal either to zero or to $-\delta$, depending on whether t is associated with an even or odd multiple of the grid size δ. The induction hypothesis implies the two inequalities $|D_R(G, t - \delta)| \leq 2N\delta$

and $|D_R(G',t)| \leq 2(N-1)\delta$. If we consider as well the two inequalities $R_{t-\delta}(G) - L_t(G) \leq 0$ and $R_t(G') - t - L_t(G) \leq 0$, which follow from the properties of the thermograph described in Chapter 24, then one obtains

$$D_L(G,t) \leq \max\left(2N\delta, 2(N-1)\delta\right) = 2N\delta.$$

It remains to establish the inequality $D_R(G,t) \geq -2N\delta$. To do so, we distinguish the two cases $t \geq t(G) + \delta$ and $t \leq t(G) + \delta$. In the first case, we have $R_{t-\delta} - L_t(G) = 0$, and thus

$$D_L(G,t) \geq R_{t-\delta}(G) - L_t(G) + D_R(G,t-\delta) \geq -2N\delta.$$

For $t \leq t(G)$ left has a move to a configuration G' with $R_t(G') - t - L_t(G) \geq 0$. If one takes on the existing possible move for $t = t(G)$ for t values with $t(G) \leq t \leq t(G) + \delta$ as well, then one sees that for all $t \leq t(G) + \delta$ there exists a move to a configuration G' with $R_t(G') - t - L_t(G) \geq -\delta$. Finally, we obtain, in this case as well,

$$
\begin{aligned}
D_L(G,t) &\geq R_t(G') - t - L_t(G) + t + 2R_0(t) + D_R(G',t) \\
&\geq -\delta - \delta - 2(N-1)\delta \\
&= -2N\delta.
\end{aligned}
$$

26

Misère Games: Loser Wins!

The rules of Conway games can be altered so that the player to move last loses instead of winning. For such inverse versions can simple criteria for winning moves be found like those obtained for normal versions using Grundy numbers?

Back in 1902, in his first analysis of nim, Charles Bouton analyzed the inverse version of standard nim.[1] In an inverse game of nim, generally called a *misère* version, each player attempts to move in such a way that he would lose according to the usual rules. Thus he must attempt to force his opponent to move to an end configuration.

Bouton's result for the inverse of standard nim is remarkably simple: the player in a winning position moves as in normal nim to a configuration with nim sum zero, except when the move would result in a configuration in which all remaining piles consist of a single stone. In this exceptional case he moves instead to a configuration that consists of an odd number of single-stone piles. Thereafter, the game proceeds according to mutual zugzwang, until the game ends with the victory of the player who used the winning strategy.

In view of other nim variants that we would like to investigate, we shall express Bouton's result another way: we first define the notion of *exceptional configuration*: these are all configurations that offer the players differing winning prognoses between the two versions. In standard nim, these are the configurations cited already in which the remaining piles

---

[1]Charles L. Bouton, Nim, a game with a complete mathematical theory, *Annals of Mathematics* **Series II, 3**, 1901/02, pp. 35–39.

consist of exactly one stone. In the exceptional configurations it is useful to distinguish them according to their winning prognoses in the normal version. Then on the one hand, one obtains the configurations of the form 1^{2k}, that is,

- 0,

- 1, 1,

- 1, 1, 1, 1, etc.

and on the other hand, configurations of the form 1^{2k+1}, namely,

- 1,

- 1, 1, 1, etc.

Based on this division, the losing configurations in inverse standard nim correspond to the configurations with nim sum 0, except for the exceptional configurations 1^{2k}, but including the exceptional configurations 1^{2k+1}. Analogously, the winning configurations of inverse standard nim include the configurations with positive nim sum, except for the exceptional configurations 1^{2k+1}, but including the exceptional configurations 1^{2k}.

Such a division of the configurations in standard nim exists in principle, of course, for other nim variants. There, too, there exists, if one considers the winning prospects of the normal and misère versions in parallel, a division of the configurations into four classes. If one denotes winning configurations by W and losing configurations by L,[2] then each of the four configuration classes can be represented by a pair of letters, where the first letter is for the normal version, and the second for the misère version:

- W; a WW configuration is a winning configuration in both versions.

- LL; an LL configuration is a losing configuration in both versions.

- WL; a WL configuration is a winning configuration in the normal version, but a losing configuration in the misère version.

- LW; an LW configuration is a losing configuration in the normal version, but a winning configuration in the misère version.

---

[2]Since we are speaking here of winning and losing configurations, the notation W and L is perhaps more suggestive than the notation that is more usual in the literature, namely, N for next player wins, and P for previous player wins.

The exceptional configurations, that is, those that offer different out-comes for the different versions, consist of the WL and LW configurations. In the case of standard nim, the WL configurations are all those of the form 1^{2k+1}, while the LW configurations are those of the form 1^{2k}. In general, this means that in an arbitrary nim variant, for every exceptional configuration there exists a move that leads to another exceptional config-uration. Namely, if there is no such possible move in a given configuration, then the equal winning prospects between the normal and misère versions from the subsequent configurations can be transferred to the given config-uration itself.

To analyze a single version of a nim variant completely, a decomposition of all configurations into winning and losing configurations must be found. In the case of the normal version, this is always possible using Grundy values. In general, if a conjectured decomposition into winning and losing configurations is to be verified, the following properties must be satisfied. They relate to the minimax principle in the sense of Zermelo's theorem and were established for the normal version of standard nim in Chapter 21. A winning move must always be offered to the player in a winning position, but his opponent must not be able to find a move that turns the tables:

- from every W configuration there is a move to an L configuration.

- from an L configuration, every move leads to a W configuration.

In the misère version, the end configuration must be excluded in the first property, which indeed can be viewed as a W configuration—after all, the player who just moved has lost—but there is no longer any possible move.

Both versions of a nim variant have been completely analyzed when the totality of all configurations has been decomposed into the four indicated classes. To confirm a conjectured configuration, the conditions given in the following table must be verified:

Normal	LL, LW	The Grundy value of the LL and LW configurations is 0.
Version	WL, WW	The Grundy values of WL and WW configurations are nonzero.
Misère	LL	From an LL configuration there is no move to WL.
	WL	From a WL configuration there is no move to LL or WL.
Version	WW	For every move from a WW configuration to an LW configuration there exists an alternative to LL or WL.
	LW	For every LW configuration (other than the end con-figuration) there is always a move to WL.

Only three of the last four conditions, those relating to the misère version, must be clarified a bit:

- for the requirement on LL configurations, in addition to moves to WL, moves to LL should also be excluded. Such moves are impossible in any case, due to the properties of the normal version.

- from the viewpoint of the misère variant, it is required of a WW configuration that there always be a move to WL or LL. The Sprague–Grundy theory for the normal version guarantees at least one move to LW or LL. If one achieves an LL configuration in this way, then all is well. Then the requirement formulated in the table is completely adequate, namely, that for every move from WW to LW there exists an alternative to LL or WL.

- from the exclusive viewpoint of the misère version, there must exist a move to LL or WL for every LW configuration. The first possibility is eliminated, however, for the normal version from the Sprague–Grundy theory.

The actual use of these conditions can best be understood with the help of an example. It is natural to check the configuration decomposition of standard nim as outlined above. The necessary considerations are not particularly difficult, but in their totality, they are anything but obvious (see Note 1 at the end of the chapter).

Unfortunately, things get much more complex with other nim variants. The reason is that configurations in the misère version, in contrast to the normal way of playing, are no longer necessarily equivalent to piles in standard nim. In general, in the misère version there are many fewer configurations that are mutually equivalent. This begins already with the fact that the doubling of a configuration W, that is, the disjunctive sum $W + W$, always leads to a losing configuration and is therefore equivalent to the end configuration 0. On the other hand, in the misère game, the standard nim configurations $1, 1$ and $2, 2$ already exhibit different winning prospects. All in all, this leads to the situation that among the configurations from which a game would last at most six moves, there are about $2^{4171780}$ inequivalent configurations.[3] It is thus hardly likely that one would be able to characterize completely this astronomical number of configurations using easily calculable data such as the Grundy values used for analysis of the normal version. In spite of this pessimistic prognosis, there are a few bright spots

---

[3]John H. Conway, *On Numbers and Games*, second edition, Natick, MA 2002, Chapter 12; E. Berlekamp, J. Conway, R. Guy, *Winning Ways*, second edition, Natick, MA 2003, volume 2, Chapter 13.

to be found for the misère games. For example, there are nim variants in which for the misère game as well, all configurations are equivalent to nim piles. The recursive criterion is as follows:

> If all possible moves in a given configuration lead to piles in standard nim, of which at least one pile contains at most one stone, then this configuration in the misère game is itself equivalent to a nim pile. Its size is equal to the smallest natural number that is not represented as the size of one of the piles reachable in one move.

This theorem of Conway is based on the fact that all moves that go beyond the possibilities of standard nim can be dispensed with (see Note 2 at the end of the chapter). A consequence of the theorem is that some inverse nim variants can be reduced recursively to the inverse standard nim; that is, within every possible course of the game, each configuration is replaced by an equivalent nim pile. For example, this works for all subtraction games, which were introduced in their normal form in Chapter 22. For these, Ferguson discovered in 1974 the decisive property that every pile with Grundy number 0 permits a move resulting in a pile with Grundy number 1.[4] Thus in subtraction nim, for every nonempty pile there is a move that leads to a pile with Grundy number 0 or 1, and that is precisely the property that the theorem requires for its hypothesis.

In the same sense, Lasker nim is *tame*, as Conway calls the misère version that is reducible to standard nim. As a consequence of Conway's theorem, the truth of this fact is at once apparent, since, as we have seen in Chapter 22, in Lasker nim there is only one pile with Grundy number 0, namely, the empty pile. By a somewhat different route, Ferguson solved in 1974 the misère version of Lasker nim together with the subtraction games.

Based on positional equivalence, tame nim games in the misère version are won the same way as in standard nim. Their exceptional configurations comprise the configurations all of whose nim piles have Grundy values 0 and 1:

- a WL configuration is a nim configuration with nim sum 0 all of whose piles have Grundy value 0 or 1.

- an LW configuration is a nim configuration with nim sum 1 all of whose piles have Grundy value 0 or 1.

---

[4]T. S. Ferguson, On sums of graph games with last player losing, *International Journal of Game Theory* **3**, 1974, pp. 159–167. See also "Nim Variants en Masse" in Chapter 22.

Other nim games, like bowling nim, are not tame at all; that is, not all of their configurations are equivalent in the misère version to a single pile of standard nim. However, such a misère game can be investigated purely theoretically in the following way, or at least one can make the attempt: starting from a complete analysis of the normal version, which is always relatively easy using Grundy values, one searches for all exceptional configurations, that is, all LW and WL configurations. If all of these have been found, then one can immediately determine the winning prospects for the misère version from those of the normal version. Since a configuration can be an exceptional configuration only if another exceptional configuration is reachable from it in a single move, it is possible in principle, starting from the end configuration as the first exceptional configuration, to determine all the others recursively. Unfortunately, there are generally infinitely many exceptional configurations, so that one must attempt to guess the entire range of exceptional configurations based on empirical data and then to confirm that guess. This path has shown itself useful especially in the case of bowling nim. It was first used by William Sibert in the mid-1970s. His extremely complex results were published only in 1992,[5] in the same year in which Ranan Banerji and Charles Dunning carried out an independent analysis of bowling nim.[6] Also in the same year came the significant refinement of the *Sibert–Conway decomposition* by Thane Plambeck. Plamback succeeded in obtaining a complete analysis of the misère versions of other octal games. His classification of configurations is based on a weight function, in addition to the Grundy values, chosen carefully for each nim variant. Here each pile size has a particular weight. In a configuration consisting of several piles, these weights are added.

Let us look now at the results on bowling nim obtained by Sibert and Conway. Their starting point is an analysis of normal bowling nim based on the Grundy values, repeating from 72 with period 12, as shown in Table 26.1.

We let $E(a, b, \dots)$ denote the configurations that contain an even number of piles of sizes $a, b, \dots$. For example, $2, 2, 2, 3$ and $2, 2, 3, 3$ are configurations in $E(2, 3)$. Similarly, we let $O(a, b, \dots)$ denote the set of configurations with an odd number of piles of sizes $a, b, \dots$. For example, $1, 2, 2$ is a configuration in $O(1, 2)$. With this notation we can now list all the exceptional configurations of bowling nim as follows, according to the following example:

---

[5] W. L. Sibert, J. H. Conway, Mathematical Kayles, *International Journal of Game Theory* **20**, 1992, pp. 237–246.

[6] Ranan B. Banerji, Charles A. Dunning, On misère games, *Cybernetics and Systems* **23** 1992, pp. 221–228.

n	$g(n)$	$g(n+12)$	$g(n+24)$	$g(n+36)$	$g(n+48)$	$g(n+60)$	$g(n+72)$
0	0	4	4	4	4	4	4
1	1	1	1	1	1	1	1
2	2	2	2	2	2	2	2
3	3	7	8	3	8	8	8
4	1	1	5	1	1	1	1
5	4	4	4	4	4	4	4
6	3	3	7	7	7	7	7
7	2	2	2	2	2	2	2
8	1	1	1	1	1	1	1
9	4	4	8	8	4	8	8
10	2	6	6	2	2	6	2
11	6	7	7	7	7	7	7

Table 26.1. Bowling configurations with one pile.

- LW configurations, that is, misère winning configurations with Grundy value 0:

$$E(5)E(4,1) \quad \text{such as} \quad 5,5,4,4,4,1,$$
$$E(17,12,9)E(20,4,1) \quad 17,9,9,9,20,1,$$
$$15E(17,12,9)E(20,4,1) \quad 25,17,9.$$

- WL configurations, that is, the misère losing configurations with nonzero Grundy value:

$$O(5)O(4,1) \quad 5,5,4,1,1,$$
$$E(5)O(4,1)) \quad 4,$$
$$O(9)E(4,1) \quad 9,1,1,$$
$$12E(4,1) \quad 12,4,1,$$
$$E(17,12,9)O(20,4,1) \quad 17,9,20,4,4,$$
$$25O(9)O(4,1) \quad 25,9,1.$$

In particular, there are no exceptional configurations among those containing at least one pile of 26 or more stones. This makes it somewhat easier to verify the given listing. To do this, we first look only at configurations each of whose piles contains at most 53 stones. From another configuration, that is, from one with at least one pile of 54 or more stones, there is no possible move that leads to one of the listed configurations. Therefore, exceptional configurations with such large piles are impossible. In an actual proof, the configurations whose piles contain at most 53 stones are characterized by the frequency with which the individual pile sizes appear. It turns out that one can determine for each of these frequencies, except from the starting values 0 and 1, whether it is even or odd. One then proves the periodicity by considering finitely many cases, where no number larger than 4 needs to be considered.

A quite different approach for the investigation of such periodicities in the winning prospects of configurations of misère variants, more suitable as the basis for extensive computer calculations, was published by Dean Allemang in 2001.[7] Allemang's results are based on the following basic idea, which actually is an analogue of the approach of Lasker that was described in Chapter 22. It is true that the general equivalence of misère configurations in reference to summands with arbitrary nim configurations cannot be practically simplified due to the already mentioned astronomical variety, even with piles with a small number of moves. However, for the analysis of a particular nim variant it suffices fully if a weaker equivalence related to the configurations of this one quite special nim variant is investigated. And in fact, there are frequently rather drastic simplifications that can be made. What is studied is piles or sums of such piles that can be replaced by smaller piles or sums of a smaller number of piles within configurations of the misère variant under consideration, particularly if the other piles of the given configuration satisfy certain properties, such as in relation to their size. The complete solution of octal games found by Allemang such as, for example, 0.53, 0.54, and 0.72 is based on two theorems that make it possible to prove (infinite) periodicities by checking finitely many periodicity conditions, and indeed, without reference to the normal rules of the nim variant being studied. Allemang's first periodicity theorem relates to the frequency with which a pile of a particular size occurs within a configuration, while his second periodicity theorem relates to the size of the pile.

However, the number of conditions to be checked using Allemang's procedure can be so large that a complete analysis cannot be practically carried out in this way. Thus, for example, Allemang gives 3^{143} for the number of conditions to be checked in bowling nim, which in decimal notation is a 69-digit number.

[7]Dean T. Allemang, Generalized genus sequences for misère octal games, *International Journal of Game Theory* **30**, 2001, pp. 539–556. The results presented stem from the author's master's thesis of 1984.

Nimbi

A nim variant that always begins with the same initial configuration is Nimbi. It was invented by the Dane Piet Hein, the
coinventor of Hex, around 1950, and it was marketed commercially for several years. Nimbi is played with 12 stones, which at
the start of the game are positioned on the intersection points
of the same number of lines, as in the figure below:

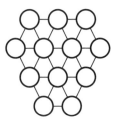

As usual, the players alternate moves. A move consists of the
removal of at least one stone. However, a player may remove
more than one stone, on the condition that the stones lie in
unbroken sequence on one of the 12 lines. The player who is
forced to take the last stone is the loser; that is, the game is
played with misère rules. Of course, one may also play by the
"normal" rules, by which the last player to move wins.

Since Nimbi always begins in the same configuration, and so
there are only $2^{12} = 4096$ possible configurations, a computer
analysis is not too difficult to carry out. The winning prospects
of all configurations are determined by a minimax process that
examines the moves in reverse order and then stores them for
further computation (see Note 3 at the end of the chapter). The
first analysis, published by Avierzi Fraenkel and Hans Herda in
1980, gave the result that the starting configuration is an LL
configuration; that is, in both versions it is a losing configuration.[8] This is noteworthy in that losing configurations in nim
games become increasingly rare as the length of the game increases; on the one hand, the first player has a stronger influence
than his opponent on the shape of the game, and on the other

---

[8]Avierzi S. Fraenkel, Hans Herda, Never rush to be the first in playing Nimbi, *Mathematics Magazine* **53**, 1980, pp. 21–26.

hand, a configuration is a losing one only if *every* configuration achievable in one move is a winning one.[9]

Chapter Notes

1. Here are the conditions:

   - a WL configuration, that is, one of the form 1^{2k+1}, can arise by a move only if a pile is cleared down to a single stone or completely removed. In the first case, the configuration before the move has the form $1^{2k}m$ with $m > 1$, while in the second case, it is $1^{2k+1}m$ with $m > 0$. Configurations of the form LL, that is, those with nim sum 0 that are not of the form 1^{2k}, are not to be found here. Moves from LL to WL are thus impossible.

   - from a WL configuration, that is, a configuration of the form 1^{2k+1}, there are only moves to configurations of the form 1^{2k}, which are exclusively LW configurations. Thus LL and WL configurations are unreachable from WL in a single move.

   - an LW configuration, that is, one of the form 1^{2k}, can arise in one move only if a pile is removed completely or down to a single stone. In the first case, the configuration before the move has the form $1^{2k-1}m$ with $m > 1$, while in the second case it is $1^{2k}m$ with $m > 0$. Except for the configurations $1^{2k}m$ with $m = 1$, these are all configurations of type WW. Each of them offers an alternative move to a configuration of the form 1^{2k+2}, that is, to WL.

   - from an LW configuration that is not the end configuration, that is, a configuration of the form 1^{2k+2}, the only possible move is to 1^{2k+1}, that is, to a WL configuration.

2. We start with the piles of standard nim, $*0, *1, *2, *3, \ldots$, which are determined formally from the set of possible moves

$$*0 = \{\ \},$$
$$*1 = \{\, *0 \,\},$$
$$*2 = \{\, *0, *1 \,\},$$
$$*3 = \{\, *0, *1, *2 \,\}, \ldots.$$

---

[9] David Singmaster, Almost all games are first person games, *Eureka* **41**, 1981, pp. 33–37; David Singmaster, Almost all partizan games are first person and almost all impartial games are maximal, *Journal of Combinatorics, Information and System Sciences* **7**, 1992, pp. 270–274.

The theorem then states that by adding certain additional possible moves, there arise misère-equivalent configurations. In detail, we have the following equivalences:

$$*0 = *1, *a, *b, \ldots \quad \text{with} \quad a, b, \ldots > 2,$$
$$*1 = *0, *a, *b, \ldots \quad \text{with} \quad a, b, \ldots > 2,$$
$$*2 = *0, *1, *a, *b, \ldots \quad \text{with} \quad a, b, \ldots > 3,$$
$$*3 = *0, *1, *2, *a, *b, \ldots \quad \text{with} \quad a, b, \ldots > 4, \ldots.$$

Each completed move leads to the affected nim pile being enlarged. Here the case of the empty nim pile $*0$ is an important special case, since the theorem depends on the assumption that one of the possible moves leads to $*0$ or $*1$. To that extent, in the first case, the move to the one-stone pile $*1$ must be present among the added moves. The equivalence expressed by the theorem means concretely the following: if any nim pile $*m$ is a summand of a configuration presented as a disjunctive sum, then the winning prospects do not change in the misère game if the nim pile $*m$ is replaced by a configuration with an appropriately increased set of possible moves. That is, the player who was in a winning position continues to possess a winning strategy. That is what we now wish to prove:

- the player in a winning position simply ignores the additional possible moves, except in the case $m = 0$, when $*0$ is the only remaining pile. In this case, the player moves to $*1$, so that he wins after the opponent's forced move to $*0$.

- on the other hand, if the player in the losing position takes one of the additional moves at any time in the course of the game and thereby increases the affected nim piles, then the player in the winning position simply plays back to $*m$.

This construction is a special case of generating equivalent configurations using reversible moves. This procedure works for all Conway games. In misère versions, however, a final clause is necessary that ensures a reply leading to a losing position for the case that no further component is available.

We note finally that configurations such as $\{ *2 \}$ and $\{ *2, *3 \}$ not only do not satisfy the conditions of the theorem, but are actually not equivalent to a nim pile in the misère game.

3. The winning prospects for all configurations can be stored in a Boolean array of 4096 values. The numbering of the configurations is best done in binary. Thus each binary digit represents a "square" of the game board; in a given configuration, a stone on a square is represented by a binary 1, while a stone that has been removed is represented by a 0. In this way, moves can be encoded. The following figure gives an example of an opening move:

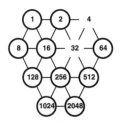

After move number 36,
the configuration 4059
results from starting
configuration 4095.

For the minimax analysis, the configurations can be investigated in the normal order $0, 1, 2, \ldots, 4095$, since this order ensures that a configuration is investigated only after all of its subsequent configurations have been analyzed.

To store the winning prospects of all the configurations, in the era of megabyte-sized storage one could just as well use a table in a relational database rather than binary encoding, in which case the configurations can be stored in just about any way desired. Rapid access is accomplished with an index in the database. In particular, this possibility is of interest for games whose configurations possess a more complex structure.

27

The Computer as Game Partner

What is going on inside the "mind" of a chess computer?

There is a subfield of computer science called artificial intelligence, but let there be no illusion that a computer thinks the way human beings do. Nonetheless, a computer can be programmed to behave as though it were thinking. A good example of this phenomenon is that of chess computers and programs. But how can one program a computer so that it can win at chess even under the time constraints of tournament rules? The difficulties to be overcome are principally of a quantitative nature, since the minimax principle offers at least a theoretical possibility of calculating all the winning prospects algorithmically. However, due to the enormous number of possible moves that arise from a given configuration, such an analysis must be relegated to the world of theory except for certain manageable endgame situations. Of course, the minimax principle is nothing to be sneezed at. Indeed, it represents a player's strategy of playing it safe so that his opponent can do him the least harm. But how can minimax techniques be simplified so that in practice, within the capabilities of computer hardware and software, acceptable results can be achieved in a limited amount of computation?

In the autumn of 1977, at the Berlin radio and television exhibition, the "Chess Challenger 3," the first of a series of chess computers, was presented. Equipped with an 8-bit microprocessor and only a few kilobytes of program space and even less RAM, its play was rather poor. Every year some new version of a chess computer would appear around Christmastime, always with some new feature—sensor board, printer, speech capability. As hardware improved, the machines' playing strength improved as well, especially with the shift to PC programs. Nowadays, one is quite well accustomed to playing games against computers, and one is also accustomed to getting soundly beaten. Good chess programs play easily at a level that offers a tournament player little chance of success. And even the reigning world chess champion Garry Kasparov was beaten in a tournament by the computer Deep Blue. A match in 2002 against Kasparov's successor, Kramnik, and the chess program Deep Fritz ended in a draw.

Aside from the fraudulent chess automaton mentioned in Chapter 18, the first chess-playing machine was invented in 1890 by the Spaniard Torres y Quevedo (1852–1936). It was capable of playing endgames consisting of king and rook against a solitary king. The electromechanical construction was designed specifically for this special situation, and thus contained no universal computational elements, in contrast to the "analytical engine" designed by Charles Babbage some 60 years earlier. A second version, from the year 1920, is on display at the University of Madrid.[1]

The two true pioneers of computer chess were the American Claude Shannon (1916–2001) and the Englishman Alan Turing (1912–1954). In the middle of the 20th century, these two men independently, and on an entirely theoretical level, thought about how a computing machine might play chess. It would appear that the time was ripe for the development of universally programmable computers: in 1936, Konrad Zuse (1910–1995) began experiments aimed particularly at chess,[2] while in the United States there appeared between 1939 and 1944 the Mark I, which operated with relays, and then in the years 1943–1945 the first electronic computer, with 17 000 vacuum tubes: the ENIAC. The first von Neumann machine, that is, the first computer with storage for data and programs, was the EDSAC computer, which was completed in England in 1949.

Computers: What They Can Do

To give an idea of how a game program works, it would not do here to explain the workings of a computer in the minutest

---

[1] A picture can be seen in Dieter Steinwender, Frederic A. Friedel, *Schach am PC*, Haar 1995, p. 32.

[2] In a document from the year 1945, *Das Plankalkül*, as Zuse called his symbolic language for carrying out computations, the last chapter is devoted to the theory of chess. There, moves and tests of certain positional properties are represented in Zuse's notation. Zuse's writing became better known after it was reprinted in 1972. As Zuse then noted, he learned to play chess in order to carry out his investigations. K. Zuse, *Das Plankalkül, Kommentierter Nachdruck der Fassung von 1945, Gesellschaft für Mathematik und Datenverarbeitung*, Sankt Augustin 1972, pp. 35 f., 235–285.

detail. Therefore, we shall carry on our discussion at a level above a number of lower system levels, namely, that of a modern programming language such as Pascal, C, C++, Fortran, and Basic.[3] These languages make it possible to describe mathematical algorithms step by step, that is, broken down into elementary arithmetic and logical operations, in a formulaic way so that they can be carried out by a computer using compilers and other system programs.

Intermediate results can be stored for later use, which is done using program variables, which can be given names with mnemonic values. These are certainly much easier to manage than the numbered storage cells of the computer's memory. Each variable can store an integer up to a certain size, which depends on the programming language and the computer's capacity. For example, if one wished to store the result of the calculation $234 \times 123 - 34 \times 91$ under the name Alpha, then one would write (more or less, depending on the details of the programming language)

```
Alpha = 234 * 123 - 34 * 91
```

With instructions like

```
Beta = 2 * Alpha + 15
Alpha = Alpha - 1
```

the values stored under the name Alpha can be read and further processed, where in the second example, the value of Alpha is altered. The last instruction shows that variables have a character different from that of mathematical symbols, whose value within a mathematical statement is always fixed.

In addition to integers, one can use variables to store floating-point numbers, Boolean values, and text characters. Another convenience of variable storage is that logically connected data can be processed together. For example, a variable named Board can contain all the information, translated into numbers, about the constellation of figures in a chess position. With the instruction

```
NewBoard = Board
```

---

[3]Though much has happened since then, Niklaus Wirth's 1975 book, *Algorithms and Data Structure*, still has much to recommend it.

the entire board configuration is copied, for example, for the purpose of making the changes required by a move in a chess game. One can also apply coordinates to such variables in order to index the various parts of a variable that may represent an array. Thus white's rook can be moved to square a4 with the following command:

```
NewBoard(1, 4).Piece = Rook
NewBoard(1, 4).Color = White
```

It is assumed here that the variables `Rook` and `White` already possess characteristic values that allow them to be distinguished from other colors, namely `Black` and `Undefined`, and these values do not change during the entire calculation. For such particular use, most programming languages permit the definition of constants.

Normally, the instructions of a computer program are carried out sequentially. However, the order of execution can be made dependent on intermediate results. For example, we test the number of squares `s` that a white rook can advance from the square $(2, 3)$, that is, b3:

```
FOR s = 1 TO 7

    IF 3 + s > 8 THEN EXIT FOR
    GoalSquare = Board(2, 3 + s)
    IF GoalSquare.Color = White THEN EXIT
    FOR
```
If the white king is not in check, then the move is legal and will be processed
```
    IF GoalSquare.Color = Black THEN EXIT
    FOR

NEXT s
```

In a real game program it would hardly be worthwhile to investigate such a special case. However, it is clear that the length that a white rook can travel is limited. It can move only until it reaches the edge of the board, the square before that occupied by another white piece, or a square occupied by an enemy piece, which it can then capture, and each of these three contingencies is noted by an `EXIT FOR`, which terminates the loop.

Shannon was a researcher at Bell Laboratories. He is best known as the founder of information theory, which goes back to a 1948 article.[4] A year later, he gave a lecture in which he presented his ideas on computer chess, which he later published in two articles.[5] Focusing on the technical aspects, Shannon briefly sketched the principles of computers and programming. Then he made a suggestion as to how a chess position might be stored in a computer. Each type of piece is given a unique identification number: a white pawn has the value 1, a knight 2, bishop 3, rook 4, queen 5, and king 6. For black, the corresponding negative numbers are used, and 0 corresponds to an empty square. For each square of the board there is a cell in computer memory available, which makes it possible to store an entire chess position. That is, other than such additional information as whose turn it is, whether castling is still permitted, en passant capture, and the 50-move rule, 64 cells of memory suffice to store a configuration.

Shannon describes the minimax principle as fundamental for a chess program. That is, the two players generate lists of possible moves, and the best move is taken, on the assumption that the opponent will respond with what he thinks is his best move. Shannon notes that a complete analysis will be impossible. As a way out of the difficulty, he suggests that only the first few moves of possible variants be investigated and the winning prospects of the configurations reached be evaluated. For this, Shannon uses the usual unit of measure for the worth of the pieces, where a pawn is worth one point, knight and bishop three, rook five, and queen nine. Since the king must not be captured, it is given a value of 200, so that its loss will never be weighed against other advantages. The evaluation is refined by giving positive and negative points for positional considerations. Thus, for example, Shannon counted negative 0.5 points for an isolated pawn, backward pawn, or doubled pawn, while mobility was rewarded with 0.1 points for each possible move. The difference between white's and black's point score was used to evaluate each side's winning prospects.

To evaluate a variant realistically, it must become stabilized, according to Shannon. Thus it would make no sense to evaluate positions in the

---

[4]Less well known are Shannon's later ideas, which he worked out with the blackjack expert Edward Thorp, on irregularities in roulette. Anecdotes on this theme appeaer in the novel *The Eudaemonic Pie*, by Thomas A. Bass (1985). See also Chips im Schuh, *Der Spiegel* **30**, 1990, pp. 152–154.

[5]C. E. Shannon, Programming a computer for playing chess, *Philosophical Magazine* **41**, 1950, pp. 256–275, reprinted in in David N. L. Levy, *Compendium of Computer Chess*, London 1988, pp. 2–13; C. E. Shannon, A chess-playing machine, *Scientific American* **182**, February 1950, pp. 48–51, reprinted in David N. L. Levy, *Computer Games I*, New York 1988, pp. 81–88. Both articles were also reprinted in Claude Elwood Shannon, *Collected Papers*, New York 1993, pp. 637–656, 657–666.

midst of an exchange of pieces, since even the stupidest capture of a pawn by a queen would seem favorable, since it apparently brings material gain. Such shortcomings are overcome if winning prospects are evaluated only for "quiescent" positions, meaning positions in which the opponent cannot effect large changes in the evaluation with his next move. This *quiescence search* is still today a significant part of every chess program.

Shannon describes two different approaches to investigating moves, which he called A and B strategies, terminology in use to this day. The difference is that either all moves, with the exception of quiescence search, are investigated to a certain level, or else only selected variants, which are investigated even more deeply. There are sharp limits placed on the A strategy because of the enormous number of possible moves; consider, for example, that there are 20 possible first moves. Using that figure, there are about $20^4 = 160\,000$ variants of only two double moves. A preselection of plausible moves, such as those made naturally by an experienced chess player, would thus be extremely useful. This would prevent the computer from wasting most of its time exploring senseless variants, by which is meant variants that contain at least one move that is clearly not the most promising for the player in question. For example, it makes no sense for white to investigate the opening move a2–a3. On the other hand, as one knows from key moves in chess problems, an unusual move can introduce a surprising turn of events.

The selective B strategy is available to a good chess player through long experience. It allows him to recognize typical patterns in a position quickly, patterns that in fact have names: double pawn, free pawn, isolated pawn, backward pawn, linked pawns, open lines, block, protection, sacrifice, tempo, fork, exchange, quality, zugzwang, discovered check, double check. Each of these terms is linked in the player's mind with a library of experience: which squares and pieces are threatened, which pieces are important, and what should one do about it? And if the player overlooks a trap in the guise of what seems an insignificant variant, it can prove disastrous in a game. On the other hand, such an experience will be added to the player's storehouse of knowledge. Without having to change his "program," a chess player can learn something new. With a great amount of practice and practical experience, he can eventually improve his play to the level of a master.

A translation of this human approach to a static program adaptable in at most a few parameters has never been achieved. Therefore, most chess programs use strategy A; that is, there are no a priori absurd moves that are excluded. The good results that are nonetheless obtained are due in large measure to the progress in hardware. Even a modest personal computer is much superior in both speed and storage capacity to a mainframe computer of the 1960s. Thus in 1995, the benchmark of evaluating 100 000 positions per second using off-the-shelf hardware and software was achieved in 1995. From a purely pragmatic viewpoint, the inelegant brute-force method of the A strategy can be used without further ado.

In 1936, Alan Turing, the second pioneer of computer chess, came up with the idea of a theoretical computer, now called a Turing machine, in order to investigate the limits of algorithmic computation, a topic to which we shall return in the next chapter. After the outbreak of World War II, Turing worked in a department of the British intelligence service on the decryption of the German ENIGMA cipher machine. The success of this special scientific unit, to which belonged two well-known chess players we might add, is often viewed as decisive in the outcome of the war, and a reason that much of their work is still classified. Turing, who was said not to have been a good chess player, probably began around that time to search for an algorithm that would find a more or less acceptable move from any given chess position. To that end, he sought a way of evaluating positions and their consequences in a way that was simple yet led generally to correct results.

Turing's approach is very similar to that of Shannon; if Turing's reflections were perhaps not so all encompassing as Shannon's, they were somewhat more concrete. Like Shannon, he assigned numerical values to the pieces: $1000, 10, 5, 3.5, 3, 1$ for king, queen, rook, bishop, knight, and pawn. Like Shannon, he refined these basic values on the basis of positional properties by measuring their mobility. For each piece he also counted the number of possible moves, with special valuations for capturing moves and those that offered check. Finally, to obtain a total value of a player's position, he added the material and mobility valuations. Unlike Shannon, Turing used the quotient of the two players' values to compare the positions of the two players.

Turing's algorithm generally operated at a search depth of two moves. That is, in the role of white, all of white's moves are evaluated together with all of black's possible replies. Then, within the context of a quiescence search, an analysis was made of all subsequent moves for which on every move:

- a piece captured in the previous move is itelf captured;

- a piece captures a higher-valued piece;

- an unprotected piece is captured;

- the opposing king is checkmated.

white: Turing's algorithm

black: Alick Glennie

Manchester 1952

White moves Qd3xd6, and black answers with Rc8-d8, winning the white queen. White resigns.

Figure 27.1. Before Turing's algorithm made a bad move.

Although Turing's algorithm was never actually programmed, it makes possible a formal, purely mechanical game. In a game played in 1952, Turing played using his algorithm against the amateur player Alick Glennie for 29 moves.[6] The result was a threatened block that was overlooked when Turing's algorithm captured an only indirectly protected pawn with the queen. The loss of the queen one and one-half double moves later was beyond the search depth—a quiescence search is not provided for such situations. Figure 27.1 shows the game before the fatal move.

Although Turing's algorithm had made errors before that, for example, when it failed to encircle a bishop with pawns, and although his opponent did not make use of all the opportunities provided him, it is nonetheless noteworthy how much Turing's algorithm accomplished with limited means, roughly equivalent to the power of a casual player. Even the inglorious end of the game exposed a significant problem, today called the horizon effect. No matter how refined a quiescence search is, a program will always be too "nearsighted" at the end of its search depth for certain developments, such as when a human opponent sacrifices material for a positional advantage with attack opportunities, even though a direct checkmate is not apparent either to man or machine.

Although today's chess programs operate according to the A strategy, they do not do so precisely in the way that Shannon envisioned. Namely, the different move variants are not given equal weight in searching, and there are exceptions not only for the quiescence search as suggested by Shannon. In the analysis, those moves are left unconsidered that demonstrably do not influence the outcome. But which moves are these, and how can they be determined? Let us put ourselves in the situation of a chess player whose turn it is, who is currently examining a possible move that

---

[6] Alan Turing, Digital computers applied to games, in B. V. Bowden, *Faster Than Thought*, London 1953, pp. 286–295; reprinted in David N. L. Levy, *Compendium of Computer Chess*, London 1988, pp. 14–19; the game itself is printed in Frederic A. Friedel, *Schach am PC*, Haar 1995.

appears to be a good one. If we ourselves find a reply by our opponent that would be bad for us, we will classify our plan as "refuted," and discard it at once. In particular, it would be completely senseless to investigate other possible replies by our opponent. Whether the opponent possesses even better refutations and how bad the move that we are considering actually is have no practical meaning.

What does this model tell us? Moves that do bring us sufficiently satisfactory results due to a discovered refutation do not need to be considered further. What "satisfactory" means in this regard can vary. Usually, it is the case that the winning prospects of another move have already been investigated sufficiently precisely and have met the threshold for further investigation. Moves that do not meet these requirements are relatively refuted and and not investigated further. One might also have the case that the minimum requirements are formulated at the start of the game as an absolute threshold without the knowledge of whether such a move actually exists. In this approach as well, refuted moves are not considered further. However, in contrast to the first case, it can happen that all moves are rejected if there is no move that is sufficiently good. In this exceptional case the analysis must be repeated with a lowered threshold.

The technique of no longer considering refuted moves can also be used at a later stage in the analysis of move variants. Here both players can focus on the results of variants with several branches that have already been investigated. For organizing the minimal requirements on such variants two parameters are used, whose values are constantly updated as the variant is explored. Since these values are generally denoted by the Greek letters α and β, the procedure has taken on the name *alpha–beta algorithm*. The alpha value describes the minimal requirement of white for a configuration. Whenever that requirement is not met, white seeks success along a different path; that is, the position in question is never reached. On the other hand, the beta value holds the minimal requirements of the opponent black. That is, if the position makes possible a move with which white can be ensured more than the beta value, then black will try to hinder the arrival at this position. Together, the parameters give a region of acceptance encompassing all numbers that are at least α and at most β. All variants that lead to positions whose value is outside this region can be prevented by one or another of the players. Since every branching and already investigated variant can bring further restrictions, the acceptance region can never get larger as the analysis progresses; it can only get smaller or remain the same.

How the alpha–beta algorithm works can best be understood with the help of an example. The two simple games that we wish to investigate are represented as tree graphs, which we first saw in Chapter 18, where the

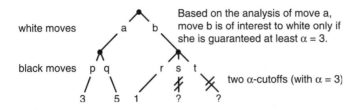

Figure 27.2. Two α cutoffs.

variants are investigated in order from left to right. In the first example, which is depicted in Figure 27.2, the positions indicated with question marks instead of a win level are insignificant for the determination of the minimax value.

If white has investigated the left move a and thus realized that it guarantees her a score of 3, then she will be interested in another variant only if she is guaranteed a score of at least $\alpha = 3$. However, with move b, the first reply r shows that white can expect no such guarantee from that move. Therefore, further moves such as s and t need not be considered, leading to α *cutoffs*.

Conversely, if black's minimal requirements are not satisfied, then the result is a β *cutoff*. To obtain an example, let us modify the game we have just been studying so that with move q no end position is reached. The resulting position is shown in Figure 27.3.

In that figure as well, the scores that have been replaced with question marks are insignificant, since the move q is bad for black in any case. Since black can limit white's score to 3 with move p, black need grant his opponent no score higher than 3 within the variant beginning with move a. The alternative move q thus fails on account of the reply y.

Although the principle of alpha and beta cutoffs is very plausible and is implicitly used by every good chess player, it took almost a decade from the first work of Shannon and Turing before it was recognized in chess programming. The first approaches are contained in the 1958 description of a conceptual program by Allen Newell, J. C. Shaw, and H. A. Simon,[7] who organized their minimax search on the basis of a one-sided acceptance threshold. The first move that exceeded this threshold was chosen, where cutoffs in later moves were not yet mentioned explicitly. It took another several years before the alpha–beta algorithm was developed in

---

[7]Allen Newell, J. C. Shaw, H. A. Simon, Chess-playing programs and the problem of complexity, *IBM Journal for Research and Development* **2**, 1958, pp. 320–335; reprinted in David N. L. Levy, *Computer Games I*, New York 1988, pp. 89–115; David N. L. Levy, *Compendium of Computer Chess*, London 1988, pp. 29–42.

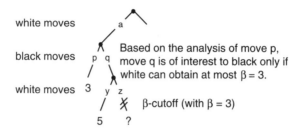

white moves

black moves

Based on the analysis of move p,
move q is of interest to black only if
white can obtain at most β = 3.

white moves

β-cutoff (with β = 3)

Figure 27.3. A β cutoff.

its full form.[8] This is noteworthy in that the alpha–beta algorithm is significantly faster than the normal minimax procedure, so that in the same amount of computing time, the search depth can generally be almost doubled (See Note 1 at the end of the chapter). In contrast to the normal minimax search, the requisite expense of searching in the alpha–beta algorithm depends on the order in which the moves are investigated. Namely, if one investigates a good move first, whether due to a clever choice or mere chance, the result will be a large number of cutoffs, which greatly reduces the computation time. But how can such promising moves be recognized? In practice, a number of approaches have been tried. Since their efficiency is never ensured in every particular case, but has nonetheless been proven in practice, one speaks of *heuristic methods*:

- variants that have been found to be good in a search with limited depth, say in the investigation of the previous move, are certainly promising.

- capture moves, particularly when they involve a return capture, are often advantageous.

- also promising are moves that have been shown to be good in parallel variants. The technique used here is called the *killer heuristic*. Here the best moves are registered statistically so that they can be selected first for other variants.

- it seems intuitively plausible that a move can be good only if it opens up opportunities in such a way that another move made by the same player would provide a measurable improvement in the position. Here

---

[8] A standard reference on the alpha–beta algorithm that also considers the historical development is Donald E. Knuth, Ronald W. Moore, An analysis of Alpha–Beta pruning, *Artificial Intelligence* **6**, 1975, 293–326. See also Alexander Reinefeld, Spielbaum-Suchverfahren, *Informatik Fachberichte* **200**, Berlin 1989, pp. 21 ff.

the imagined denial of a turn to the opponent has given this approach the name *null move*. The null move technique is used by some programs such as the PC program Fritz, often for *forward pruning* in the sense of a B strategy. However, this type of action can lead to quite delicate situations, especially in the endgame, since the right to move, such as in zugzwang situations, is not always an advantage.

Furthermore, the number of cutoffs can be increased by a "hopefulness" principle, in which a priori requirements are formulated. Here, on the basis of an analysis with limited search depth an alpha–beta search window is specified, for which then a move variant must be found that lies within the window.

Null Window Search

Generally, the numbers used for the evaluation of positions on the search horizon are normalized so that they are always integers. In this case, the minimax values that are calculated from these are also integers. An alpha–beta search window that lies in the space between integers cannot then contain a minimax value, and one therefore speaks of a *null window* search. Of course, the calculation of a minimax value using such an approach is not to be expected. However, as we shall see, the alpha–beta algorithm is usually programmed in such a way that in a complete cutoff at least the information is provided as to whether the minimax value lies above or below the search window. Thus the null window search gives a relatively efficient answer to the question of whether the minimax value exceeds a specified bound.

The null window search is used in a variety of ways: in addition to heuristic use for presorting possible moves, one might mention the so-called L improvement of the alpha–beta algorithm and above all the Negascout procedure.

With the *L improvement* of the alpha–beta algorithm, also called *last move improvement*, for every position, the previously investigated move is evaluated in a null window. The result for this last investigated move is "only"—sufficient at least at the first move level—the statement as to whether this move is better than the previously investigated move.

The *Negascout procedure* is based on the hope that the first investigated move for a position is already the best; with a clever heuristic presorting of moves, this hope is by no means necessarily in vain. After the investigation of the first possible move, every additional move is checked using a null window to see whether it might be better than the first move. If the hope is justified, then thanks to many cutoffs, one has obtained an efficient proof that the first move is in fact the best. In the opposite case, the procedure is broken off as soon as a better move is found, so that the procedure can be started again on the currently evaluated position, but now with a shorter list of moves and a new candidate as the hoped-for best move.

The search can be speeded up even more if transpositions are taken into account, that is, if positions that appear in two or more variants are evaluated only once. In that case, some of the intermediate results, as in the analysis of a variant for one move, must be stored to be used in the evaluation of other variants. This assumes the availability of a large amount of storage, and so such concepts were actualized only in the 1970s. However, it is not enough merely to store intermediate results. The results stored must be able to be quickly retrieved. To this end, in 1980 Joe Condon and UNIX coinventor Ken Thompson used *hash tables* in their special-purpose computer Belle, where every position possessed an index number, for example, between 0 and $2^{32} - 1$. It was possible for different positions to have the same index number, though collisions were rare. Once a position within a move variant has been thoroughly investigated, the result, together with data about the position, is stored in the associated hash index, making it available for the analysis of further variants. To make the procedure as practical as possible, the formula for the hash index was created in such a way as to make collisions extremely rare, and a hash index can be continually updated; that is, on each move the new hash index is computed from the old one. Common are binary summations without carry, denoted by XOR for "exclusive or," which we saw when we were studying nim addition. For each combination of piece and square a binary random number is determined at the start of the program, for example,

♔	a1:	1101001010011101
♔	a2:	0011101011101011
♔	a3:	1011110100100101
...		
♖	c4:	1011001010011010
...		
♚	b7:	0011001110011010
...		

The hash index of a position is now an XOR sum of the associated individual values:

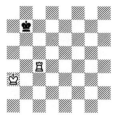

In order to update the hash index after a move, the data for the pieces and squares involved in the move must be updated. Since the XOR of a number with itself gives zero, one obtains a new hash index from the old by adding all the changed data on the pieces. Especially in the case of positions with many pieces, such an update can be much quicker than a complete calculation from scratch. For example, if the white king moves from a3 to a2, then one obtains

old value:		0011110000100101
♔	a2:	0011101011101011
♔	a3:	1011110100100101
	XOR	_____
new value:		1011101011101011.

Moreover, hashing collisions are usually ignored in chess programs, so that positions with the same hash index are given the same evaluation, even if that is not desirable. The resulting error is simply accepted, since it only seldom has a serious negative effect. The advantage is that by not paying attention to collisions, only the important evaluation data, and not the complete characterization of a position, need to be stored in the hash table.

Hash tables can also be used for the evaluation of endgame situations, which are hopeless using the normal alpha–beta procedure. For the analysis of special endgame situations, such as king with bishop and knight against king, it is better to adapt the process and storage management to the special circumstances (see "Endgame Databases").

Endgame Databases

Endgames belong to classical chess theory. Beginning with the easily won endgame of king and rook against king, and then more complicated endgames such as king with bishop and knight against king, by the 19th century, even more complex constellations such as king with two bishops against king and knight had been analyzed.

It is fun to check these classical endgame analyses with a computer and extend them if possible. To investigate a constellation of pieces completely, one generates a database that contains the winning prospects for every possible configuration of the given type. This corresponds to the technique that we employed in Chapter 26 in our investigation of Nimbi. In comparison to Nimbi, however, chess endgames are much more complex, and not only on account of the large number of possible moves. On the one hand, in contrast to Nimbi, chess is not impartial. Thus one must consider whose turn it is in addition to the disposition of the pieces. In addition, positions in the endgame can repeat themselves after a number of moves.

For the actual investigation of a special type of endgame, all possible positions up to symmetry are generated, and using a position index, storage space is reserved, for each position, for the as yet unknown result. For all positions it is assumed that it is white's move. All positions that are impossible based on the rules of chess are removed, for example, one in which black is in check. One asks whether white can force checkmate, and if so, in how many moves.

The actual analysis can take place by searching all positions in succession with a search depth of $1, 3, 5, \ldots$, where the positions on the search horizon are distinguished by whether or not white can force checkmate. If at every search depth the results of the previous search are considered, then effectively only the minimax values of the two additional (half) moves need be investigated.

The following method is faster but more complicated: one begins with the "one-move" positions in which white can checkmate in one move. With a backward-running minimax process, the "two-move," "three-move," and so on positions are constructed, where reference is made to the results already obtained. Thus to find all two-move positions in which white can checkmate her opponent in three (half) moves, starting with the one-move checkmates, one goes back one double move: first, every possible black move is reversed, so that every position is generated that could lead in one move to one of the one-move positions. Such a precursor position, which has the additional property that *all* its possible black moves lead to one-move positions, forms an intermediate position between the one-move and two-move positions. A two-move position is then obtained by backing up, from one of the intermediate positions, one move that white could have made previously, unless for such a precursor position there is already a faster checkmate, namely, checkmate in one move.

Regardless of how the actual investigation is carried out, in each case one obtains a database that contains information on every position as to whether white can force mate and how many moves are necessary to do so. The first computer endgame analyses were carried out in 1970 by Thomas Ströhlein in connection wih his dissertation at the Munich Technical University.[9] In the 1980s, Ken Thompson investigated many of the more complicated endgames. His extensive results were published in compressed form on CD-ROM.[10] In the following table we have collected some of the results of particular endgames. What is tabulated there is the maximal number of double moves until the game is "decided," that is, the number of double moves

---

[9]Thomas Ströhlein, *Untersuchungen über kombinatorische Spiele*, Munich 1970. In Chapter 9, the endgames KR–K, KQ–K, KR–KB, KR–KN, and KQ–KR are investigated. Using the computer TR4 from AEG-Telefunken with about 114 Kilobyte of memory, the analysis of the rook endgame KR–K took nine minutes. See also Gunther Schmidt, Thomas Ströhlein, *Relationen und Graphen*, Berlin 1989, pp. 199–202.

[10]See also C. Posthoff, G. Reinemann, *Computerschach—Schachcomputer*, Berlin 1988, pp. 123 ff., as well as Dieter Steinwender, Frederic A. Friedel, *Schach am PC*, Haar 1995.

that white needs in particular positions of this type to checkmate black, promote a pawn, or capture a piece in order to
arrive at a simpler type of endgame.

White	Black	Maximal Number of Double Moves to a "Decision"
KQ	K	10
KR	K	16
KP	K	19
KQ	KR	31
KR	KB	18
KR	KN	27
KBB	K	19
KBN	K	33
KBB	KN	66
KRB	KR	59
KRN	KR	33
KQ	KNN	63
KQ	KNB	42
KQ	KBB	71

Aside from those positions that permit black a quick decision in
his favor, be it checkmate or stalemate, white can always win in
the given positions. This is noteworthy for two reasons. On the
one hand, it is in opposition to classical theory. For example,
in the endgame king and bishop pair against king and knight,
certain positions, the so-called Kling–Horowitz positions, were
considered defensible by black. However, and this is the second
point, white can actually force a win if the 50-move rule is
appropriately modified for such positions.

Using an idea with which endgames without pawns can be analyzed in parallel, Lewis Stiller managed in the early 1990s to
settle even more complex constellations of pieces.[11] The object
of his investigations were the various endgames in which four
figures plus the two kings, but no pawns, were in play. His noteworthy discovery is the pictured KRN–KNN endgame position,
in which white is able only at the 243rd double move to force
the capture of a black knight.

[11] Lewis Stiller, Multilinear algebra and chess endgames, in: R. J. Nowakowski (ed.),
Games of No Chance, Cambridge 1996, pp. 151–192; Lohn der Geduld, *Spektrum der
Wiss.* **4**, 1992, pp. 22–23.

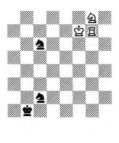

Although we have described here some of the main ideas and techniques of chess programming, we have not discussed how they can actually be programmed. To give at least an impression of this, we will sketch how one might program the main part, namely, the minimax procedure and the alpha–beta algorithm. We shall not go into other parts of a chess program, such as quiescence search and hashing, in order to keep the exposition simple.

Subprograms and Recursion

In order to make large computer programs manageable, it is advisable to package individual tasks into independent subprograms. Such subprograms use their own local variables and are accessible by the main program and other programs only through a well-defined interface of variables, both for input and output.

All modern programming languages permit recursive subprograms, with which one can often simplify the expression of a complex algorithm. Thus for example, one can compute the factorial function, say 5!, by the assignment `fac5 = Factorial(5)`, which calls the following subprogram:

```
FUNCTION Factorial(n)
    IF n = 0 OR n = 1 THEN
        Factorial = 1
    ELSE
        Factorial = n * Factorial(n - 1)
```

```
END IF
END FUNCTION
```

The subprogram works because with each call to the function, a new copy of the variable n is generated, each copy being stored on a *stack*, a particular type of memory organization. Following ideas developed at the end of the 1950s, the stack manages all the variables of subprograms that have begun but have not yet terminated, and without regard to duplication of names among subprograms or for multiple calls to the same subprogram. That is, all the variables that appear in the subprogram, including those for input, output, intermediate results, and the state of the program, are created anew for each call of a subprogram, and these variables are maintained until the program terminates. The stack is organized on the "last in, first out" principle, in analogy to a desk on which new documents to be acted on are placed on the top of the pile, without regard to whether there are unprocessed papers below. When a piece of work is completed, be it a document or a subprogram, then whatever was being worked on at the time of interruption is picked up where it was left off and work is continued on it. This means that in the original call of Factorial(5), the subprogram is processed until it encounters the line

```
Factorial = 5 * Factorial(4)
```

at which point the variable n and the internal variables for intermediate results and the program state are stored on the stack. Before the multiplication, the calculation is interrupted, and the program proceeds with a new call to the Factorial subprogram, using a new set of variables, consisting of the variable n, this time with the value 4, and the internal variables for intermediate results, final results, and the process state. This continues until the value 1 is reached. Only at the end, when Factorial(1) has returned its results via its internal variables to Factorial(2), are the subprograms Factorial(2), Factorial(3), Factorial(4), and finally Factorial(5) continued where they were broken off and brought to an end.

We begin by formulating the minimax algorithm according to the principles of common programming languages, as was done in the previous

sidebars. In particular, a program called `Minimax` will be created that calculates the minimax value as a function of the search depth **n** and the starting position `Position` of a suitable variable type. This can be done relatively simply using a recursive procedure (see "Subprograms and Recursion"). We do not show the determination of the most interesting result, namely, the best move, but it can be easily added for the first search level.

```
FUNCTION Minimax(n, Position)
  IF n = 0 THEN
    Minimax = EstimateValue(Position)
  ELSEIF Position.OnMove = White THEN
    (determine positions P(1), ..., P(s), to which white can
move from Position)
    IF s = 0 THEN
      Minimax = Win(Position)
    ELSE
      MaxValue = -infinity
      FOR j = 1 TO s
        MaxValue = max(MaxValue, Minimax(n - 1, P(j)))
      NEXT j
      Minimax = MaxValue
    END IF
  ELSE
    (determine positions P(1), ..., P(s), to which black can
move from Position)
    IF s = 0 THEN
      Minimax = Win(Position)
    ELSE
     MinValue = infinity
      FOR j = 1 TO s
        MinValue = min(Minvalue, Minimax(n - 1, P(j)))
      NEXT j
      Minimax = MinValue
    END IF
  END IF
END FUNCTION
```

The program can be easily explained. The function subprogram `Minimax` is based on the following:

- the function subprogram *Win*, which returns the game value for white for the end positions.

- the function subprogram `EstimateValue`, that estimates the winning prospects for white in a certain position based on the value of the pieces and positional considerations.

- the move generator, which generates the subsequent positions $P(1),\dots,$ $P(S)$ from a given position.

- the `Minimax` subprogram itself, but now with a search depth reduced by 1 and applied to the subsequent positions.

Depending on the current case, namely,

- upon reaching the search horizon,

- upon reaching an end position,

- when white is to move,

- when black is to move,

the minimax value is calculated, generally by maximizing or minimizing the recursively determined minimax values of the subsequent positions.

For programming the alpha–beta algorithm we use a modified minimax function. If $v(P)$ is the usual minimax value of a position P, then instead of that, one calculates a function $u(P, \alpha, \beta)$, whose value agrees with that of the minimax value $v(P)$ within the acceptance interval determined by α and β, and otherwise, takes on a value on the "right" side of the acceptance interval:

$$u(P, \alpha, \beta) \begin{cases} \leq \alpha & \text{for } v(P) \leq \alpha, \\ = v(P) & \text{for } \alpha \leq v(P) \leq \beta, \\ \geq \beta & \text{for } \beta \leq v(P). \end{cases}$$

```
FUNCTION AlphaBeta(n, Position, Alpha, Beta)
  IF n = 0 THEN
    AlphaBeta = EstimateValue(Position)
  ELSEIF Position.OnMove = White THEN
    (generate Positions P(1),..., P(s), to which white can
move from Position)
    IF s = 0 THEN
      AlphaBeta = Win(Position)
    ELSE
      MaxValue = Alpha
      FOR j = 1 TO s
        CurrentValue = AlphaBeta(n-1, P(j), MaxValue, Beta)
```

```
      MaxValue = max(MaxValue, CurrentValue))
      IF MaxValue >= Beta THEN EXIT FOR
    NEXT j
    AlphaBeta = MaxValue
  END IF
ELSE
  (generate Positions P(1),..., P(s), to which black can
move from Position)
    IF s = 0 THEN
      AlphaBeta = Win(Position)
    ELSE
      MinValue = Beta
      FOR j = 1 TO s
       CurrentValue = AlphaBeta(n-1, P(j), Alpha, MinValue)
       MinWert = min(MinWert, WertAkt)
       IF MinValue <= Alpha THEN EXIT FOR
      NEXT j
      AlphaBeta = MinValue
    END IF
  END IF
END FUNCTION
```

In contrast to the minimax program, here we have the variables alpha and beta, whose values, beginning with the original input, are adapted to the results obtained during the optimization process. Thus, for example, a win that was assured with the move to position P(j) becomes the minimum requirement for the moves P(j+1), ..., P(s) yet to be investigated. If it turns out during the optimization process that the current player can attain a result that is favorable with respect to the acceptance region, the process will be terminated at once with a cutoff. The beginning of the alpha–beta procedure is also important: to obtain the desired minimax value, the acceptance region must be made large enough at the beginning. Namely, for an arbitrary position P, one has

```
Minimax(n, P) = AlphaBeta(n, P, -infinity, infinity)
```

Equally unattractive in both subprograms is that black's moves and white's moves are handled in two almost identical program segments. However, with a bit of work this drawback can be overcome if the game is always viewed from the point of view of the player whose turn it is trying to maximize his own game value. The corresponding variants of the two algorithms are called *negamax* versions.

In practical chess programming one wants absolutely to avoid the position variables P(1), ..., P(s), for whose initialization many bytes have to be copied. More effective is to change the stored data directly in the variable Position. However, in this case the original data have to be recreated after a move has been analyzed. That is, the move must be "taken back."

Chapter Notes

1. In a game tree of depth d and n possible moves per position, there are n^d end positions to evaluate in a normal minimax search. With the alpha–beta algorithm, there are significantly fewer, where the actual number depends on the order in which the moves of a particular position are evaluated. If the analysis of each position begins with the best move, then only $2n^{d/2} - 1$ end positions need be evaluated for even depth d, and $n^{(d+1)/2} + n^{(d-1)/2}$ for an odd depth. As an order of magnitude, one can take the number $2n^{d/2}$, so that for the same search time, about double the depth can be achieved.

   The number of cutoffs achieved is plausible if one imagines the time it takes for a typical situation in alpha–beta search, and the as yet unknown result at this point in time gives some idea of how good the individual moves are:

   - white is about to investigate a move that in comparison to the move already analyzed will turn out to be worse. Then the remaining moves are investigated by white.

   - for black a reply is investigated that will turn out to be a refutation of white's first move. The other alternatives to this move can therefore be ignored.

   - for white, a defense against the refutation is sought, but this search will not meet with success.

   - a refutation for black will thus be sought.

   And so forth. Since for each player it suffices to find a sufficiently good refutation, generally only a few moves need to be searched at the second level, in any case, many fewer than there are altogether.

Further Literature on Chess Programming

[1] David Levy, Monty Newborn, *How Computers Play Chess*, New York 1991.

[2] Rainer Bartel, Hans-Joachim Kraas, Günter Schrüfer, *Das grosse Computer-schachbuch*, Düsseldorf 1985.

[3] Hans-Peter Ketterling, Frieder Schwenkel, Ossi Weiner, *Schach dem Computer*, Munich 1980.

[4] Feng-hsiung Hsu, Thomas Anantharaman, Murray Campell, Andreas Nowatzky, A Grandmaster Chess Machine, *Scientific American* **263:4**, 1990, pp. 18-24.

28

Can Winning Prospects Always Be Determined?

Two mathematicians play the following game: they alternate turns, and the game lasts exactly five moves. A turn consists in choosing an arbitrary nonnegative integer and making it known to the opponent. After the five moves have been taken, and the integers x_1, x_2, x_3, x_4, x_5 chosen, the first player wins if and only if

$$x_1^2 + x_2^2 + 2x_1x_2 - x_3x_5 - 2x_3 - 2x_5 - 3 = 0.$$

Which player has a winning strategy available?

That the players are mathematicians is certainly not of importance for the existence of a winning strategy, but then again, who else would play such a weird game? Of course, the purpose of the game is not for it to be played, even by mathematicians. Rather, we shall use it as preparation for the next chapter, in which we promise the reader to return to "real" games.

The first thing that we notice in looking at the above equation is that x_4 does not appear at all. Furthermore, we can transform the equation into the following form:

$$(x_1 + x_2)^2 + 1 = (x_3 + 2)(x_5 + 2).$$

In this form it becomes apparent how the two players should develop their strategies. The first player can win precisely when the result of the first two moves yields an integer $(x_1 + x_2)^2 + 1$ that is not a prime. Then and only then can the first player choose x_3 and x_5 in such a way that the equation is satisfied. To be sure of winning, the second player must attempt to choose x_2 in such a way that $(x_1 + x_2)^2 + 1$ is prime. Independent of the opening move x_1, this will be possible only if there are infinitely many prime numbers of the form $n^2 + 1$. Whether that is the case is something that we are not going to answer here, since it is an unsolved problem in mathematics.[1] If there are only finitely many primes of the given form, then the first player needs only to choose the first number x_1 large enough to ensure a win.

In sum, we know that one of the two players has a winning strategy. However, we don't know which player it is!

This game originated with James Jones, who also constructed an entire class of similar games of much greater interest.[2] Again the two players alternate in choosing numbers, this time over 17 moves. The first player wins if the expression

$$
(n + x_5 + 1 - x_4)
$$

$$
\times \left[\left((x_5 + x_7)^2 + 3x_7 + x_5 - 2x_4 \right) + \left[\left((x_{12} - x_7)^2 + (x_{14} - x_{11})^2 \right) \right. \right.
$$

$$
\times \left((x_{12} - x_5)^2 + (x_{14} - x_9)^2 \left((x_4 - n)^2 + (x_{14} - x_{11} - n)^2 \right) \right)
$$

$$
\times \left((x_{12} - 3x_4)^2 + (x_{14} - x_9 - x_{11})^2 \right)
$$

$$
\left. \times \left((x_{12} - 3x_4 - 1)^2 + (x_{14} - x_9 x_{11})^2 \right) - x_{15} - 1 \right]^2
$$

$$
\left. \times \left((x_{14} + x_{15} + x_{15} x_{12} x_3 - x_1)^2 + (x_{14} + x_{17} - x_{12} x_3)^2 \right) \right]
$$

is equal to 0. The parameter n is not chosen by the players, but belongs to the rules of the game. That is, there is a different game for each value $n = 0, 1, 2, \ldots$. The question, then, is this: for which games does the first player have a winning strategy, and for which games does the second player have one?[3] We must decide how the winning prospects of the game associated with the parameter n can be determined.

---

[1] See Paulo Ribenboim, *The Book of Prime Number Records*, New York 1988, p. 322 (6. III. A. Conjecture E). There is a formula for the approximate number of such primes less than a given value based on experimentation and probability arguments. The formula suggests that there are infinitely many prime numbers of the form $n^2 + 1$, but there is no known proof that this is so.

[2] J. P. Jones, Some undecidable determined games, *International Journal of Game Theory* **11**, 1982, pp. 63–70.

[3] The existence of winning strategies is assured, even though the assumption of Zermelo's theorem, namely, the finiteness of the number of possible moves, is not satisfied.

It is not difficult to see that a normal minimax procedure will not work here. After all, for every move there are infinitely many possibilities, which gave us difficulties even in the first game that we considered here. For how long should we search for a good move? Can a search that has not met with success be broken off because no winning move will be found? Or is there some other way that the infinitude of possibilities can be evaluated in a finite number of calculational steps?

It gets even worse, as one may readily imagine: the problem is undecidable; that is, one can prove that there is no algorithm that can determine the winning prospects for every game of this class. Thus it is impossible to program a computer so that for each value of n, the associated winning prospects can be calculated. No mathematician, no matter how clever, will ever find a general solution of the question of which games can be won by the first player and which by the second.

How can such statements be explained? They seem at first astounding, even unbelievable. To understand what is going on here, we need to introduce some ideas from the foundations of theoretical computer science and mathematical logic, subjects that often lead to the limits of human reasoning ability, which perhaps explains why such a comprehensive and difficult book like *Gödel, Escher, Bach*,[4] which deals with such topics, became a bestseller in the early 1980s.

We begin our discussion of the foundations of theoretical computer science and mathematical logic with the notion of *computability*: even before the existence of universal, that is, freely programmable, computers, there were a number of approaches in the 1930s to defining a formal notion of computability. Of course, such a definition should agree with existing notions and experience related to computation. Thus, for example, arithmetic operations and calculational methods should be included. A most suggestive definition is due to Alan Turing, based on the model of a seemingly primitive, yet universally programmable, computer, later called a Turing machine.[5] Other approaches were purely arithmetic, though they turned out to be, together with other approaches, equivalent to Turing's definition. In 1936, Alonzo Church (1903–1995) formulated a theory that later became known as Church's thesis, according to which everything that we think of intuitively as computable can be computed with a Turing machine. For our purposes, let us agree on the following definition of computability:

---

[4] Doulgas R. Hofstadter, *Gödel, Escher, Bach*, New York 1979.

[5] Alan M. Turing, On computable numbers, with an application to the Entscheidungsproblem, *Proceedings of the London Mathematical Society* (2) **42**, 1936, pp. 230–265, (2) **43**, 1937, pp. 544–546.

Everything that can be programmed on a normal, that is, deterministically programmable, computer will be said to be computable. We limit ourselves to programs that for a given input always produce the identical output. We deviate from practical reality only in permitting an unlimited amount of computer memory.

Although the idea of computability is generally associated with numbers, our definition does not limit us in this regard. For example, the object of our computation can be text or game configurations. For ease of programming we shall again, as we did in the previous chapter, program our fictive model computer using a high-level modern programming language.[6] That is, with the aid of system programs such as compilers and interpreters, the high-level commands of a program can be translated level by level into machine commands and then processed; this translation is nothing more than a special series of calculations in the form of transformation of text. Thus a compiler is a form of universal program that for a given input, namely, a program in a particular programming language, can carry out all possible computations.

Let us remain a bit on the topic of compilers. During the translation to machine code the compiler may detect syntactic errors, that is, violations in the program against the rules of the programming language. These must be corrected before a successful compilation can take place. Yet, as every programmer has experienced, a program free of syntactic errors that has been successfully compiled will not necessarily run free of errors. Logical errors in the program can cause it to compute something other than what the programmer had in mind. And things can get even worse: the program might end up in an infinite loop and never terminate.

It would therefore be highly practical if a compiler could be improved in such a way that it would automatically detect infinite loops. This does not seem a difficult assignment, as the following example illustrates:

```
N = 1
WHILE N > 0
  N = N + 1
WEND
```

But what about the following program, whose execution depends on an integer N that is to be input before the program runs?

---

[6]A reduction of programming languages similar to Pascal to a universal minimal code corresponding to the programming of Turing machines is given in Jürgen Albert, Thomas Ottmann, *Automaten, Sprachen und Maschinen für Anwender*, Zurich 1983, pp. 274 ff.

```
INPUT N
WHILE N <> 1
  IF (N MOD 2) = 0 THEN N = N/2 ELSE N = 3 * N + 1
WEND
```

For which values of N does the program halt, and for which values does in land in an infinite loop? Beyond this specific question, it would of much greater interest to consider the general question, called the *halting problem*. For a given program and its input can it be determined whether it will halt, and if so, how; that is, can one create a computer program that will check whether the program will halt in a finite amount of time and produce a unique output?

Such a program would be of great interest, and not only to programmers. Even mathematical problems could thereby be solved. For example, if we wished to check whether the equation

$$x^n + y^n = z^n$$

has a solution for positive integers n, x, y, z with $n \geq 3$, then a program could be developed along the following lines: with the help of subprograms for adding and multiplying arbitrarily large numbers, the program would systematically test all combinations of possible values (n, x, y, z) and check whether they satisfy the equation. The search will continue until the program finds a solution and then terminate. If there is no solution, then of course the program will never halt. If one could determine independently whether this search program will ever halt, that is, whether it will ever find a solution, then the solubility of the equation would also be determined. In the same way, the question posed earlier whether there are infinitely many prime numbers of the form $n^2 + 1$ could also be answered analogously. Essentially, one would test the halting behavior of a search program that beginning with an input N, tests all larger numbers in succession as to whether they are of the form $n^2 + 1$ and halts on the first success. That value is then used as the new input N, and the process is continued until an input is found that leads to an input that can be shown to result in an infinite loop. Then there are infinitely many primes of the form $n^2 + 1$ precisely if this program never halts.

The high quality of the two example problems leads to serious doubts about the solvability of the halting problem; namely, as we have already mentioned, the answer to the second problem is unknown, while the first deals with the famous Fermat conjecture, which required more than 350 years since its formulation by Pierre de Fermat before a solution was found (see Note 1 at the end of the chapter). In fact, the halting problem is undecidable; that is, it is a task that has no algorithmic solution. In other words, there is no possible computer program that would be a solution to the halting problem. This impossibility can be proved indirectly. To this end, one assumes that such a test program existed, extends the program in a suitable way, and thereby obtains a contradiction, namely, that the extended program is able to test itself. See "The Halting Problem."

The Halting Problem

We make the following assumptions in our investigation of the finite or infinite running time of computer programs:

- the programs run on an imaginary computer that possesses unlimited storage capacity.

- the programs are formulated in a particular programming language, such as Pascal, C, or Basic, that is strong enough so that a compiler for the language itself can be written in that language.

- the program has input in the form of text, where numerical input is suitably transformed and multiple inputs are separated by a unique separator, such as $, that is used for nothing else.

The statement of the halting problem is this: does there exist a program, we shall name it STOPTEST, that given as input a program together with its text input determines whether the given program with the given input will ever halt? That is, the program STOPTEST itself should always halt after a finite amount of time and then output either "INFINITE" or "HALTS" according to whether the input program run on its text input runs forever or halts in a finite amount of time. Finally, in the special case in which the input program is not syntactically correct, the program STOPTEST will output "HALTS." This last test can be made by a normal compiler.

The proof that no such program STOPTEST with the required characterstics can exist is an indirect one. That is, we assume that we indeed have such a program and attempt to derive a logical contradiction from that assumption. To this end, we extend STOPTEST to a program that we will call DIAGONAL, which consists of the following three program steps:

1. first, the input text is duplicated, and the two copies are separated by the special separator symbol used in STOPTEST.

2. then STOPTEST is launched on the duplicated input; that is, the original input to the DIAGONAL program is used twice, once as the program to be tested, and once as the input to that selfsame program.

3. the further course of the DIAGONAL program depends on the result produced by the subprogram STOPTEST:

   - if the result is "HALTS," then DIAGONAL will run in an infinite loop.
   - if the result is "INFINITE," then DIAGONAL stops at once.

The properties of our program DIAGONAL can be immediately derived from those of STOPTEST. If the input is not a syntactically correct program, then DIAGONAL runs without halting. Of greater interest are of course the other cases, in which the input corresponds to a syntactically correct program:

- if the input program eventually halts upon being launched with itself as input, then DIAGONAL runs forever with this program as input.

- if the input program does not halt when run with itself as input, then with this input, DIAGONAL halts.

And now for the punch line: what happens when the program DIAGONAL receives its own program text as input? Does it eventually halt, or does it run forever? Each of the two possibilities leads to a contradiction: given itself as input, DIAGONAL halts if it doesn't halt, and conversely. This result is similar to the fact that the proposition "This sentence is false" contains a logical contradiction. The cause is a negating self-reference, which was known in antiquity by the Cretan scholar Epimenides as the "All Cretans are liars" paradox. In the case of the program DIAGONAL, the only way of resolving the contradiction is by abandoning the assumption that the program STOPTEST can be created in the first place.

It is not only in the halting problem that contradictory self-references play a role in mathematical thought. The first to use such diagonalization procedures was Georg Cantor (1845–1918), the founder of set theory. Using such a technique, he proved in 1874 that the set of real numbers, unlike the set of rational numbers, cannot be presented in the form of an infinite list. (Said another way, there is a one-to-one correspondence between the rational numbers and the natural numbers, but not between those two sets and the real numbers.) This was the first proof that not all infinite sets have the same "size."

Less pleasing to Cantor was another application of the diagonalization procedure by Bertrand Russell (1872–1970). In 1906 he proved that Cantor's set theory contains a logical contradiction by considering the set of all sets that contain themselves as an element and asked whether that set contained itself as an element. In 1931, Kurt Gödel (1906–1978) used a much more complex diagonalization argument to prove his famous incompleteness theorem. According to this theorem, any mathematical theory such as arithmetic or classical geometry can never be decided on the basis of a finite number of axioms. That is, no matter what set of axioms you choose for your theory, and regardless of whether they are free of contradictions, there will always be propositions in your theory that are true but are nonetheless unprovable. Thus one can formulate propositions such that neither the proposition nor its negation can be proven. Yet one of the two assertions must be true, and thus there always exist true propositions that cannot be proven.

Gödel's incompleteness theorm marked the end of an era in mathematics that began at the end of the 19th century, which was marked by the goal of putting the foundations of mathematics on a secure footing. It is clear that a secure foundation for mathematics was desirable. After all, unlike the natural sciences, mathematics is based not on experiment and observation of the natural environment, but on sets of axioms. Recall the difficulties, discussed in Chapter 8, of creating a mathematical foundation for probability theory.

Therefore, mathematicians sought a system of unprovable axioms that would serve for the logical derivation of all known mathematical laws while never leading to any logical contradictions. And mathematical logic must be included as well: what logical rules of derivation are permitted for proving facts from the axioms and additional facts from the facts already proved?

The first axiomatization of classical geometry occurred in antiquity, namely, in the third century B.C.E., with Euclid's *Elements*. In particular, Euclid formulated the fundamental relationships between objects such as "point" and "line." For example, "For every pair of distinct points P and Q there exists precisely one line ℓ on which both P and Q lie."

Euclid's attempts to clarify just what the objects of geometry are, along the lines of, "A point is that which has no part," are essentially worthless for modern mathematics, which insists that a set of axioms describe the properties of the relations between objects, such as between a line and a point lying on that line, so precisely that the consequences of such relationships remain clear even if any attempt at visualization, even one borrowed from everyday experience, is done without. We can state this in the stark words of David Hilbert: "One must be able at any moment to replace the words 'points, lines, planes' with 'tables, chairs, beer mugs.'"

Euclid's axiom about parallel lines aroused controversy even in ancient times. One form of the axiom is this: for every line ℓ and point P that does not lie on ℓ, there is precisely one line h in the plane determined by P and ℓ on which P lies and that has no points in common with ℓ. It was long questioned whether the parallel axiom could be derived from Euclid's other axioms. After all attempts at a proof failed, two mathematicians, János Bolyai (1802–1860) and Nikolai Lobachevski (1792–1856), succeeded independently around 1930 in showing that such a proof was impossible. The reason is both simple and ingenious. One gives an example of a geometric system that satisfies all of Euclid's axioms except the parallel axiom and such that the parallel axiom is violated. That is, one defines notions such as point, line, and what it means for a point to lie on a line in a special way; for example, one could consider a line to be a great circle on a sphere and define a point in the usual way. As simple as this idea sounds, it required a courageous break with the interpretation of Euclid's axioms as something God-given. In other words, while axioms may attempt to describe the objects of our existence, and while the results derived from those axioms might thus be of practical use, it is also possible to create axioms that partake of an entirely different interpretation, one that may reflect little or nothing of the world around us, for example, a non-Euclidean geometry, as such a system is called that satisfies all the Euclidean axioms except for the parallel axiom (See Note 2 at the end of the chapter).

To prove his theorem, Gödel constructed an arithmetic proposition whose interpretation amounted to a declaration of its own unprovability. If the proposition were false, it would have to be provable. Therefore, the proposition must be true, but then, if it were true, it would have to be unprovable. Even this hint at Gödel's proof suggests the great subtlety required in his argument in order not to entangle himself in a web of contradictions. We can approach Gödel's theorem a bit less abstractly by looking at the theorem's close connection with the halting problem. (See "Gödel's Incompleteness Theorem.")

Gödel's Incompleteness Theorem

There are two facts that underlie the close connection between Gödel's incompleteness theorem and the halting problem:

- if an arithmetic proposition is provable from the axioms of the system, then a proof can be found—at least theoretically—by a computer program.

  To this end, all proofs that can be formulated using the finitely many axioms and finitely many logical derivation rules are listed by a computer program. If a proof is found, then the program stops. This makes possible a process by which all arithmetic propositions including the axioms can be translated into a purely formal language, such as could be created using the symbols

  $$+, \times, =, (,), \neg, \wedge, \vee, \Rightarrow, 0, S, \forall, \exists, x, y, \ldots.$$

  The permitted logical proof steps involve rules by which strings of these symbols can be transformed. It can then always be determined in a strictly formal and unique manner whether a proof presented in the form of a character string is correct, that is, whether it is the result of permitted textual transformations of the axioms or other propositions whose proofs have already been established. It is not necessary, nor even desirable, to supply a semantic interpretation to these strings. However, even "obviously true" propositions such as $0 = 0$ and $SSS0 = SSS0$, which we might interpret as $3 = 3$, require a proof. And that is not always a simple matter, even for "theorems" like $S0 + SSS0 = SSS0$, which stands for $1 + 3 = 4$.

- the proposition that a given pair of program and input will or will not eventually halt is of a purely arithmetic nature: in a computer program, the stepwise changing of values of the variables, including those of the internal variables that reflect the program state, can be described by arithmetic formulas. Thus for every concrete halting problem there corresponds a string of arithmetic symbols, where the necessary transformation of the program and input into a string can be calculated.

If every arithmetic proposition could be decided on the basis of the axioms, that is, if one could always prove or disprove the proposition, then the halting problem would always be decidable. Therefore, there must be true arithmetic propositions that cannot be derived from the axioms. Gödel gave a concrete example (see Note 3 at the end of the chapter).

Of course, one could consider enlarging the axiomatic system. But even then, there would remain undecidable propositions in the system. Moreover, every extension of the axioms contains within it the danger of a logical incompatibility among the axioms. That no such contradiction exists among the axioms of arithmetic was shown by Gödel in 1931 to be undecidable.[7]

The propositions that were constructed for Gödel's incompleteness theorem are quite artificial, and thus do not seem particularly relevant to the parts of mathematics that deal with the real world. But "real" mathematical problems are not exempt from the possibility of being algorithmically undecidable. A famous example is Hilbert's tenth problem, the tenth problem in the list of the 23 most significant open problems in mathematics presented by David Hilbert at the Second International Congress of Mathematicians in Paris in 1900.

Hilbert's tenth problem deals with *Diophantine equations*, that is, equations based on polynomials with integer coefficients, such as the Fermat equation

$$x^{11} + y^{11} = z^{11}$$

or even

$$y^z = x^3 - 3x + 5,$$

for which one seeks integer solutions. The name "Diophantine" goes back to the Greek mathematician Diophantus, who lived in Alexandria in the third century C.E. and investigated special types of such equations. Hilbert asked for a procedure by which "using a finite number of operations it can be decided whether the equation is solvable in rational integers." In a more modern formulation, Hilbert's problem asks, Is it decidable whether a Diophantine equation is solvable? That is, can a computer be programmed so that for every text input such as x**11+y**11=z**11 or y**2*z=x**3-3*x+5 it can be determined after a finite amount of processing time whther the

---

[7]Since such a contradiction allows every propostion to be proved, the lack of contradiction is equivalent to the statement that there can be no proof of the statement $0 = 1$. This is related to Hilbert's second problem from the year 1900.

equation has a solution in integers? Such a program could be written for solvable equations, since one could, at least theoretically, try all possible solutions one after the other. If we are guaranteed a solution, then this process must eventually terminate. However, it can be much more difficult to prove a Diophantine equation unsolvable. Of course, in special cases, such as the equation $x^2 + y^2 = 3$, one needs to investigate only a limited number of cases either on the basis of size (we must have $|x| \leq 1$ and $|y| \leq 1$) or divisibility (the sum of two squares always has remainder 0 or 1 when divided by 4, and so cannot equal 3). However, a general procedure that would work for all Diophantine equations cannot be found in this way.

A conclusive solution to Hilbert's tenth problem was found only in 1970, by the 22-year-old Russian mathematician Yuri Matiasevich. Building on earlier partial results, he proved that the solution of Diophantine equations is undecidable. In other words, as with the halting problem, there can be no computational procedure, and therefore no computer program, that will establish for every Diophantine equation in finite time whether it is solvable (see Note 4 at the end of the chapter).

Algorithmically undecidable problems can also be used to construct games with indeterminable winning prospects. The first to do so was M. O. Rabin in 1957 with a not very concrete example.[8] In 1982, James Jones presented the examples that we saw earlier of a series of games each of which is defined on the basis of a Diophantine equation with 17 variables. Although one of the two players possesses a winning strategy in each game, there is no general procedure for determining which of the players is in a winning position.

But of course, even for games whose winning prospects can be theoretically calculated, in practice such a calculation can be very difficult. How difficult this can be will be examined in the next chapter using the "real" games go-moku and Hex.

Chapter Notes

1. The history of the Fermat conjecture has been repeated in a variety of venues and at a variety of levels. Here are some references: Harold M. Edwards, Das Fermatsche Theorem, *Spektrum der Wisschenschaft* **12**, 1978, pp. 38–45; Ferne Zukunft, *Der Spiegel* **28**, 1983, p. 146; Süsses Gift, *Der Spiegel* **12**, 1988, pp. 272–275; Griff nach dem Gral, *Der Spiegel* **26**, 1993

---

[8]M. O. Rabin, Effective computability of winning strategies, in: H. W. Kuhn, A. W. Tucker (eds.), *Contributions to the Theory of Games III, Annals of Mathematics Studies* **39**, 1957, pp. 147–157.

pp. 203 f.; Christoph Pöppe, Der Beweis der Fermatschen Vermutung, *Spektrum der Wissenschaft* **8**, 1993, pp. 14–16; Thiagar Devendran, Der Widerspenstigen Zähmung, *Bild der Wissenschaft* **4**, 1994, pp. 42–44; M. Ram Murty, Reflections on Fermat's Last Theorem, *Elemente der Mathematik* **50**, 1995, pp. 3–11; Jürg Kramer, Über die Fermat-Vermutung, *Elemente der Mathematik* **50**, 1995, pp. 12–25; René Schoof, Fermat's Last Theorem, *Jahrbuch Überblicke Mathematik*, 1995, pp. 193–211; Christoph Pöppe, Die Fermatsche Vermutung ist bewiesen—nun auch offiziell, *Spektrum der Wissenschaft* **8**, 1997, pp. 113–116; Simon Singh, Kenneth A. Ribet, Die Lösung des Fermatschen Rätsels, *Spektrum der Wisschenschaft* 1998/1, pp. 96-103; Simon Singh, *Fermat's Last Theorem*, New York 1997.

2. In the sense of object-oriented programming, one can imagine mathematical objects not only as belonging to an axiomatic system, but also as instances of particular classes whose structure is hidden by strong encapsulation. That is, one cannot tell, for example, whether instances of two classes `point` and `line` refer to points and lines on a plane or points and great circles on a sphere. All that can be checked, using appropriate methods, are the relations between the given instances; for example, it can always be determined for a point and a line whether the point lies on the line. Other methods can be used to establish equality between two instances or to generate additional instances.

3. Setting out from the ideas on the halting problem that we have been examining, let us consider a program that given a program A as input, searches, possibly without ever halting, for a proof of the proposition "Program A does not halt on input A" and halts if it finds such a proof and exhibits the proof as its output. Such a program is actually realizable as a combination of two programs:

   - the first part of the program translates a concrete halting problem, including program and input, into a string of arithmetic symbols.

   - the second part searches for a proof of the string that was thus generated.

Once our program, which we will call program D, has been constructed, we run it with program D itself as input. What happens? Does the program halt on this input? We would like to investigate the halting behavior on two levels, namely, on the level of symbolic arithmetic and on that of semantic interpretation. Within the semantic interpretation—and within it only—the propositions involve the properties "true" and "false" in just such a way as the parallel axiom does not represent a true or false proposition in and of itself, but only when interpreted within the context of a particular geometry. In connecting the two levels, we begin with the correctness of the system comprising the axioms and rules of deduction; that is, every formally decidable (symbolic) proposition in the system should be able to be interpreted as a proposition about numbers:

- first, it turns out that the proposition "Program D does not halt on input D" cannot be decided at the level of symbolic arithmetic; if the proposition were decidable, then the proof could be found with the proof-search program D. The necessary input is program D itself. Since program D halts as soon as the proof has been found, the statement "Program D does not halt on input D" is false, and therefore cannot be proved.

- on the other hand, the proposition "Program D does not halt on input D" is true: if it were false, the program D would have to halt given itself as input and thus return a proof of the proposition "Program D does not halt given itself as input." Therefore, in contradiction to the hypothesis that it is false, the proposition must be true.

Therefore, "Program D does not halt on input D" is an undecidable yet true proposition.

It remains to note that the argument just presented is not a proof in the strict sense. It cannot be, since the truth of a proposition is not an object of formal proofs carried out on a symbolic level. However, that changes nothing in the legitimacy of accepting the conclusions of the argument and thus accepting the proposition as true, just as one does for axioms.

4. In general, every set of natural numbers that can be listed by a computer program, such as the even integers, prime numbers, or numerical encodings of programs that halt, can be characterized as a Diophantine equation. In particular, for each such set S there exists a Diophantine equation $Q_S(y, x_1, \ldots, x_k) = 0$ such that a natural number n is in S if and only if the equation $Q_S(n, x_1, \ldots, x_k) = 0$ is solvable in natural numbers. If one chooses for the set S an undecidable set such as can be derived from the halting problem, then the solvability of the associated sequence of Diophantine equations $Q_S(1, x_1, \ldots, x_k) = 0$, $Q_S(2, x_1, \ldots, x_k) = 0$, … is undecidable. That is, there can be no computer program that recognizes the decidability or undecidability of every equation in finite time.

Since solvable and provably unsolvable equations are always recognizable in finite time, the sequence must contain unsolvable equations whose unsolvability is unprovable within the axiomatic theory of arithmetic. Such a Diophantine equation can be explicitly given. It is so constructed that each of its solutions would correspond to the encoding of a proof of its unsolvability. Thus the equation is unsolvable, but this fact cannot be derived from the axioms.

A more precise overview of Hilbert's tenth problem can be found in Die Hilbertschen Probleme, *Ostwalds Klassiker der exakten Wissenschaften* **252**, 1976, pp. 53, 177–195; Martin Davis, Reuben Hersh: Hilbert's 10th problem, *Scientific American* **229:5** 1973, pp. 84–91; Keith Devlin, *Mathematics: The New Golden Age*, London 1988; Martin Davis, Hilbert's tenth problem is unsolvable, *American Mathematical Monthly* **80**, 1973, pp. 233–269 (contains a complete proof).

Further Literature on Computability

[1] Herbert Meschowski, *Lust an der Erkenntnis: Moderne Mathematik*, Munich 1991. A multifaceted overview is offered.

[2] John E. Hopcroft, Turing Machines, *Scientific American* 1984/5, pp. 70-80. A survey article.

[3] A. K. Dewdney, *The (New) Turing Omnibus*, New York 1993, Chapters 31, 51, 59, and 66. Various aspects of computability are considered.

A more thorough mathematical treatment is offered by the following books:

[4] Hans Hermes, *Aufzählbarkeit, Entscheidbarkeit, Berechenbarkeit*, Berlin 1971.

[5] Wolfgang Paul, *Komplexitätstheorie*, Stuttgart 1978.

[6] Uwe Schöning, *Theoretische Informatik kurz gefasst*, Mannheim 1992.

29

Games and Complexity: When Calculations Take Too Long

Does there exist for the game Hex, as there does for many nim variants, a "formula" that permits a rapid calculation of winning prospects?

The border-to-border game Hex was introduced in Chapter 19. Its configurations are characterized by two subsets of squares, one containing those with white stones, the other those with black. The totality of all configurations is thus structured relatively simply, and it seems highly plausible that one should be able to find, as in nim, criteria for winning configurations. In the case of Bridge-it, a game similar to Hex, which we also met in Chapter 19, this can in fact be done, as was shown by Alfred Lehman in 1964 (see "Bridge-It and Shannon's Switching Game").

Bridge-It and Shannon's Switching Game

In order to determine the winning prospects for a Bridge-it configuration simply, Lehman generalized the game[1] so that the object of this new game, dubbed Shannon's switching game, is

---

[1] Alfred Lehman, A solution of the Shannon switching game, *Journal of the Society for Industrial and Applied Mathematics (SIAM Journal)* **12**, 1964, pp. 687–735.

a graph. Such a graph consists of a set of vertices and a set of edges, where in this case, an edge is not necessarily associated with two distinct vertices. One can imagine the vertices as points, which are partially, possibly multiply, connected by undirected paths, namely, the edges. It is possible for an edge to connect a vertex with itself. In the graphs used by Lehman for his game, there are two special vertices, identified by + and − in the 5 × 5 Bridge-it graph depicted below:

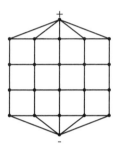

Now players black and white take turns. A move consists in selecting an edge that has not been selected in a previous move. White plays by "erasing" the edge she has chosen, that is, removing it from the set of edges. Black colors his edge black, thus making it untouchable by white. The position arising from a black move can be represented graphically by uniting the two vertices in the edge selected into a single vertex. The following figure shows a Bridge-it configuration after moves by white and black; the right-hand figure shows the vertices of the edge selected by black (the heavy line in the left figure) joined into a single vertex.

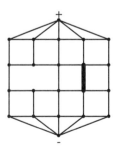
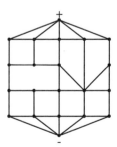

Black wins if he succeeds in linking the special vertices + and − with the edges that he has selected. Graphically, this means that the two vertices merge to a single vertex. Otherwise, white wins.

Since going first can never be disadvantageous, there are three distinct classes of configurations:

- white to move second possesses a winning strategy.
- black to move second possesses a winning strategy.
- the player to move, whether black or white, possesses a winning strategy.

The criterion found by Lehman is based on the special kind of graph called a *tree*. A tree is a graph in which every pair of vertices is linked by precisely one path, perhaps consisting of several edges. Equivalently, a tree is an acyclic connected graph; that is, every two vertices are connected by some path (connected), and no vertex is connected to itself by a path (acyclic). According to Lehman, black moving second possesses a winning strategy precisely when two trees can be formed from the edges of the graph, with both trees containing the same vertices, including + and −, but with no edges in common. For example, if black moves in the configuration depicted above directly below his first move, then he can now force a win as the player to move second. The following figure shows the resulting configuration and two trees that satisfy Lehman's criterion for the set of remaining nodes.

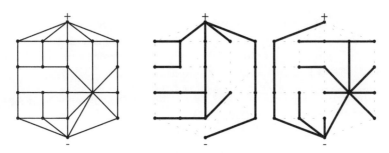

If, as in the depicted example, Lehman's criterion is satisfied, then black can refute every one of white's moves: if white moves so that neither of the two trees is broken into two parts, then black can move as he pleases. If white erases an internal edge from one of the two trees, then black must select an edge from the other tree in such a way that the divided tree is made whole.[2]

In comparison to extensive analyses of long sequences of moves, Lehman's criterion is significantly simpler to apply. In particular, one can immediately confirm whether two given trees satisfy the required conditions. Moreover, there exist algorithms that complete the search for such trees relatively quickly.[3] Lehman's criterion can also be modified to analyze the other two classes of configurations (see Note 1 at the end of the chapter).

At the end of his investigations into Bridge-it and related games, Lehman discussed the game of Hex, to which he was unable to generalize his technique. That Lehman's inability to do so was not due to a mere oversight was demonstrated in 1979 by Stefan Reisch, of the University of Bielefeld, in his noteworthy bachelor's thesis *Die Komplexität der Brettspiele Gobang und Hex*.[4] Reisch proved that every general procedure for determining the winning prospects of Hex configurations on large playing boards exceeds all currently practicable computational capacity. The argument depends on a conjecture that has not been proved, but is generally accepted as most likely true.

How are such propositions possible, and what exactly has Reisch proved? As in the previous chapter, here, too, we must make a detour into theoretical computer science, this time into the area of complexity theory. Whereas in the last chapter we were investigating the limitations on what is theoretically computable, now we will focus on what is computable in practice, that is, on the minimum amount of computation necessary to solve a given problem.

---

[2]The converse part of Lehman's proof, showing that every winning strategy for black to move second implies the existence of two trees with the given properties, is much more complex.

[3]Harold M. Gabow, Herbert H. Westermann, Forests, frames and games: algorithms for matroid sums and applications, *Algorithmica* **7**, 1992, pp. 465–497.

[4]These results were eventually published: Stefan Reisch, Go-bang ist PSPACE-vollständig, *Acta Informatica* **13**, 1980, pp. 59–66; Stefan Reisch, Hex ist PSPACE-vollständig, *Acta Informatica* **15**, 1981, pp. 167–191. The game gobang is better known under the name go-moku. The name gobang is used for the variant in which stones can be captured.

We begin with some simple examples. If two integers in their decimal representation are to be added or multiplied, there are well-known procedures—algorithms—for carrying out these operations. Clearly, addition is simpler than multiplication, since for the sum of two n-digit numbers there are n pairs of digits to add, while in the usual multiplication algorithm of two n-digit integers there are n^2 multiplications of pairs of digits, which results in n integers that themselves must be added. With very large numbers with their correspondingly long representation as sequences of digits, the complexity of addition grows more slowly than that of multiplication. As a function of the length of the input, that is, the total length of the two decimal representations, the complexity of addition grows in proportion to the length of input, while that of usual multiplication grows quadratically. Such tendencies of the computational complexity, and thereby the computational time for a programmed algorithm, are a good measure of the complexity of an algorithm. In particular, such asymptotic results are independent of the encoding of the input. Thus the growth rates for a given algorithm are the same for other numerical representations, such as the binary system, or two-byte and four-byte encodings used by the computer for its internal operations.

The approach of describing the *computational complexity*, as the asymptotic cost for ever longer inputs is called, is useful in many situations. In particular, the efficiency of an algorithm can be compared with the efficiency of other algorithms. Thus, for example, the usual method of multiplication is anything but efficient on large inputs of hundreds or more decimal digits. Its complexity of $O\left(n^2\right)$, which indicates a quadratically growing upper bound for an input of n characters, can be improved using a simple idea to $O\left(n^{1.585}\right)$ (see Note 2 at the end of the chapter).

In practice, it often suffices to use an algorithm that is fast *on average*, since it then operates efficiently in most, but not necessarily in all, cases. However, great demands can be placed on an algorithm when results are required in real time, such as in encrypted data transfer, controlling a manufacturing process, or analyzing a game configuration under tournament conditions. In such cases it is required that the time required to solve a problem be guaranteed in advance. In regard to such absolute requirements one generally applies, as in the minimax procedure, the principle of the worst case; that is, the measure of the time complexity is always the worst possible result for input of a particular length. The classification is done using a coarse measure: algorithms whose computing time can be bounded by a polynomial, that is, $O\left(n\right)$, $O\left(n^2\right)$, $O\left(n^3\right)$, ..., are considered *efficient*; that is, they are considered amenable to practical calculation, though a bound of $O\left(n^{1000}\right)$ would seem to be stretching the point. In contrast, worst-case computing times that cannot be polynomially bounded, such as growth of order 2^n, rapidly approach astronomical orders of magnitude, and are impractical for large inputs (see Note 3 at the end of the chapter).

This asymptotic growth of the computational demands required by an algorithm makes it possible as well to predict the future rate of technological progress. Thus in recent years the speed of computers has doubled every one to two years. The utility of an algorithm of type $O(n)$ therefore is increased to inputs of double the previous size every year or two, while an algorithm with quadratic complexity can handle inputs of length increased by only 41%, namely, in the ratio $\sqrt{2} : 1$. On the other hand, for algorithms with exponential growth, the increase is only by a fixed number of digits. Instead of individual processes, one can also investigate particular problems. Their computational complexity can be characterized by identifying the algorithm that guarantees the most rapid solution for large inputs; this is no easy matter in practice, since of course one must include algorithms that may not yet have been discovered. This difficulty can be overcome in part by finding some measuring rod for the difficulty of solving a problem. Thus, for example, we can compare the complexity of games like nim, Bridge-it, and Hex among themselves: what is the minimal computational requirement in such a game in comparison to the input length, that is, the length of an encoded configuration, that guarantees a determination of the winning prospects of an arbitrary configuration?

- This is simple in the case of standard nim. With respect to the numerical encoding of a configuration, the complexity is at most $O(n)$. The same holds for nim variants whose Grundy values are periodic or grow periodically. In these cases the Grundy values can be calculated with linear computational cost. That holds as well for the subsequent nim addition.

- The case of Bridge-it is somewhat more complex. Thanks to Lehman's criterion and the relevant algorithms for graphs, the winning prospects of Bridge-it configurations can be calculated with a computational cost that grows less than quadratically. Thus even relatively large game boards do not cause a significant problem.

- For Hex, on the other hand, a simple criterion for winning has remained elusive. If one accepts long computation times, one can, of course, carry out a complete minimax analysis. Then every variant is determined by the order in which the squares of the player board are occupied alternately with black and white stones. However, the computational cost of calculating all these variants is immense. In comparison to input length, it grows for large boards like the factorial function, that is, exponentially.

Nonetheless, this analysis will not fail for lack of storage space, whose requirement remains modest, so long as the moves are investigated depth first. Then for each point in time in the analysis there is only one move to be stored per move level. Even if all the moves arising from a given move are investigated, the next move is generated on the level in question. With this depth-oriented search (which is hardly ever used in the analysis of chess positions in practical chess programs, say to sort the moves according to cutoffs, due to run-time considerations) the storage requirements can be polynomially bounded.

Complexity Theory: P–NP–PSPACE–EXPTIME

Each of the abbreviations displayed in the title of this box stands for a class of decision problems. The restriction to such problems prevents the complexity reflecting merely an amount of routine work, rather than an actual difficulty. For example, to output the number of 1s equal to the value of an input decimal number requires work proportional to the length of the output, which is of exponential size in relation to the input, even though the task is quite simple.

However, the limitation of the field of application to decision problems is not so restrictive as one may at first imagine. In particular, every optimization task corresponds to a class of decisions for which one asks whether a particular specified upper or lower bound is achievable.

The four classes of decision problems are bounded as follows:

- the class P contains all decisions that can be computed in polynomially bounded time. With respect to the calculation of winning prospects, nim and Bridge-it belong to this class.

- the class NP comprises those decision problems for which there is an efficient procedure by which every yes decision can always be *confirmed* with the help of suitable supplementary information. Thus the assertion that a number is composite can be confirmed quickly if one is given one of its factors and then carries out a single long division (see Note 4 at the end of the chapter).

Many combinatorial problems for which an efficient solution procedure is unknown belong to the class NP. The best known of these is the *traveling salesman problem*, which seeks the shortest route for a traveler who has to visit a specified number of cities. In the associated decision problem one asks whether a route shorter than a specified maximum can be found. Positive decisions can be found simply by confirming that a given route satisfies the condition. To be sure, without any additional information one can examine, in the case of n cities, all $n!$ routes. However, for a large number of cities the task is enormous, and certainly not polynomial.

- the class PSPACE consists of all decision problems that can be solved in unrestricted time with a memory requirement that grows at most polynomially in the size of the input. To this set belong questions about the winning prospects in games such as Hex that end after a fixed number of moves. To achieve polynomial space, the minimax search is conducted depth first, so that for each move level only a single configuration needs to be stored.

- the class EXPTIME contains all decision problems that can be solved in exponential time.

These four classes are related as sets according to the following hierarchy:

$$P \subseteq NP \subseteq PSPACE \subseteq EXPTIME.$$

The second inclusion follows from the fact that one can check all inputs that might serve as supplementary information one after the other. Since furthermore, only a limited storage area can be read and written in polynomially restricted time, the class NP must be a subset of the class PSPACE. The third inclusion is based on the fact that the number of internal storage states is limited exponentially in relation to the storage size.

Which of these inclusions are proper, that is, which of these classes are actually larger than the next smaller class, is an open problem. However, it is conjectured that all four are of

different sizes. All that is certain is that the class EXPTIME
contains problems that do not belong to the class P.

How complex is it, then, to compute the winning prospects in games like
Hex, go-moku, go, checkers, and reversi? That is, given a playing board of
arbitrary size and the appropriately generalized version of the game, one
asks what complexity an algorithm would have to possess in order to work
for all board sizes. Can one find efficient algorithms, those with polynomial
complexity, for determining the winning prospects for an arbitrary configuration? The answer is no, even if such were proved for all the games named
here. For the games Hex, go-moku, and reversi,[5] which always end with
the board completely filled, only a relative proof can be given, according to
which the task of determining the winning prospects of these games is at
least as difficult as all the problems of a large class for which there appears
no hope of ever finding efficient algorithms. It is doubtless inefficient to
compute the winning prospects of go and checkers.[6] The same holds for
chess and its Japanese counterpart shogi, whose generalizations to larger
boards are rather forced.[7]

The arguments that form the basis for such results are very complicated,
and we are unable to discuss them here. For every game one conceives of
a "construction kit" containing subconfigurations from which one can assemble configurations that satisfy certain conditions. In particular, inputs
of other decision problems are transformed into generally very large configurations of the game under study, so that the original decision as to the
winning prospects is a consequence of the configuration thus arising. The
configurations are constructed so that in the further course of the game,
carefully calibrated multiple threats move over the playing board along prescribed paths. Thus in the case of go-moku, open triple chains are strung

---

[5] Shigeki Iwata, Takumi Kasai, The Othello game on an $n \times n$ board is PSPACE-
complete, *Theoretical Computer Science* **123**, 1994, pp. 329–340.

[6] J. M. Robson, N by N checkers is EXPTIME complete, *SIAM Journal on Com-
puting* **13**, 1984, pp. 252—267; J. M. Robson, The complexity of Go, in: *Proceedings
Information Processing* 1983, pp. 413–417. The construction configurations contain kos.
The corresponding result without kos is weaker: David Lichtenstein, Michael Sipser, Go
is Pspace hard, in: *Proceedings 19th Annual Symposium on Foundations of Computer
Science*, Ann Arbor 1978, pp. 48–54; also in *Journal of the Association for Computing
Machinery* **27**, 1980, pp. 393–401.

[7] Aviezri S. Fraenkel, David Lichtenstein, Computing a perfect strategy for $n \times n$
Chess requires time exponential in n, *Journal of Combinatorial Theory* **A 31**, 1981,
pp. 199–214; H. Adachi, H. Kamekawa, S. Iwata, Shogi on an $n \times n$ board is complete
in exponential time, *Transactions IEICE* **J70-D**, 1987, pp. 1843–1852 (in Japanese).

together; in Hex, almost closed linked paths are constructed with holes only at some carefully chosen branch points.

Which decision problems can be transformed in this way? That is, which decisions can be reduced to the analysis of a Hex configuration? If one makes use of a multistage chain of proof as is done in complexity theory, then the following picture emerges: what are transformed are not the individual decision problems, but the representatives of an entire class all at once. To obtain such a universal transformation one begins not with the decision problem itself, but with a computer program that computes the decision in question. The construction of the configurations is supported by the mechanism by which the program alters the internal states of the processor and memory within the computer, from the various inputs step by step until the decisive output bit. In particular, in the cases under consideration the sequential alteration of the bits is characterized so cleverly via Boolean formulas that the equivalent configurations can be generated from the original inputs in polynomially bounded time.

With Hex, but also with the other games mentioned, one can carry out this construction for every decision problem that can be decided with a program whose memory requirement is bounded by a polynomial function of the input length (see Figure 29.1).

With the represented transformation a Hex configuration is generated for each input of a given decision problem whose winning prospects are reflected by the originally posed decision and whose board size grows at most polynomially as a function of the input length. That is, every problem that can be decided with polynomially bounded storage space can be reduced to the analysis of Hex configurations for which the size of the playing board is not "essentially" larger than the input that was originally to be investigated. Hex thus belongs among the most difficult problems in its class, since the following statement holds: if there exist decision problems that can be solved in polynomially bounded storage space but not in polynomi-

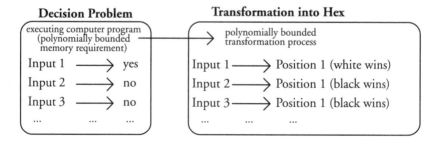

Figure 29.1. Transformation of a decision problem into Hex.

ally bounded time, then the analysis of Hex configurations is an example of a problem of such complexity.

Not only does all this sound highly theoretical and inapplicable in the real world, that is, in fact, the case. But what was to be expected? For how else could such far-reaching statements be justified that exclude the possibility of the existence of simple formulas and procedures like those for nim and Bridge-it for these other games?

There are additional consequences that affect the game directly: Hex, reversi, and go-moku are polynomially transformable one into the other. Thus at least in theory there exist efficient algorithms by which arbitrarily large Hex configurations can be transformed into equivalent go-moku configurations, and conversely. That is, white is in a winning position in the one configuration precisely when she is in a winning configuration of the other. A perfect Hex player, one who can always find an optimal move in every configuration on an arbitrarily large playing board, would be able to use the algorithm to become a perfect go-moku player, and vice versa.

In sum, the uncertainty of combinatorial games can be completely overcome only in very simple cases, like nim and Bridge-it. Therefore, our playing strategy must always remain inaccessible to codification, and we are required always to attempt to overcome this deficit. There is plenty of room for this, such as in the search for almost perfect algorithms whose runtimes or results are very good as often as possible. A good example is the approximation algorithm thermostrat presented in Chapter 24, with which the complexity of special endgame situations in go can be drastically reduced. Finally, we must not neglect to mention that complexity theory makes statements only about the asymptotic complexity for large inputs. How difficult go-moku and Hex are when played on a normal playing board is a different question altogether.

NP and PSPACE Complete Problems

Although efficient algorithms were discovered for a wide variety of problems up to the beginning of the 1970s, other problems proved intractable. Among these included primality tests, factorization algorithms, linear optimization problems,[8] and many combinatorial problems such as the traveling salesman problem.

---

[8]Linear optimization will be discussed in depth in Part III of this book. The simplex algorithm, used widely in practice, leads usually, but not always, to a solution. However, since 1979 there have been known efficient, that is, polynomially bounded even in the worst case, algorithms (see Chapter 36).

Attempts to demonstrate that some of these problems were inherently difficult remained fruitless at first. The question whether efficient solution algorithms existed remained open. A great step forward was made in 1971 by Stephen Arthur Cook (1939–), who came up with a completely new approach. Cook proved that every decision problem in the class NP can be solved by reducing it to a particular decision problem called the *satisfiability problem* (see Note 5 at the end of the chapter). That is, every instance of the original problem is transformed into an equivalent instance of the satisfiability problem, so that an efficient solution procedure of the satisfiability problem, if it exists, can be used to obtain an efficient solution of the original decision problem. It follows that the satisfiability problem cannot be essentially easier to solve than any other problem in the class NP. One therefore calls it NP hard. With this result an entire avalanche was let loose, for the satisfiability problem can be reduced in turn, up to a polynomially bounded transformation, into many other problems, including the traveling salesman problem. These problems are also NP hard.

There are thus two possibilities:

- the two classes P and NP are equal.

  In this case, one has simply not been clever enough to find an efficient algorithm for the satisfiability problem or the traveling salesman problem, even though such algorithms exist.

- the class NP contains problems that are not in the class P.

  All NP complete problems, as NP hard problems lying in the class NP are called, are then examples of such difficult problems. In particular, there do not exist efficient algorithms for the satisfiability problem and the traveling salesman problem.

Which of these two possibilities actually obtains remains an unsolved problem. Because of its great significance, the P = NP problem was named one of the seven "millennium problems" by the Clay Mathematics Institute in Cambridge, Massachusetts, with a one-million-dollar reward for its solution.

In 1973, Stockmeyer and Meyer translated Cook's method to a class of conjecturally difficult problems. They showed that there are also PSPACE hard problems and PSPACE complete

problems. Hex, go-moku, and reversi are PSPACE complete.[9]
Checkers, go, chess, and shogi are more difficult, namely, EX-
PTIME complete, which excludes the possibility of an efficient
solution procedure, since the class P is a proper subset of EX-
PTIME.

Even sums of short Conway games are very difficult, which is
shown by the limits of approximately good procedures like thermostrat, described in Chapter 24. Thus Morris proved in 1981
that relatively simple Conway games suffice for the construction
of sums whose winning prospects can be computed only with
PSPACE complete cost.[10] In the configurations used by Morris, after three moves at the latest an integer is obtained. Even
simpler configurations, namely, those of the form $\{\,a\mid\{\,b\mid c\,\}\,\}$
with three integers a, b, c, possess sufficient difficulties. With
these one can construct sums whose analysis is NP hard.[11]

Chapter Notes

1. It would not do simply to negate the condition, since one would have find a
   nonconstructive criterion to prove the requisite unsatisfiability. Better are
   the following variants, which are built on the winning criterion for black
   that we have described:

   - the configurations in which white to move second possesses a winning
     strategy can be characterized in Bridge-it by switching the colors of
     the two players before the transformation into Shannon's switching
     game. Lehman found a corresponding construction directly on the
     level of the switching game.

   - the configurations of the third class can be recognized by specifying a
     winning move for black and one for white and confirming the winning
     character of the configurations that arise using the criteria for the
     other two classes.

---

[9] A unified representation of PSPACE hard problems together with their application
to go can be found in Karl Rüdiger Reischuk, *Einführung in die Komplexitätstheorie*,
Stuttgart 1990, Section 7.4.

[10] F. L. Morris, Playing disjunctive sums is polynomial space complete, *International
Journal of Game Theory* **10**, 1981, pp. 195–205.

[11] See Elwyn Berlekamp, David Wolfe, *Mathematical Go*, Wellesley 1994, pp. 109–111.
The result is due to Yedwab and Moews.

The generation of winning strategies can also be reduced to the case in which black to move second is assured of a win.

The most effective of the known procedures is a method in which the graph—at first without considering the $+$ and $-$ vertices as special—is decomposed in a particular way. See the cited work of Gabow et al.

2. Quadratic complexity means a quadrupling of the computational cost as the length of input is doubled. Using the standard procedure, one can see this from the equation

$$(aB + b)(cB + d) = acB^2 + (ad + bc)B + bd.$$

Here the numbers to be multiplied are broken into two pieces, each half the length of the original number. One requires four multiplications of numbers of this size. On the other hand, the additions and multiplications by B and B^2 are significantly simpler, particularly if B is a power of the base of the number system used, in which case B^2 is computed by shifting B to the left. If one rewrites the equation, using a 1962 idea of Karatsuba and Ofman, to

$$(aB + b)(cB + d) = acB^2 + [ac + bd - (a - b)(c - d)]B + bd,$$

then one needs only three multiplications of numbers with half-sized representations, namely, ac, bd, and $(a - b)(c - d)$. That is, the computational complexity triples when the input length doubles. Since $2^{1.585} \approx 3$, we conclude that the complexity is $O\left(2^{1.585}\right)$. The growth of the complexity for very long numbers can be further reduced. See A. K. Dewdney, *The (New) Turing Omnibus*, New York 1993, Chapter 25; Gilles Brassard, Paul Bratley, *Algorithmics: Theory and Practice*, Engelwood Cliffs, NJ 1988.

3. Theory and practice can be far apart in such cases, in particular, when the computational demands rise significantly only with very long inputs. An example of this phenomenon is the 1980 primality test of Adleman and Rumely, whose computational complexity for an n-digit input is bounded by $O\left(n^{c \ln \ln n}\right)$ for a constant c. Although there is no polynomial bound, that makes little difference in computations of orders of magnitude that a computer can handle.

The computational complexities of various primality tests and algorithms for factorization have been well studied. Aside from the purely mathematical interest, cryptographic applications have provided a strong motive for such analyses. For example, the security of the 1978 RSA public key encryption system of Rivest, Shamir, and Adleman is based, among other things, on the belief that factorization is much more difficult, from the point of view of computational complexity, than primality testing. Otherwise, it would be possible to compute the secret key for decryption using the public key for encryption.

More on this topic can be found in Carl Pomerance, Primzahlen im Schnelltest, *Spektrum der Wissenschaft*, February 1983, pp. 80–92; Jürgen Wolfart, Primzahltests und Primfaktorenzerlegung, in: *Jahrbuch Überblicke der Mathematik* 1981, pp. 161–188; John D. Dixon, Factorization and primality tests, *American Mathematical Monthly* **91**, 1984, pp. 333–352; Paulo Ribenboim, *The Book of Prime Number Records*, New York 1988, Chapter 2; Martin E. Hellmann, Die Mathematik der Verschlüsselungssysteme, *Spektrum der Wissenschaft* 1979/10, pp. 93–101; Albrecht Beutelspacher, *Kryptologie*, Braunschweig 1987.

4. The decision whether a number is prime belongs to the class co-NP. This class is defined analogously to the class NP, except that here, every no decision must be able to be confirmed with the help of suitable additional information.

One can convince oneself relatively easily that the decision whether a number is prime is also in the class NP. The confirmation of the fact that a given number n is prime can always be carried out by producing an integer a such that the numbers $a^1, a^2, a^3, \ldots, a^{n-1}$ are representatives of distinct remainder classes upon division by n. (If b is in the remainder class of c for division by n, then $b = mn + c$ for some integer m. We say that b is congruent to c modulo n and write $b \equiv c \pmod{n}$. For example, for the integer 7, the powers of $a = 3$ are $3^1 = 3$, $3^2 = 9 \equiv 2 \pmod{7}$, $3^3 = 27 \equiv 6 \pmod{7}$, $3^4 = 81 \equiv 4 \pmod{7}$, $3^5 = 243 \equiv 5 \pmod{7}$, and $3^6 = 729 \equiv 1 \pmod{7}$, giving the six distinct representatives $\{3, 2, 6, 4, 5, 1\}$. This test, based on Fermat's little theorem, is very easy to carry out when the prime factorization of $n - 1$ is available as additional information. One checks that for no divisor t of $n - 1$ is the expression $a^{(n-1)/t} - 1$ divisible by n. That the given factors of $n - 1$ are actually prime can be checked recursively in a like manner (for Fermat's little theorem see Chapter 15, "The Generation of Random Numbers").

It was long conjectured that the problem of primality testing belongs to the class P, though no proof could be found. What has been known since 1976 is a technique based on Fermat's little theorem that operates in polynomially bounded time on the assumption of the extended Riemann hypothesis. However, even in its standard version this conjecture has remained unproved for over 100 years. It was given by David Hilbert as his eighth problem in his 1900 address to the International Congress of Mathematicians. (See Stan Wagon, Primality testing, *The Mathematical Intelligencer* **8/3**, 1986, 58–61.) Moreover, Riemann's conjecture belongs among the seven unsolved problems that in honor of the hundredth anniversary of Hilbert's address were endowed with a prize of one million dollars each for their solution by the Clay Mathematics Institute in Cambridge, Massachusetts.

Without recourse to the Riemann hypothesis, the efficiency of a procedure was proved in 2002 by the Indian mathematicians Agrawal, Kayal, and Saxena. This algorithm always returns in polynomial time an answer to the question whether a given number is prime. (See Folkmar Bornemann, Primes in P: Ein Durchbruch für "Jedermann," *Mitteilungen der DMV* **4**, 2002, pp. 14–21.)

5. The satisfiability problem asks whether the variables $x_1, \ldots, x_n$ of a given Boolean formula can be instantiated with the values true and false in such a way that the result is a true value. For example, the expression

$$x_1 \wedge (\neg x_1 \vee x_2)$$

is satisfiable with the values $x_1 = 1$ and $x_2 = 2$. Clearly, the satisfiability problem belongs to the class NP, since a yes decision can be efficiently checked with the supplementary information of the values with which the variables are to be instantiated.

To reduce an arbitrary problem in the class NP to the satisfiability problem via an efficient transformation, Cook proceeded as follows: the starting point is a computer program reflecting the associated decision problem that is capable of confirming every yes decision with the help of suitable supplementary information. A Boolean variable is now assigned to every bit in active memory and the processor. In particular, there are variables for the ultimately valid decision bit, the bits of input, and the bits of the supplementary information. Since such a bounded amount of memory can be processed in polynomially bounded time, the number of variables is also polynomially bounded as a function of the input size. Moreover, based on the computer architecture, for every variable there is set up a transformation equation arising from the variable values of the previous step. Finally, using logical AND operations one obtains a single equation.

In this way, for every input length a single Boolean formula can be efficiently obtained, that is, with polynomially bounded cost, that characterizes the total behavior of the computer program. Every input that is decided with the help of suitable supplementary information leads to a satisfiable equation, and this correspondence is reversible, since a satisfied instantiation of variables always contains the supplementary information for a confirmation. In this way, the existence of suitable supplementary information is reduced to the satisfiability of the generated Boolean expression.

Further Literature on Complexity Theory

[1] Edmund A. Lamagma, Infeasible computation, NP-complete problems, *Abacus* **4**, 1987, Book 3, pp. 18–33.

[2] Harry R. Lewis, Christos H. Papadimitriou, The efficiency of algorithms, *Scientific American* 1978/1, pp. 96–109.

[3] John E. Hopcroft, Turingmaschinen, *Spektrum der Wissenschaft* **7** 1984, pp. 34–49.

[4] A. K. Dewdney, *The New Turing Omnibus*, New York 1993, Chapter 54.

[5] D. B. Shmoys, É. Tardis, *Computational Complexity, Handbook of Combinatorics* (R. L. Graham, M. Grötschel, L. Lovász, eds.), Amsterdam 1995, volume 1, pp. 1599–1645.

[6] Michael R. Garey, David S. Johnson, *Computers and Intractabilitiy: A Guide to the Theory of NP-Completeness*, San Francisco 1979.

[7] Wolfgang J. Paul, *Komplexitätstheorie*, Stuttgart 1978.

[8] Gilles Brassard, Paul Bratley, *Algorithmics: Theory and Practice*, Engelwood Cliffs, NJ 1988.

[9] Aviezri S. Fraenkel, Complexity of games, in: *Combinatorial Games*, Richard K. Guy (ed.), *Proceedings of Symposia in Applied Mathematics*, AMS Short Course Lecture Notes **43**, Providence, RI 1991, pp. 111–153.

30

A Good Memory and Luck: And Nothing Else?

To win at the game memory, one needs a good memory and a bit of luck. But are there additional strategic considerations that could improve one's chances of winning?

The game memory is a well-known children's game. It is eternally fascinating to young and old alike. Its most salient feature is that it requires concentration.

The game appeared in Germany in 1959. Its creator is said to be Heinrich Hurter, who had played it among his family since 1946. However, the game has many precursors, such as the English card game concentration, whose roots can be traced to the 19th century.[1]

Memory is usually played with a special deck of cards. Different pictures are printed on the face side, with each picture appearing on two cards. The game begins with the cards being distributed face down on the table. The players take turns turning over first one card, and then a second, so that all the players can see the cards. If a matching pair is revealed, then the player takes the pair and draws again. Otherwise, the cards are returned to their original positions face down. When all the pairs have been found and removed, the player who has found the most is the winner.

Here we shall restrict our attention to the two-person version of the game. We begin with the question of how the game might be approached

---

[1] Erwin Glonnegger, *Das Spiele-Buch*, Munich 1988, pp. 106 f.

Figure 30.1. White to move: which card does she turn up second?

mathematically. What are the random elements? Is memory a game with perfect information? Fortunately for us, these two questions are not too difficult to answer:

- random elements of the game are characterized by their expectations. That is, if a player uncovers a card whose value he does not know, then all possible resulting game variants with their probabilities and corresponding results are considered. The score, positive or negative, of a player is the number of pairs that he has taken in comparison to those of his competitor.

- memory is a game with perfect information, since all players always have the same information. We shall assume that the players' memories are perfect, as it would be if the overturned cards remained face up.

On this basis we may now apply Zermelo's theorem; that is, for every game configuration we can calculate the winning expectation by means of minimax optimization. But what is to be optimized? Are there different strategies? Let us first consider an example in which three pairs of cards, labeled 1, 2, and 3, are lying face down on a table, and the location of a single card, a 1, is known. The player whose turn it is, let us call her white, turns over one of the unknown cards, resulting in the situation of Figure 30.1.

What should she do? Which card should white turn up second? There are two possibilities:

1. It appears most profitable for white to turn over one of the four still unknown cards.

    - With probability 1/4 the matching card to the exposed 2 will be turned, so that white obtains one point. Then she will have another turn, knowing exactly one of the remaining four cards.

    - With the same probability of 1/4 white will expose the matching card to the known 1. This new information will give white's

opponent black a certain point at the start of his turn, after
which black would move in the situation of knowing exactly one
of the remaining cards.

- Finally, with probability 1/2 white will expose one of the two
  3s. In this case, black will be able to take all three pairs on his
  next turn.

Since the expectations cancel out in the first two cases, this entire
variant gives white an expectation of

$$\frac{1}{2} \times (-3) = -\frac{3}{2}.$$

2. It must seem much less advantageous for white simply to expose the
   already known 1. However, such a move is a legitimate means of
   denying black any additional information. Therefore, let us consider
   in detail what results from this move. The starting point for black's
   next move is the situation in which two different cards are known
   among the three pairs, as shown in the following figure:

We assume that black begins his turn by exposing one of the four
unknown cards:

- with probability 1/4 black turns over a 1, which he can immediately match. This wins him a point and another turn. With his
  next move black can assure himself of the two remaining pairs
  unless he draws a 3 and then a 2, so that the two remaining pairs
  go to white; the probability of black's second move turning out
  that way is

$$\frac{2}{3} \times \frac{1}{2} = \frac{1}{3}.$$

3. With probability 1/4 a 2 will be chosen. Regarding expectations, this
   case is equivalent to the first.

4. With probability 1/2 the first card drawn will be a 3. Despite the
   unlucky beginning, it makes no sense for black to draw a known card
   as his second selection, since then white will be assured of winning all
   three pairs. Therefore, black goes for the small probability of drawing
   another 3.

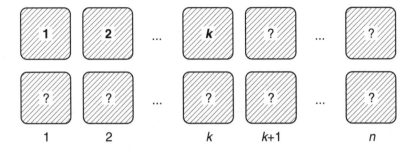

Figure 30.2. The configuration $P_{n,k}$: k of the total of $2n$ cards are known.

- With net probability of $(1/2) \times (1/3) = 1/6$ black has good luck and finds the 3. Then he collects the remaining two pairs.
- With net probability of $(1/2) \times (2/3) = 1/3$ black has bad luck and draws a 1 or a 2. This allows white to win all three pairs.

Altogether, white therefore has in this variant a winning expectation of

$$\frac{1}{4} \times \left[\frac{2}{3} \times (-3) + \frac{1}{3} \times (-1+2) \right] +$$

$$\frac{1}{4} \times \left[\frac{2}{3} \times (-3) + \frac{1}{3} \times (-1+2) \right] - \frac{1}{6} \times 3 + \frac{1}{3} \times 3 = -\frac{1}{3}.$$

This shows that it is to black's advantage to make a nontrivial move (that is, not simply "pass" by turning over two known cards). Avoiding the entire move, which black can manage by turning over two already known cards, is therefore not to be recommended.

With respect to our initial question, which card white should expose second, it has become clear that the second variant, as destructive as it may seem, greatly reduces white's expected loss. This shows as well that with strategic skill one can increase one's winning prospects in the game memory; this seemingly simple game contains some hidden surprises. Such results were published first by Uri Zwick and Michael Paterson, who in 1993 offered a complete analysis of memory,[2] in which they determined recursively the optimal strategy for all possible memory configurations. The

---

[2]Uri Zwick, Michael S. Paterson, The memory game, *Theoretical Computer Science* **110**, 1993, pp. 169–196; see also Ian Stewart, Mathematische Unterhaltungen, *Spektrum der Wissenschaft* **6** 1992, pp. 12–15; David Gale, Mathematical Entertainments, *The Mathematical Intelligencer* **15/3**, 1993, pp. 56–60. In a postscript to their publication, Zwick and Paterson mention that after they had completed their work they became

Figure 30.3. The configuration $P_{2,0}$.

Figure 30.4. The configuration $P_{3,1}$.

basis of their work is a recursion formula that allows the computation of the winning expectations of all possible memory configurations in the case of error-free memory of the players. Every such configuration is characterized by two numbers, namely, the number n of remaining card pairs and the number k of already known individual cards, where k can take on all values between 0 and n (see Figure 30.2). These numbers do not take into account the special case in which a player knows the locations of complete pairs. But this poses no problem, since in such cases the optimal behavior is clear.

Each such configuration, which we denote by $P_{n,k}$, corresponds to a minimax value of $v_{n,k}$, which corresponds to the expected pair surplus accruing to the player whose move it is under mutually optimal play. As an example, we consider the configurations $P_{2,0}$ and $P_{3,1}$.

In the configuration $P_{2,0}$ depicted in Figure 30.3, two pairs remain on the table in which no card has yet been identified. The next move involves no skill. Luck alone determines whether the two of the four cards drawn will match. If a player draws, with probability $1/6$, one of the two pairs, then he wins both pairs. In the other cases the opponent can win both pairs:

$$v_{2,0} = \frac{1}{3} \times 2 + \frac{2}{3} \times (-2) = -\frac{2}{3}.$$

For the configuration $P_{3,1}$ shown in Figure 30.4 we return to our original analysis. Its starting position is achieved with probability $4/5$, namely, whenever a 2 or 3 is chosen as the first card. Otherwise, that is, with probability $1/5$, a 1 is uncovered, which permits the completion of a pair.

aware of an analysis of the game memory in Dutch by S. H. Gerez. This work, which was done in 1983 at the University of Twente, contains some significant results on memory strategies and their optimization.

The next move is from configuration $P_{2,0}$:

$$v_{3,1} = \frac{4}{5} \times \left(-\frac{1}{3}\right) + \frac{1}{5} \times (1 + v_{2,0}) = -\frac{1}{5}.$$

The general situation for an arbitrary configuration $P_{n,k}$ can be investigated analogously. The player has up to three significantly different move options, ignoring obviously bad moves such as beginning a turn with a card that is already known and then searching for its mate. According to the number of newly exposed cards, the move types are designated as types 0, 1, and 2:

- a 0 move amounts to passing. One simply turns over two known cards. Such a move is possible in cases of $n \geq k \geq 2$. In such cases a 0 move prevents the winning expectation from becoming negative, though at the cost of terminating the game if both parties make such a move, even though there are cards remaining on the table.

- with a 1 move, first an unknown card is exposed. If it does not form a pair with a known card, then no attempt at forming a pair is made. That is, apparently destructively and with no chance at winning a point, a known, nonmatching, card is turned over. This move is possible in cases $n - 1 \geq k \geq 1$.

- a 2 move is the usual type of move. One first turns over an unknown card, and if it does not match a known card, an additional unknown card is exposed. A 2 move is always possible except in the cases $P_{0,0}$, $P_{1,0}$, and $P_{1,1}$.

The winning expectations of the two last move types can be determined recursively:

$$v_{n,k}^{(1)} = \frac{k}{2n - k}(1 + v_{n-1,k-1}) + \frac{2(n - k)}{2n - k} \cdot (-v_{n,k+1}),$$

$$
\begin{aligned}
v_{n,k}^{(2)} &= \frac{k}{2n - k}(1 + v_{n-1,k-1}) \\
&+ \frac{2(n - k)}{2n - k} \cdot \frac{(1 + v_{n-1,k} + k(-1 - v_{n-1,k}) + 2(n - k - 1) \cdot (-v_{n,k+2})}{2n - k - 1} \\
&= \frac{k}{2n - k}(1 + v_{n-1,k-1}) \\
&- \frac{2(n - k)}{2n - k} \cdot \frac{(k - 1)(1 + v_{n-1,k}) + 2(n - k - 1)v_{n,k+2}}{2n - k - 1}.
\end{aligned}
$$

k\n	0	1	2	3	4	5	6	7	8	9	10	11	12	13	14
0	0.000	1.000	-0.667	-0.200	-0.114	-0.029	0.002	0.053	-0.033	0.035	-0.038	0.014	-0.020	0.017	-0.016
1		1.000	0.667	-0.200	0.114	-0.029	0.002	0.053	0.033	0.035	0.038	0.014	0.020	0.017	0.016
2			2.000	0.333	0.267	0.143	0.095	0.026	0.111	0.024	0.097	0.033	0.067	0.023	0.056
3				3.000	0.000	0.543	0.124	0.229	0.046	0.176	0.020	0.153	0.028	0.118	0.024
4					4.000	0.000	0.771	0.095	0.367	0.058	0.246	0.024	0.207	0.025	0.167
5						5.000	0.000	0.984	0.065	0.504	0.064	0.326	0.030	0.261	0.023
6							6.000	0.000	1.190	0.035	0.642	0.057	0.412	0.034	0.318
7								7.000	0.000	1.394	0.005	0.766	0.046	0.499	0.036
8									8.000	0.000	1.596	0.000	0.883	0.034	0.586
9										9.000	0.000	1.797	0.000	0.997	0.023
10											10.000	0.000	1.998	0.000	1.109
11												11.000	0.000	2.199	0.000
12													12.000	0.000	2.399
13														13.000	0.000
14															14.000

Table 30.1.   The winning expectations $v_{n,k}$ with optimal play: k individual cards among n pairs are known.

The formula is relatively easy to verify for a 1 move: first, one of the $2n - k$ unknown cards is turned over. In k cases the player finds a card that can be converted at once into a pair, and the next move take place from the configuration $P_{n-1,k-1}$. In the remaining $2(n - k)$ cases the player has bad luck and so turns over a known card, so that his opponent is now in configuration $P_{n,k+1}$.

For the second formula we have still to deal with the case that the first card does not lead to an immediate pair. The probability of that is $2(n - k)/(2n - k)$. Then together with the chosen card there are $k + 1$ cards known of the n pairs. In a 2 move the player, trusting to luck, turns over one of the additional $2n - k - 1$ unknown cards. One of these cards is a match for the first, which would bring a point and another turn in the configuration $P_{n-1,k}$. For the remaining $2(n - k - 1)$ cards neither player wins a point, and the opponent draws from configuration $P_{n,k+2}$.

In reverse chronology, starting from $v_{n,n} = n$, $n \geq 0$, and $v_{1,0} = 1$, all winning expectations $v_{n,k}$ with $n > k \geq 0$ can be calculated. The selection of the optimal strategy for $n \geq 2$ is made using the formulas

$$v_{n,0} = v_{n,0}^{(2)},$$

$$v_{n,1} = \max\left(v_{n,1}^{(1)}, v_{n,1}^{(2)}\right),$$

$$v_{n,k} = \max\left(0, v_{n,k}^{(1)}, v_{n,k}^{(2)}\right) \quad \text{for } 2 \leq k \leq n.$$

The last equation allows for the possibility of a 0 move in order to avoid a negative expectation. The winning expectations thus calculated for $0 \leq k \leq n \leq 14$ are shown in Table 30.1, with the associated optimal moves in Table 30.2.

$k\backslash n$	0	1	2	3	4	5	6	7	8	9	10	11	12	13	14
0			2	2	2	2	2	2	2	2	2	2	2	2	2
1			2	1	2	1	1	1	2	1	2	1	2	1	2
2			1	2	1	2	1	2	1	2	1	2	1	2	1
3				1	0	1	2	1	2	1	2	1	2	1	2
4					1	0	1	2	1	2	1	2	1	2	1
5						1	0	1	2	1	2	1	2	1	2
6							1	0	1	2	1	2	1	2	1
7								1	0	1	2	1	2	1	2
8									1	0	1	2	1	2	1
9										1	0	1	2	1	2
10											1	0	1	2	1
11												1	0	1	2
12													1	0	1
13														1	0
14															1

Table 30.2.   The winning expectations $v_{n,k}$ with optimal play: k individual cards among n pairs are known.

It remains to remark that in the case of configuration $P_{4,3}$, in addition to the tabulated 0 move, the 2 move is equally optimal.

Within the confines of our discussion, we see that the game memory can be calculated with a spreadsheet program. Paterson and Zwick, on the other hand, went far beyond the level presented here in their publication in order to obtain general results on optimal strategies for arbitrary configurations. To that end, they compared the various strategies on the basis of more complex estimates. Thus they could confirm that hidden behind the regularity of Table 30.2, in which the 1 and 2 moves alternate in checkerboard fashion, lies a general law, by which the following strategy is optimal:

- if $n + k$ is odd with $k \geq 2(n + 1)/3$, then make a 0 move. If the opponent does likewise, the game is over.

- in the case of an even value $n + k$ with $k \geq 1$, as well as for the exceptional configuration $P_{6,1}$, make a 1 move.

- in all other situations make a 2 move.

31

Backgammon: To Double or Not to Double?

If a backgammon player believes himself to be sufficiently ahead, then he has the option of doubling the bet. His opponent must either accept the double or lose the amount of the current wager. Needless to say, a player who is behind has no interest in doubling. Are there thus doubling situations that arise in error-free play?

The roots of backgammon and its variants go back to antiquity.[1] For example, the Romans played a game called *ludus duodecim scriptorum*, the game of 12 lines. Later references to backgammon are found in paintings and drawings in which the characteristic backgammon board appears. One of the oldest such representations is a miniature from the Middle Ages, appearing in the Mannessischen Handschrift of 1330.

Except in the countries of the eastern Mediterranean, where the backgammon variants Plakato and goul continued to be widely played, backgammon

---

[1] More information on the rules, history, and variants can be found in David Pritchard, *The Family Book of Games*, Brockhampton Press 1983, pp. 22–27; Rüdiger Thiele, *Das grosse Spielevergnügen*, Leipzig 1984, pp. 182–184; Erwin Glonnegger, *Das Spiele-Buch*, Munich 1988, pp. 31–37; R. C. Bell, *Board and Table Games from Many Civilizations*, New York 1979, volume I, pp. 23–46 and volume II, pp. 12–23; L. U. Dikus, Black Mammon, *Spielbox* **2**, 1986, pp. 14–16; Wir sind die Clochards ohne Durst und Hunger, *Der Spiegel* **49**, 1987, pp. 244–250. Greater detail can be found in backgammon books such as Oswald Jacoby, John R. Crawford, *The Backgammon Book*, New York 1970; Tim Holland, *Beginning Backgammon*, New York 1973; Charles H. Goren, *Goren's Modern Backgammon Complete*, New York 1974; Bill Robertie, *Backgammon for Winners*, New York 1993.

declined in popularity over time. It saw a renaissance only in the 20th century. The first phase began at the end of the 1930s, when the game was rediscovered by London's intellectuals. In the 1970s came the great breakthrough, when it became fashionable in the USA to play the game.

On its face, backgammon is a race governed by the luck of the dice, similar to pachisi. The two players attempt to get their own pieces to the goal. On the way, opposing pieces can be hit under certain conditions, namely, if a single such piece lies on a field that the player can reach. The distance that one is allowed to move is determined by the throw of two dice. A player moves a piece by the amount shown on one die, and then another or the same piece the amount shown by the second die. If doubles are thrown, the player moves the number shown on a die four times.

Backgammon is much more complex than the children's game pachisi. First of all, there are many more pieces, and up to two of them can be moved on a single turn, even four in the case of doubles. Thus backgammon acquires a combinatorial character above and beyond the element of chance, which is quite strong due to the interaction between the pieces on account of the small number of fields and the fact that the players move in opposite directions. Whoever believes that all one needs to win in backgammon is a bit of luck is sorely mistaken. Let him try. He will rarely succeed.

Like chess, backgammon is played in international tournaments. The highest honor that one can achieve is to be crowned world champion. Competition from computers came much earlier to backgammon than it did to chess. Thus, for example, already in 1979 the then reigning world champion Luigi Villa was defeated 7 : 1 by a computer program designed by Hans Berliner.[2]

The quiet that prevails at a chess tournament is hardly imaginable at a backgammon tournament. The reason for this is the dice. They are always making noise, for the pace is rapid: there is no point taking a great deal of time in analyzing a position. One recognizes certain patterns and quickly makes a realistic risk assessment. Both of these can be accomplished quickly by an experienced player.

One of the most attractive elements of backgammon is the possibility of doubling. A special die is used, called the doubling cube, which displays on its six faces the numbers $2, 4, 8, 16, 32, 64$. It was introduced in the 1920s and established the custom of playing backgammon for money. Since

---

[2]Hans Berliner, Ein Computer spielt Backgammon, *Spektrum der Wissenschaft* **8**, 1980, pp. 53–59; Hans Berliner, BKG: A program that plays backgammon, Computer Science Department, Carnegie-Mellon University, Pittsburgh 1977; Hans Berliner, Backgammon computer program beats world champion, *Artificial Intelligence* **14**, 1980, pp. 205–220. The last two publications are reprinted in David N. L. Levy, *Computer Games I*, New York 1988, pp. 3–28, 29–43.

money is involved, the doubling rule makes sense, and it is used even in tournaments.

In order to prevent a player from continually doubling the wager, no player who doubles is allowed to double again before his opponent doubles. When the game is played, the right of the next redouble is indicated by placing the doubling cube on the side of the board of the player who has the next right to double. After the first double, the face labeled 2 is placed upward. For the remainder of the game, the only player permitted to double is the one with the doubling cube on his side of the board. If the opponent accepts a redoubling, he receives the doubling cube, which is turned so that the next higher level is displayed.

Of course, a player whose double or redouble has been accepted can make better use of his positional advantage. However, he loses some strategic potential, namely, the option of doubling at a more suitable time. This loss of initiative is particularly apparent in the case of a redouble: if a player can redouble, but does not do so, he can wait calmly for a better opportunity, while his opponent, should the tide turn in his favor, will certainly regret not having the opportunity to redouble. In contrast, passing up an opportunity to make the first double leaves a player exposed to the possibility that his opponent will double if the tide shifts in his favor. Thus not redoubling can have more advantages than not doubling. Therefore, for a redouble, the current position must be relatively more advantageous than is required for the first double.

Let us introduce an example to show why in error-free play doubling and redoubling can take place. Consider the situation shown in Figure 31.1, in which white is to play.

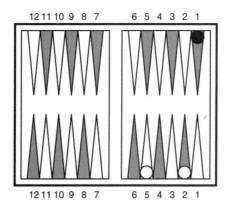

Figure 31.1. Should white double?

What are the chances that the player to move will bear all his pieces off the board in one turn?

- White wins on her first move by rolling any of the following combinations (where order does not matter): 2-2, 3-3, 4-4, 5-2, 5-3, 5-4, 5-5, 6-2, 6-3, 6-4, 6-5, 6-6. Therefore, the probability that white will win in one move is 19/36.

- If white does not succeed in bearing off both of her pieces on her turn, then black wins on his next turn. Thus the probability that black will win is 17/36.

Thus without doubling, white's winning expectation is $19/36 \times 1 + 17/36 \times (-1) = 1/18$. And if she doubles? Let us see first how black should deal with the situation:

- if black does not accept a double, then the game ends, and white wins one point.

- if black accepts, then white's expectation is $1/18 \times 2 = 1/9$.

Thus it is black's interest to accept the double despite his positional disadvantage, for even after a double his expected loss is much less than an entire point, which he would lose by accepting.

The last argument applies not only to this special situation, but to every doubling: a player possessing odds of winning of at least 1/4 should accept a double, since against a certain loss of one full point acceptance represents the lesser of two evils, with a probability of at most 3/4 of losing a doubled bet, but in return the probability of at least 1/4 of winning a doubled bet. And of course, the player will not have to fear the consequences of a redouble, since having accepted the double, he now has control over the doubling cube.

The situation of our first example, in which the game will end in at most two moves, can be generalized. We assume, then, that it is white's turn and that she will win on that turn with probability p. Thus black will win with probability $1 - p$. We now calculate the minimax value for white. We compute the winning expectations in the three possible doubling situations:

- without a double: $p \times 1 + (1 - p) \times (-1) = 2p - 1$;

- with an accepted double: $p \times 2 + (1 - p) \times (-2) = 4p - 2$;

- with a declined double: 1.

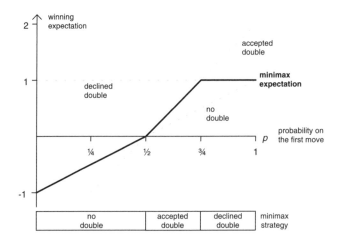

Figure 31.2. Two-move game: when should white double?

Altogether, this yields a minimax value for white of

$$\max(2p - 1, \min(4p - 2, 1)).$$

The significance of this formula can be seen in Figure 31.2. There one can see clearly that white should double in the position under consideration as soon as she has the smallest advantage, that is, in the case $p > 1/2$. Black should accept if white's winning expectation does not exceed $3/4$.

In the positions that we have been considering, black also has the option of doubling if he gets a turn. White would decline the double, of course, and therefore lose the current bet, just as if no doubling had taken place. However, if a game can last more than two moves, then, of course, the possibility of multiple doublings must be taken into account. Let us consider the position shown in Figure 31.3, where again it is white's move.

This configuration differs only slightly from that of the first example; a second black piece has been added, so that white's chances of winning are even greater than before. Let us see how that game can progress, first without regard to doubling. To make things clearer, in Figure 31.4 all the positions offering the same probabilities are combined into a single node. Along the edges are the conditional probabilities corresponding to the results of the dice.

Here is an explanation of the possible courses of the game as shown in Figure 31.4:

- as in the first example, white wins on the first move with probability $19/36$. The other dice combinations, those that do not lead to an im-

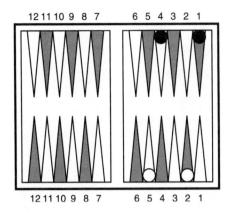

Figure 31.3. Black's situation is even worse than that depicted in Figure 31.1.

mediate win, must be subdivided into groups, depending on whether white can bear off all her pieces on her second move. This is always the case, in fact, except for the roll 2-1. If white rolls 2-1 twice, then two moves do not suffice.

- if black gets a turn, then he can bear off both of his pieces with conditional probability 29/36, namely, with every roll except 1-1, 1-2, 1-3, or 2-3. If one move does not suffice, then a second move will.

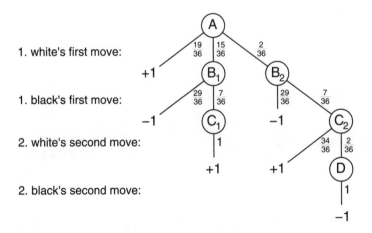

Figure 31.4.

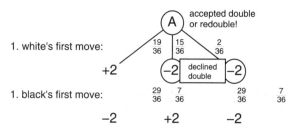

Figure 31.5.

And how does doubling affect the winning expectations? We begin with the case shown in Figure 31.5, in which white, in position A, doubles or redoubles before her turn. We assume that the basic amount wagered is one unit; that is, all amounts are to be understood as multiples of the doubling level reached at this point in the game. Black will accept the double in any case, since his winning probability exceeds 1/4. If white fails to win on her first move, then black will redouble, forcing white to decline, since her chances in both B positions are too small: in position B_1, which is a two-move position of a type already considered, the winning probability of $7/36$ is less than 1/4, and in position B_2 the situation is even worse, though it could grow to $7/36$ through a further redouble. Altogether, in the case of a double or redouble we obtain the minimax remainder of the game shown in Figure 31.5; the winning expectation is $19/36 \times 2 + 17/36 \times (-2) = 1/9$.

However, even if she is in possession of the doubling cube, white in position A can decline to redouble. Then we obtain the minimax game

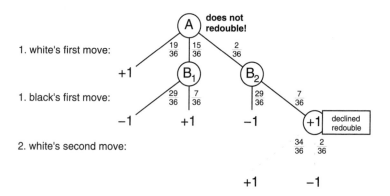

Figure 31.6.

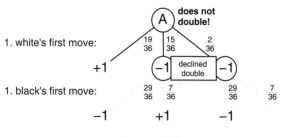

Figure 31.7.

shown in Figure 31.6, in which white can expect a winning amount of $19/36 \times 1 + 17/36 \times (29/36(-1) + 7/36 \times 1) = 155/648$.

It is to white's advantage *not* to redouble. This appears paradoxical if one compares the situation with that of the first example. There black does not have a piece on point 4, and despite white's worse prognosis, it made sense to redouble. How is this phenomenon, known as the *Jacoby paradox*, to be explained? All we need to do is to remind ourselves of what we have already learned. Redoubling permits, to be sure, a better exploitation of an existing advantage, but at the same time, it gives a bit of initiative to the opponent, and this initiative has a higher value in the second example than in the first, where this initiative cannot be profitably implemented by black. For this explanation we also have the fact that a double, unlike a redouble, is absolutely to be recommended: in declining to double, white cannot block later doublings by her opponent. But let us look again in detail at the course of the game shown in Figure 31.7: the winning expectation for white is $19/36 \times 1 + 17/36 \times (-1) = 1/18$, which is less than for a double.

Taken all together, the position that we are considering offers the winning prospects tabulated in Table 31.1.

If one wishes to investigate more complex positions, then the method used in this example of representing the course of the game graphically will become unmanageable. But even a computer-supported calculation will soon reach its limits due to the large number of positions that have to be considered. Therefore, we need to make some simplifying assumptions. To this end, various mathematical approaches have been tried in recent decades. They are based almost exclusively on the so-called running game, in which the opposing armies have already passed each other, and thus no further stones can be hit. Furthermore, many wins based on a gammon are also not considered:

- as we did in Chapter 14, a model is created in which each player has only a single piece, which has to cover a distance of, say, 60, 80, or

The doubling cube belongs to...	white...	black...	winning expectation	
...white	...redoubles	...accepts	0.11111	
	...redoubles	...declines	1.00000	← 1
	...does not redouble	-	0.23920	×2 dice and move analysis
...no one yet	...doubles	...accepts	0.11111	
	...doubles	...declines	1.00000	← 1
	...does not double	-	0.05556	
...black	-	-	0.05556	dice and move analysis
for comparison: game without doubling			0.23800	

Table 31.1. Winning expectations for white for the position shown in Figure 31.3 (white: 5-2, black: 4-1).

even 100 fields. This two-piece model has some serious shortcomings due to the dice points that are lost in bearing off.

- one obtains a simpler model when one considers only the winning probabilities of a game governed strictly by the throws of the dice. Instead of the concrete dice results and their probabilities, one considers a continuous process in which the winning probabilities change randomly. The details of the random process are ignored. One assumes only that the changes occur continuously, that is, without breaks. This assumption suffices for obtaining some quite fundamental results.

- positions with only a few pieces can be investigated explicitly. This is best done recursively, where intermediate results are stored in a sufficiently large database. In practice, such results are not very usable, since they do not cover much ground, but they would be usable in theory if, for example, an approximation model was used that gave heuristic "rules of thumb" based, for example, on point counts that indicated whether a move was sufficiently close to optimal in the majority of cases.

- investigations of positions characterized parametrically are suitable only for elementary considerations. Thus one may investigate all positions that lead to the end of the game in at most three moves. Such positions are characterized by two parameters, namely, the probability of a win for each of the two first moves.[3]

---

[3]E. O. Tuck, Doubling strategies for Backgammon-like games, *Journal of the Australian Mathematical Society* **21** (Ser. B), 1980, pp. 440–451. Tuck investigates, among

We turn now to positions in which the game lasts for a number of moves. We turn our attention to white's winning probability in a game governed solely by the dice. In the course of a game this probability changes move by move. At the start, we leave completely open the details of how these random changes occur. However, the probabilities must be compatible with the initial values for the various developments; that is, in relation to a later point in time, the average improvements and worsenings must be in balance, so that altogether the expected change is zero. One can now idealize the course of the game as a continuous process in which the winning probabilities change continuously. This backgammon model was first investigated in 1975 by Emmett Keeler and Lel spencer.[4] Figure 31.8 shows a typical game, during which the winning probabilities change continuously toward one of the two possible end values of 0 and 1. Two additional courses of the game are sketched in.

It has already been mentioned that from one point in time to another, the expected change in probability is equal to 0. However, changes about whose probability one is asking need not necessarily relate to two fixed points in time. For example, how probable is it that white starting from the 4/5 level reaches the 1 level without passing through the 1/5 level? Since one of the two levels will be passed through in any case, the sought-for probability p satisfies the equation

$$\frac{4}{5} = p \times 1 + (1 - p) \times \frac{1}{5}.$$

That is, $p = 3/4$, which is indicated in Figure 31.8 by the two arrows.

We have not yet commented on the meaning of the two levels indicated in Figure 31.8 by dotted lines, in which white wins with probability 1/5 or 4/5. We know already that a player should accept a double if his winning chances are at least 1/4. If black should play so that he declines a double precisely when white has a winning probability of at least 4/5, then white should not double before that; that is, she should not double below a winning probability of 4/5. This is so because this being a game with

---

other things, games that last at most three moves in which doubling according to the rules of backgammon is allowed. Thus, for example, the Jacoby paradox can be obtained systematically. Namely, white can double offensively (in more situations) if her probability of winning in one move increases. In the conditional probabilities that black wins on the second move the situation is reversed: with high values it is advisable for white to double less offensively. In regard to redoubling, the monotonic influence of the parameters can be violated if white wins on the first move with probability between 0.5 and 0.6 and black wins on the second move with conditional probability 0.75.

[4]Emmett B. Keeler, Joel Spencer, Optimal doubling in Backgammon, *Operations Research* **23**, 1975, pp. 1063–1071; see also **24**, 1976, p. 1179. Reprinted in David N. L. Levy, *Computer Games I*, New York 1988, pp. 62–70.

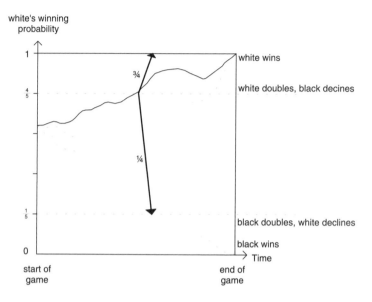

Figure 31.8.   Double or not? Accept or not? Games with continuous change in winning probabilities.

perfect information, one may assume without loss of generality that white knows black's level at which he declines a double. Thus to double below that level represents a risk for white. For in every game that she wins, the 4/5 level will certainly be passed, and in games that are lost, doubling is bad in any case.

In contrast to real backgammon, in which the winning probabilities develop discontinuously, in the model that we are considering there is no position in which the players are forced by optimal play on both sides to double and to accept that double. But what level is optimal for a doubling on the one hand, and for declining the double on the other? Is it really the level 4/5 as shown in the figure, and symmetrically the 1/5 level on the other side? The level that is best can be recognized in that white, regardless of whether black accepts or declines a double, has the same winning expectation. And in fact, such is the case for the 4/5 level and for no other:

- if black declines the double, then white wins one point.

- if black accepts the double, then the game develops at some time with probabilities 3/4 and 1/4 as determined above to the levels 1 and 1/5, respectively. In the first case, white wins a doubled bet, while in the second, black assures for himself with a redouble the same win level. In this way, white can expect altogether a win of

$$\frac{3}{4} \times 2 + \frac{1}{4} \times (-2) = 1.$$

Let us summarize: if the probabilities develop continuously during a continuously progressing game, then one should double and redouble precisely when one's winning chances reach 80%. That is also the bound above which the opponent should decline a double. Figure 31.9 shows the effect of these minimax strategies. There are shown three winning expectations depending on the current winning probability. The dotted line represents the winning expectation when no doubling is permitted at all. If white to move next is allowed to redouble, then her winning expectation increases according to the upper, heavy, line. In the reverse case the expectation worsens analogously. As one can see, there corresponds to possession of the doubling cube a constant value of half a wager, unless one player is too far ahead. Moreover, the doubling cube in any backgammon-type game can never have a value of more than half the current wager, since the advantage for any one player is limited by two factors: on the one hand, a player can at most double his winning expectation, while on the other, the achievable winning expectation can be limited by the opponent to 1.

At the end of a backgammon game, a single throw in the actual race can drastically change the odds. Such sudden changes have little to do with continuity, and so while the results that we have obtained are interesting in principle, in practice they have little to offer for endgames. Thus in a real game, the doubling cube seldom achieves its theoretical maximum value of 1/2. Much more realistic is the model in which each of the two players must move one piece over a possibly long stretch to the goal. Such model

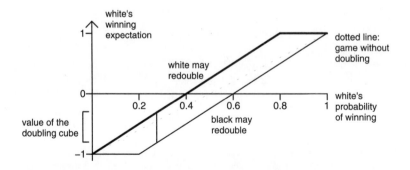

Figure 31.9. Winning expectations in a continuous game.

positions are representative for normal backgammon to the extent that one adds the fields to be traveled by the individual pieces from their original positions and then estimates how many dice points will be lost in bearing off.

In comparison to real backgammon, the two-stone model is much simpler in its combinatorial elements. The reason is the reduced number of possible positions, so that the actual race is completely removed from the players' influence. All that remains are the random influences and the decisions about doubling.

Every model position is characterized by two point numbers and the current state of the doubling cube. By symmetry, it suffices to investigate all positions under the assumption that white is to move. The winning expectations can be determined most easily recursively, where one stores the results obtained in a database[5] so that they can be accessed for use in the analysis of other positions. If one restricts consideration to the pure race part of the game, with no player allowed to double, then white's winning expectation from a position G can be derived from the formula

$$E_{none}(G) = -\sum_d P(d)E_{none}\left(\overline{G^d}\right).$$

Here G^d denotes, for a possible roll d, which appears with probability $P(d)$, the position attainable with that move, and the subscript on E_{none} indicates that none of the players doubles. To obtain again a position for which it is white's turn, the colors of the positions that arise are interchanged, which is indicated in the formula with the overbar. In changing colors, the signs of the expectations must be changed. If one is interested in the winning probabilities in the dice race, then one obtains them using the formula

$$P(G) = \frac{1}{2}(1 + E_{none}(G)).$$

To optimize the doubling strategy, each position must be analyzed three times, namely, depending on who is in possession of the doubling cube. In each case, a winning expectation is calculated based on the current level of the wager. For expectations E_{white} and E_{black} we assume that the exclusive right for the next redouble belongs to white or black, respectively. The expectation E_{both} represents the case in which the first double has not yet taken place, and both players have the right to double. If white is unable to

---

[5]With the model positions, the intermediate results are of course stored most naturally in arrays or in sets of random access files. But by the time one is investigating real backgammon positions, one should consider the use of indexed tables within relational databases. The coded positions are used as keys.

double or redouble, then her expectation is as above, where in the exchange of colors the right of the next redouble must be taken into account:

$$\text{E}_{\text{black}}(G) = -\sum_{d} P(d)\text{E}_{\text{white}}\left(\overline{G^d}\right).$$

In the other cases the corresponding formulas hold only for rhe case that white declines to double at his current turn or if the opportunity to double is unavailable:

$$\text{E}_{\text{white}}\left(G^{\text{roll}}\right) = -\sum_{d} P(d)\text{E}_{\text{black}}\left(\overline{G^d}\right),$$

$$\text{E}_{\text{both}}\left(G^{\text{roll}}\right) = -\sum_{d} P(d)\text{E}_{\text{both}}\left(\overline{G^d}\right).$$

Moreover, at the beginning of the move, the decisions by the players relating to doubling must be taken into account with regard to the minimax value:

$$\text{E}_{\text{white}}(G) = \max\left(\text{E}_{\text{white}}\left(G^{\text{roll}}\right), \min(1, 2\text{E}_{\text{black}}(G))\right),$$

$$\text{E}_{\text{both}}(G) = \max\left(\text{E}_{\text{both}}\left(G^{\text{roll}}\right), \min(1, 2\text{E}_{\text{black}}(G))\right).$$

Both cases are based on the same sequence of decisions, within which white either simply throws the dice or first doubles the wager. In the case of a double, black may choose the minimal winning expectation from his two choices. Either black declines, so that white wins the value 1, or white obtains double the bet that she could have achieved without access to the doubling cube.

Figure 31.10 shows how large white's advantage must be for doubling and redoubling to be worthwhile. Furthermore, one can see the degree of advantage at which it is better for black to decline a double or redouble.

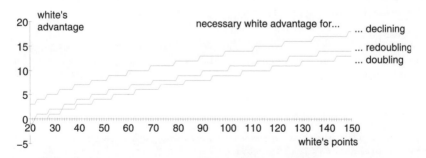

Figure 31.10. Two-stone model of backgammon: one piece per player.

Of note is the constant distance between the three curves. Although the rate of growth decreases to the right, one can take the average gradient of the curves in the region shown to be about $1/10$, where the basic values of -1, 0, and 3 are to be added. Thus for doubling in the backgammon model we have the following nearly optimal recommendations:

- with an advantage of at least 10%, that is, if the second player has at least 10% more field to traverse, then one should double or redouble.

- for doubling it suffices to have an advantage of one field fewer to traverse than the 10% value.

- if the advantage is greater than 10% plus three fields, then black should decline the double.

These laws were discovered in the mid-1970s in simulations and calculations.[6] Backgammon experts such as Crawford and Jacoby had earlier offered authoritative advantages of 7.5, 10, and 15%. Since then, the results for actual backgammon positions have found their way into the backgammon literature.[7] But how is one to determine the "correct" number of fields in a given backgammon position? That is, what position in the two-piece model reflects a particular backgammon position so that the probability distributions for further moves most nearly correspond? An approach to finding suitable model positions consists in approximating the number of lost dice points in bearing off. Edward Thorp, known for his work on blackjack, obtained the approximation formula

$$p + 2a + a_1 - b,$$

where p is the actual number of fields, a the number of pieces, a_1 the number of pieces on the first field, and b the number of occupied fields in the range 1 to 6. If one generates the model position for both players using this estimate, then good strategic recommendations within the limits of the model's criteria result.

Finally, we note that the minimal advantage numbers presented above behave much more regularly than the corresponding probabilities that would exist in a game based purely on dice. Depending on the distance to

---

[6] In addition to the work by Keeler and Spencer cited above, see Norman Zadeh, Gary Kobliska, On optimal doubling in Backgammon, *Management Science* **23**, 1977, pp. 853–858, reprinted in David N. L. Levy, *Computer Games I*, New York 1988, pp. 71–77. Additional investigations are reported in a review article by Edward O. Thorp, *Mathematical Reviews* **57**, 1979, #2594.

[7] Jeff Ward, *The Doubling Cube in Backgammon*, San Diego 1982; Bill Robertie, *Advanced Backgammon*, Arlington, MA 1983.

the goal, the following winning probabilities are necessary for a favorable redouble: 0.69 for 30 fields, 0.73 for 60 fields, 0.74 for 100 fields, and 0.75 for 150 fields.

As with the positions of the two-stone model, one may investigate, within limits, actual backgammon positions. There are additional difficulties posed by the various possible moves among which a player can choose given the values that appear on the dice.[8] Of course, white attempts to move to a position G^d that offers her maximal winning expectations, which corresponds to the following formulas, where with regard to doubling, the unrepresented minimax optimization takes place as in the two-stone model:

$$ \mathrm{E}_{\mathrm{black}}(G) = \sum_d \mathrm{P}(d) \max_{\text{choose roll } d} \left(-\mathrm{E}_{\mathrm{white}}\left(\overline{G^d}\right) \right), $$

$$ \mathrm{E}_{\mathrm{white}}\left(G^{\mathrm{roll}}\right) = \sum_d \mathrm{P}(d) \max_{\text{choose roll } d} \left(-\mathrm{E}_{\mathrm{black}}\left(\overline{G^d}\right) \right), $$

$$ \mathrm{E}_{\mathrm{both}}\left(G^{\mathrm{roll}}\right) = \sum_d \mathrm{P}(d) \max_{\text{choose roll } d} \left(-\mathrm{E}_{\mathrm{both}}\left(\overline{G^d}\right) \right). $$

To speed up the recursion it is recommended to analyze additional intermediate states beyond the positions G^{roll}. To this end, the dice results are divided into their individual values, which are then played one after the other. This is done formally by investigating positions in which white maintains a "credit balance" of one to four dice values during the current move.

Figure 31.11 shows the analysis of $207 \times 207 = 42\,849$ endgame positions. All positions are considered in which the expected number of throws to the end for white and black, though for each without considering the opponent, is at most 2.8, as it would be, for example, for a player with five pieces on the fields 1-2-3-3-4 with an expectation for the number of throws of 2.7979. Such expectations for the number of throws can be easily calculated recursively.

In the figure are shown four relative—depending on the throw expectation for white, which is exhibited on the horizontal axis, though not linearly—throw expectations for black:

- up to what point it is advantageous not to redouble;

---

[8] According to the rules, in bearing off, for each of the two (or possibly four) dice values, a piece must be moved such that the fewest possible dice points are lost. Otherwise, the order of moving is arbitrary. Therefore, a player whose three stones occupy fields 1, 3, and 5 who throws a 2-4 with the dice may first play the 2 to position 1-3-3 and then the 4 to 1-3. This offers him better chances on his next turn than first playing the four and then the two, which would by necessity result in the position 1-1-1.

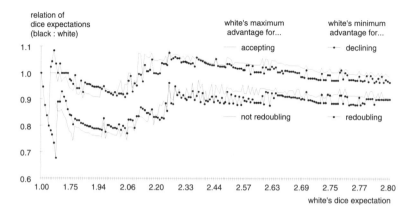

Figure 31.11. Optimal redouble: greatest and least advantage for white in relation to the expected number of throws.

- from what point a redouble can be advantageous;

- up to what point the acceptance of a redouble can be advantageous;

- from what point the acceptance of a redouble can be advantageous.

To keep things clear, for the first two points we have not shown the corresponding results for the first double.

If the optimal redoubling behavior could be determined from the two throw expectations alone, then the two curves above and the two below could be replaced by a single curve for each pair. That such simple criteria cannot be found can be seen from the right part of the figure. There we see that the curve marked with points often runs below its partner curve, and each place represents a pair of positions for which the optimal doubling strategy and the roll expectations are not in their usual relationship; that is, as in the case of the Jacoby paradox, an increase in the black roll expectation can allow a more defensive redoubling behavior to become optimal.[9] Since such occurrences are the exception, one can essentially characterize every position with the two relatively simply calculated, or through simulation approximated, roll expectations.[10] To such an extent, Figure 31.11

---

[9]However, in contrast to the Jacoby paradox, the two positions differ in relation to the placing of the white pieces; only the roll expectation must agree.

[10]Using statistical analysis, Jim Gillogly found the following asymptotic formula for the roll expectation:

$$0.603 + 0.1014(p + 2a + a_1 - b).$$

See the review article of Edward O. Thorp mentioned above.

stands in direct analogy to Figure 31.10, where corresponding assertions for the two-piece model can be made on the basis of the two numbers of fields remaining to be traversed.

Further Mathematical Investigations into Backgammon

[1] Norman Zadeh, On doubling in tournament Backgammon, *Management Science* **23**, 1977, pp. 986–993.

[2] E. O. Tuck, Simulation of bearing off and doubling in backgammon, *The Mathematical Scientist* **6**, 1981, pp. 43–61.

[3] Edward O. Thorp, End positions in backgammon, *Gambling Times*, 1978, October, November, December; reprinted in David N. L. Levy, *Computer Games I*, New York 1988, pp. 44–61.

[4] Edward Thorp, *The Mathematics of Gambling*, Hollywood 1984, pp. 83–109.

32

Mastermind: Playing It Safe

What is the quickest way to crack the code in Mastermind? How many turns are sufficient to decode an arbitrary four-place, six-color code?

The commercially marketed game Mastermind was one of the most successful games of the 1970s. Imitating the English pencil and paper game "bulls and cows," Mastermind was invented in 1973 by the Parisian Israeli Marco Beirovitz. Within a few years, over ten million of the games were sold.[1]

Mastermind is a two-person game of logic. At the start of the game, one player chooses a color code; this is his sole active decision for the entire game. In particular, he places a specified number n of colored sticks in a row, hidden behind a shield from the opponent. There are k colors from which to choose, and each color may be used more than once or not at all, so that there are k^n possible codes. In the usual variants we have $k = 6$ and $n = 4$ or $k = 8$ and $n = 5$.

The challenger attempts to break the secret code in the fewest possible guesses, where a guess consists of a code, after which the first player supplies the following information about the number of correct matches:

- first, the number of true "hits," that is, the number of sticks whose color and position are correct. For each such hit the encoder places a black stick, but without revealing which colored sticks these relate to.

[1]David Pritchard, *The Family Book of Games*, Brockhampton Press 1983, pp. 190 f.; Rüdiger Thiele, *Das grosse Spielevergnügen*, Leipzig 1984, p. 210; Erwin Glonnegger, *Das Spiele-Buch*, Munich 1988, p. 228. Practical tips for playing can be found in Leslie H. Ault, *The Official Mastermind Handbook*, New York 1976.

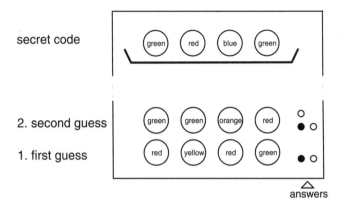

Figure 32.1. The first two guesses in a game of Mastermind.

- the second number is the number of colored sticks that would form additional hits if they were permuted to different positions. The encoder indicates this by placing a suitable number of white sticks.

Figure 32.1 should help to clarify how the game works.

In the game of Mastermind, logical conclusions must be drawn with the utmost precision: which colors does one know appear in the secret code? Which appear more than once? Which colors can be definitively excluded? Can a particular colored stick be identified both as to color and location? How should one frame one's next guess so as to obtain as much new information as possible? It is only with answers to such questions that the large number of possible codes can be addressed, so that out of the entirety of $6^4 = 1296$ or even $8^5 = 32\,768$ possible codes the correct one can be found in a small number of turns.

We would like now to attempt to perfect our Mastermind playing strategy. First, however, we must establish the boundary conditions and the criteria to be optimized. These are closely connected to the category of game to which Mastermind belongs. From a game-theoretic point of view is it a one- or two-person game. Can Mastermind be formulated as a game with perfect information? Three completely different approaches are possible. All three models agree in their determination of what it means to win, namely, that for each guess, the decoder has to pay one unit to his opponent:

- in the sense of a worst-case approach, one can seek a strategy that guarantees the smallest possible bound on the number of moves to crack any code. One might imagine that the encoder can cheat, changing the code in the middle of the game while keeping it in conformity with the answers already given. This means that the encoder can answer arbitrarily, provided that his answers do not contradict those already given. In this interpretation, Mastermind becomes a nonrandom two-person game with perfect information, whose minimax value can be calculated.

- one can also interpret Mastermind as a one-person game, in which the single decision of the encoder is replaced by a random process. Here one might decree, though it is not necessary, that each code appear with equal probability. With knowledge of the probability distribution employed, the guesser then seeks a strategy that minimizes the expected number of turns. This is an average-case optimization, which can be solved with the methods of probability theory. Minimax techniques play no role here.

- a realistic Mastermind analysis may not, in contrast to the first two approaches, overlook the role of the encoder and his strategic influence. Thus one considers Mastermind a two-person game, of course without perfect information. One seeks strategic considerations such as those known from poker: how do I size up my opponent? What guesses does he think I am least likely to make?

Common to all theoretic Mastermind analyses is that the deductive elegance of the game is dispensed with. Instead of artfully drawing logical deductions between the individual answers, one operates with a simple but universal sorting mechanism. That is, in accord with the motto "quantity instead of quality," one does precisely what a player avoids at all cost: one checks all possible codes to see whether they are in conformity with the knowledge thus far obtained. Future moves are planned in this way, in that one tests the set of remaining possible codes as to how it will grow smaller under the planned guesses and the possible replies to them.

If one collects all k^n codes into a set C_0, then every intermediate state, that is, every position, can be formally characterized by a subset $C \subseteq C_0$ that contains all codes that are in accord with the knowledge thus far obtained. In principle, Mastermind could now be investigated by "simply" examining all subsets of codes. But there are too many of them! In the case of the relatively small 6^4 Mastermind there are 2^{1296} subsets. One therefore restricts one's investigation to the subsets that actually correspond to a position. One therefore defines sets of the form $C(q, a)$, where such a set contains all codes that give the answer a to the question q. Thus every

Mastermind position corresponds to a subset of the form

$$C = \bigcap_{t=1}^{s} C(q_t, a_t).$$

This set represents the situation in which s guesses $q_1, \ldots, q_s$ have received the replies $a_1, \ldots, a_s$. At the latest, when the intersection of all remaining codes is reduced to a single element, the decoder can prepare his last move, for which he is guaranteed a complete set of black sticks.

For the decoder it is natural, even if not necessarily optimal, to make a guess in such a way that the size of the resulting code subset is as small as possible in the worst case. For 6^4 Mastermind this technique was first investigated in 1976 by Donald Knuth.[2] Before we look at his results, let us agree on a simpler notation, even though we have to sacrifice the visual aesthetic of the game: the colors are denoted by the numbers $1, 2, \ldots, k$, so that, for example, "3221" is a code of length 4. The replies consist of two numbers bw, the numbers of black and white sticks displayed. In 6^4 Mastermind there are 14 possible answers:

$$04, 03, 02, 01, 00; \quad 13, 12, 11, 10; \quad 22, 21, 20; \quad 30; \quad 40.$$

We observe that the reply 31, that is, three black and one white stick, is impossible.

Let us consider the first guess in a game of 6^4 Mastermind. Up to symmetry, there are five possible opening moves: 1111, 1112, 1122, 1123, and 1234. The left-hand portion of Table 32.1 shows how many codes remain after the first move, depending on the answer.[3]

As one sees in Table 32.1, the guess 1122 guarantees in the worst case the greatest reduction in the code set. At most 256 codes remain. "Fortunately," thus Knuth, one can continue this opening so that every code is cracked in at most five moves, being confirmed with four black sticks. Which code one should guess second is seen in the right-hand part of the table. There are tabulated both the guess and the resulting reduction in the number of codes, in the form of the number of codes that are still possible after the second guess. Knuth's description of the entire strategy takes two pages, and a program based on the above table should not be too difficult to write. Furthermore, Knuth notes that his strategy is not optimal in the average number of moves, but it is almost optimal.

---

[2]Donald E. Knuth, The computer as master mind, *Journal of Recreational Mathematics* **9**, 1976/77, pp. 1–6.

[3]The left part of the table can be found in Robert W. Irving, Towards an optimum Mastermind strategy, *Journal of Recreational Mathematics* **11**, 1978/79, 81–87.

| | First Guess | | | | | Second Guess | |
Answer	1111	1112	**1122**	1123	1234	after **1122**	
04			**1**	2	9	2211	done
03			**16**	44	136	1213	4
02		61	**96**	222	312	2344	18
01		308	**256**	276	152	2344	44
00	625	256	**256**	81	16	3345	46
13				4	8		
12		27	**36**	84	132	1213	7
11		156	**208**	230	252	1134	38
10	500	317	**256**	182	108	1344	44
22		3	**4**	5	6	1213	1
21		24	**32**	40	48	1223	6
20	150	123	**114**	105	96	1234	20
30	20	20	**20**	20	20	1223	5
40	1	1	**1**	1	1	done	

Table 32.1. Codes remaining after the first two moves.

The proof that five guesses in 6^4 Mastermind always suffice is long, but it is easily checked due to its constructive nature. But what about the negation of this statement? Is there perhaps a better strategy that would always work with four guesses? It is easy to see that this is not the case. We make use of a relatively general investigation of sequential search games, as given in 1979 by Viaud.[4] Viaud begins his estimates with the number α of different answers that can result from a guess. In the case of 6^4 Mastermind, $\alpha = 14$. Every additional guess divides the set C of still possible codes into α mutually disjoint subsets. One of these, namely, the one with the correct code, contains a single element. Therefore, at least one of the $\alpha - 1$ remaining sets contains at least

$$\frac{|C| - 1}{\alpha - 1}$$

elements. If one applies this result repeatedly, then one sees that with more than

$$(\alpha - 1)^m + \cdots + (\alpha - 1)^2 + (\alpha - 1) + 1 = \frac{(\alpha - 1)^{m+1} - 1}{\alpha - 1}$$

codes one requires at least $m + 2$ guesses to be sure of breaking the code. Therefore, with 184 or more codes one needs at least four guesses in 6^4

---

[4]D. Viaud, Une formalisation du jeu de Mastermind, *R.A.I.R.O. Recherche opérationnelle/Operations Research* **13**, 1979, pp. 307–321.

Mastermind, since $\alpha = 14$. Furthermore, we see from Table 32.1 that with the first move in 6^4 Mastermind it is impossible to make a reduction to fewer than 256 codes. Therefore, after the first move in 6^4 Mastermind one needs at least an additional four, thus five altogether. Knuth's strategy is therefore optimal in the sense of worst case.

Knuth's worst-case optimal strategy is also very good if one is seeking an *average* of very few moves to break the code, on the assumption, of course, that the encoder selects a code randomly from among the $6^4 = 1296$ possible codes. With Knuth's strategy the decoder needs on average $5804/1296 = 4.478$ guesses. That this is not optimal was shown by Irving in 1978, when he produced a strategy that reduced the expected number of turns to $5662/1296 = 4.369$. The first guess of type 1123 already differs from Knuth's strategy. From Table 32.1 one can see the plausibility of the approach: although after the 1123 opening there could remain as many as 276 codes in play, these codes are divided into 14 parts, and not the mere 13 that arise from the play 1122. The sets of remaining codes are therefore smaller on average. Like Knuth, Irving oriented his strategy toward seemingly plausible criteria. As a measure he took the expected number of remaining codes. For example, after the 1111 opening, this average is

$$\frac{625}{1296} \times 625 + \frac{500}{1296} \times 500 + \frac{150}{1296} \times 150 + \frac{20}{1296} \times 20 + \frac{1}{1296} \times 1 = 511.98.$$

A much smaller number, namely, 185.27, results when one starts with the code 1123. Irving optimizes the second turn in the same way. The additional turns are so obvious that they can be investigated explicitly and so can be directly optimized.

A few years after Irving, the Viennese statistician Neuwirth investigated, among other things, strategies by which a player always guesses only those codes that have not yet been excluded; that is, those for which one may realistically hope that the reply will be a full set of black sticks.[5] A danger arising from this limitation is that good strategies can be rejected. On the other hand, the remaining set of strategies can be more easily dealt with. Despite this limitation, it turns out to be too difficult for 6^4 Mastermind using currently available computing power, and Neuwirth thus had to restrict his attention to the 5^4 variant. Using a combination of various approaches, however, he was able to improve a bit on Irving's result, namely, to an average of $5656/1296 = 4.364$ guesses.

---

[5]Erich Neuwirth, Some strategies for Mastermind, *Zeitschrift für Operations Research* **26**, 1982, pp. B257–B278.

Finally, the search for an optimal average-case strategy was concluded in 1993 by Kenji Koyama and Tony Lai.[6] With their strategy, which was found using complete optimization, the decoder succeeds in breaking the code in $5625/1296 = 4.340$ turns on average. In the worst case, however, six guesses are necessary. A small modification of the strategy limits the number of guesses to at most five, but then the average grows to $5626/1296 = 4.341$.

But what happens if the encoder does not behave as we have specified? In order to ensure that the code is selected with equal probability from among the 6^4 choices, the encoder refrains from taking an active role in the game, even if all he can do is choose from among only five different codes, up to symmetry. Would he not be better off leaving his opponent in the greatest possible uncertainty? For such strategic uncertainty, such as is found in poker, we have not yet developed a mathematical concept. We will therefore return to Mastermind at a later point in this book.[7]

Black Box: Mastermind of Molecules

Great economic successes like Mastermind are frequently taken as cause to offer similar "successor" games. The English game Black Box was marketed in Germany under the names Ordo and Logo, where it attained little success.[8] The game, invented by Eric Solomon, contained elements of the games Mastermind and Battleship, but it is nonetheless an interesting game in its own right.

The game begins by a player "hiding" a "molecule" consisting of four or five "atoms" on an 8×8 playing board. He does this by simply making a cross in the relevant squares on a marked piece of paper, which is of course hidden from the opponent. As with Mastermind, the opponent is the only active player in the game. This player obtains information by shooting "x-rays" from the edge of the board, where such a ray can be reflected or absorbed, or else it simply exits the playing field at another point. This, and only this, information is given by the molecule

[6]Kenji Koyama, Tony W. Lai, An optimal Mastermind strategy, *Journal of Recreational Mathematics* **25**, 1993, pp. 251–256.

[7]See Chapter 44.

[8]Werner Fuchs, *Spieleführer 1*, Herford 1980, p. 101; David Pritchard, *The Family Book of Games*, Brockhampton Press 1983, p. 195.

builder. The following illustration shows on the left some paths of x-rays, and on the right the conclusions that can be drawn; in particular, it is certain that no atom is located on the squares marked with a cross (×):

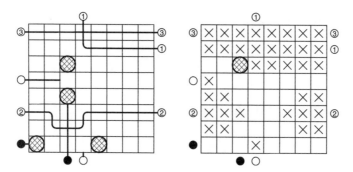

In general, an x-ray is always absorbed if it would hit an atom in the next square. Otherwise, the x-ray proceeds in a straight line or, if it encounters one or more atoms in a neighboring diagonal square, its path is altered. The direction of alteration is determined by the concentric "force field" that every atom emits, by which the angle of incidence is equal to the angle of reflection. The x-ray is directly repelled by two atoms or by a single atom on the boundary.

The player who is guessing does not, of course, see the atoms or the exact path of the x-rays. He is simply informed as to the point of exit, whether and where it occurs. This occurs in the form of the traces left on the boundary. In the diagram, black circles indicate absorption, white circles reflection, and the remainder pairs of entry and exit locations.

Black Box can be played on a checkerboard. Even better is a reversi game, since one can use both sides of the playing pieces to indicate proven or excluded locations of atoms.

Black Box can be analyzed mathematically in a manner similar to that of Mastermind. That is, one sorts out the successively refuted possible molecules. In comparison to Mastermind, Black Box is much more complex, since the molecule builder has greater scope for a larger number of asymmetric constructions. There are some quite refined constructions that can easily lead an uninitiated player to some very wrong conclusions. An example is the x-ray denoted by a circled 2 in the diagram, which is diverted several times. A further difference from Mastermind is that some molecules cannot be determined uniquely, in which case it suffices to produce an equivalent molecule.

Further Mathematical Publications on Mastermind

[1] V. Chvátal, Mastermind, *Combinatorica* **3**, 1983, pp. 325–329.

[2] D. Viaud, Une stratégie générale pour jouer au Master-mind, *R.A.I.R.O. Recherche opérationnelle/Operations Research* **21**, 1987, 87–100.

[3] Merill M. Flood, Sequential search strategies with Mastermind variants, *Journal of Recreational Mathematics* **20**, 1988, pp. 105–126, 168–181.

[4] H. P. Wynn, A. A. Zhigljavsky, J. H. O'Geran, Search methods and observer logics, *Fifth Purdue International Symposium on Statistical Decision Theory and Related Topics*, 1992, pp. 533–535.

[5] J. H. O'Geran, H. P. Wynn, A. A. Zhigljavsky, Mastermind as a test-bed for search algorithms, *Chance* **6**, 1993, pp. 31–37.

In the articles by Chvátal and Viaud, general strategies for k^n Mastermind are investigated, including the version without repetition of colors. In the two articles of Wynn et al., general search strategies are discussed with reference to Mastermind.

Part III

Strategic Games

33

Rock–Paper–Scissors: The Enemy's Unknown Plan

If two persons wish to decide who is to pick up the check at a restaurant, one option available is to play rock–paper–scissors. In this game the two players have the same moves and the same odds of winning. Unlike symmetric two-person games with perfect information, one cannot know in advance what move will lead to disaster. What is one to do?

We have already analyzed the first two among the three causes of uncertainty in games mentioned in the preface, namely, chance, combinatorial explosion, and differing levels of information. We have not yet explored the issue of the uncertainty presented to a player who does not know everything that his opponent knows. We therefore would like to look at games without perfect information, also known as games with *imperfect information*.

Rock–paper–scissors is a game without random influences whose combinatorial complexity is trivial. The entire uncertainty of the game resides in the lack of perfect information, that is, in the fact that the two players must move simultaneously, each of them without knowledge of what move the opponent has chosen. Each of the three possible moves can lead to a loss: the rock is captured by the paper, the scissors by the rock, and the paper by the scissors. The only way of being sure of avoiding defeat is to be able to guess in advance the opponent's plan. Then, of course, one would always have a winning move.

An evaluation of the opponent's psychology can be very useful in games like this one; one need think only of poker, which one knows about, if not from first-hand experience, then from any number of Hollywood films. Does the opponent really have as good a hand as it seems? Or is he bluffing? That is, can one believe that an opponent would have made his previous bets with a worse hand than the one that we possess? An important role is played by what has happened already. How risk-averse does our opponent seem? How poker-faced? How has the opponent played in previous games? The chains of tactical reasoning that even a simple game can unleash can be sampled in Edgar Allan Poe's 1845 story, "The Purloined Letter." In this story a schoolboy is depicted who achieves great success in the game "even or odd," which is similar to rock–paper–scissors:

> This game is simple, and is played with marbles. One player holds in his hand a number of these toys, and demands of another whether that number is "even or odd." If the guess is right, the guesser wins one; if wrong, he loses one. The boy to whom I allude won all the marbles of the school. Of course he had some principle of guessing; and this lay in mere observation and a measurement of the astuteness of his opponents. For example, an arrant simpleton is his opponent, and, holding up his closed hand, asks, "Are they even or odd?" Our schoolboy replies, "odd," and loses; but upon the second trial he wins, for he then says to himself, "The simpleton had them even upon the first trial, and his amount of cunning is just sufficient to make him have them odd upon the second; I will therefore guess odd"; he guesses odd, and wins. Now, with a simpleton a degree above the first, he would have reasoned thus: "This fellow finds that in the first instance I guessed odd, and, in the second, he will propose to himself upon the first impulse, a simple variation from even to odd, as did the first simpleton; but then a second thought will suggest that this is too simple a variation, and finally he will decide upon putting it even as before. I will therefore guess even." He guesses even, and wins. Now this mode of reasoning in the schoolboy, whom his fellows termed "lucky"—what, in its last analysis, is it?

The procedure by which the successful schoolboy assuredly places his opponents in various categories of simpleton to achieve victory must remain—as is typical with Poe—a mystery. There seems to be no way of transforming such a process into a mathematically formulatable algorithm. Or is there? Let us consider. In "even or odd" there exists, as with rock–paper–scissors, a certain symmetry among the various moves; that is, there are no better or worse moves. In contrast to rock–paper–scissors, however, the game "even or odd" is not symmetric, since a draw does not result if the two players choose the same strategy.

Much greater importance must be attached to the strategy by which a move is chosen than to the move itself. Poe's schoolboy takes the measure of his opponent and considers as well his actions in previous games. But is there an opposing strategy to counter this successful play? Or is one without defense against such a brilliant player, with no choice save that between ruin and quitting the game?

In fact, there is a simple method of thwarting the psychological genius. Instead of trying to figure out which of our thoughts our opponent can guess, we let lady luck choose our moves for us. For example, in "even or odd" we can cut a shuffled pack of cards and if the card is red, take one marble, and if black, two. Of course, we must keep the color of the cards hidden from the opponent, but we do not have to hide our intention of choosing our move according to this procedure.

What is achieved with this trick of leaving the decision to chance? Our opponent can psychologize as long and as cleverly as he wishes, but it will help him not at all, because when all is said and done, he is playing a pure game of chance. However he chooses his moves, he will win a marble with probability $1/2$ and will therefore break even in the long run. Our trick of randomization has met all the requirements of defensive play, though at the cost of having no advantage ourselves when playing against an opponent whom we can see through, since our strategy takes no information about the opponent into account.

To transfer this technique to other games, we will describe it more formally. We will use the normal form that we met in Chapter 18. There we saw that every two-person zero-sum game can be represented by a table, though it might be a very large table indeed. Namely, if a player decides his entire strategy for a game in advance, then the game is reduced to one gigantic double move, and the course of the game becomes equivalent to that of rock–paper–scissors or "even or odd." In particular, before the game begins, a player must decide how he will respond to every possible game configuration that may present itself. This may prove completely unrealistic in practice. However, for the theory all that is important is that the game remain substantially unchanged by such a modification. Von Neumann and Morgenstern (1902–1977), the founders of game theory, have this to say on the subject:[1]

> Imagine now that each player ... instead of making each deci-
> sion as the necessity for it arises, makes up his mind in advance

---

[1] John von Neumann, Oskar Morgenstern, *Theory of Games and Economic Behavior*, Princeton 1944.

for all possible contingencies, i.e., that the player ... begins to play with a complete plan, a plan which specifies what choices he will make in every possible situation, for every possible actual information which he may possess at that moment ... We call such a plan a strategy.

Observe that if we require each player to start the game with a complete plan of this kind, i.e., with a strategy, we by no means restrict his freedom of action. In particular, we do not thereby force him to make decisions on the basis of less information than there would be available for him in each particular instance in an actual play. This is because the strategy is supposed to specify every particular decision only [in dependence on] ... actual information which would be available for this purpose in an actual play. The only extra burden our assumption puts on the player is the intellectual one to be prepared with a rule of behavior for all eventualities—also he is to go through one play only. But his is an innocuous assumption within the confines of a mathematical analysis.

In contrast to most other games, in which the normal form is of purely theoretical interest due to the combinatorial explosion, the normal forms of rock–paper–scissors and "even or odd" are extremely simple. Let the player trying to guess his opponent's move take the role of black. Then the entries of the normal form, as shown in Table 33.1, correspond to black's score against his opponent white.

The game possesses no saddle point, which, by Zermelo's theorem, always exists for games with perfect information. Thus each player must strive to keep his or her strategy secret from the opponent. A player who knows his opponent's strategy can always win. From the point of view of white's score, in the terminology of maximin and minimax values as described in Chapter 18, this has the following significance:

- the maximin value, which is the greatest value that white can be sure of achieving through her own efforts, is equal to -1.

			Black Guesses	
			Odd	Even
			1	2
White	Odd	1	-1	1
Chooses	Even	2	1	-1

Table 33.1. The normal form of "even or odd": white chooses, black guesses.

- the minimax value, which is the lowest value to which black can be
  sure of limiting white's score, is equal to $+1$.

Since each player can force his or her value independent of the opposing
stragegy, this value is not worsened if the opponent chooses his strategy
via a random process. On the other hand, it is entirely possible to improve
one's own value with this trick. Let us see what happens if white does not
choose a fixed strategy but decides instead to make her choice randomly
in the proportion $1 : 1$. The game proceeds randomly, depending on the
opponent's strategy. That is, as with a pure game of chance there is no
fixed score but only a probability distribution for the score, where in the
mathematical analysis we treat the expected value like a fixed score in
that amount. One can thus formulate the random selection of the strategy
actually used as a new strategy, called a *mixed strategy*, and we can thus
extend the normal form to include the corresponding game values for white:

			Black Guesses Odd	Even
			1	2
White	Odd	1	-1	1
Chooses	Even	2	1	-1
	$1:1$ Random Choice	3	0	0

Of course, white could decide on another weighting of the two basic
strategies, usually called *pure strategies*. If the strategy "odd" is chosen
with probability p, and the strategy "even" with probability $q = 1 - p$,
then the following normal form results:

			Black Guesses Odd	Even
			1	2
White	Odd	1	-1	1
Chooses	Even	2	1	-1
	$p:q$ Random Choice	3	$1-2p$	$2p-1$

In this normal form as well, the additional game values are expectations,
which again are treated like fixed scores. If white decides on a mixed
strategy in the relation $p : q$, then her winning expectation is $1-2p$ or $2p-1$,
depending on how black moves. Except for the case $p = 1/2$, one of the two
values is always negative, so that white must fear a negative expectation

should black choose a winning strategy. This makes it clear that white
can defend herself against negative expectations only by choosing a mixed
strategy in the proportion 1 : 1.

Of course, black can also choose mixed strategies by making his move
according to an established probability distribution. However, black will
not succeed in attaining a positive expectation against white's 1 : 1 mixed
strategy. If one assumes no fixed strategy in advance for white, then the
1 : 1 strategy is black's only chance of avoiding the risk of an expected loss.

The 1 : 1 mixed strategies thus form a *saddle point* in the game extended to mixed strategies, as is guaranteed for every two-person zero-sum
game with perfect information even in the form of a combination of two
pure strategies. Minimax and maximin values thus agree in the extended
game and are equal to zero. The strategies themselves are called minimax
strategies or simply optimal strategies. Here the word "optimal" is applied
in the sense of the worst case, that is, the best possible defense in which
the total risk is minimized:

			Black Guesses		
			Odd	Even	1 : 1 Random
			1	2	3
White	Odd	1	−1	1	0
Chooses	Even	2	1	−1	0
	1 : 1 Random Choice	3	0	0	**0**

As we have already noted, once found, the value of a saddle point is not
altered by additional mixed strategies. To that extent, a game of "even or
odd" extended to mixed strategies has attained a certain stability, such as
we have seen with games with perfect information. Each of the two players
can make his or her mixed strategy, that is, the proportion of choices,
known in advance to the opponent without fear of being at a disadvantage.
From this point of view, the difference between such a game and chess is
only in the influence of chance; that is, as with backgammon or memory, all
the propositions apply to the winning expectations, and not to the actual
results of an individual game.

In the game we are considering the saddle point is unique. As we have
seen in white's case, every mixed proportion other than 1 : 1 represents a
risk of loss for the player if the opponent counters appropriately. A careful
player will not trust opponent error in formulating a strategy, but as with
chess will put up the best possible defense.

In the game rock–paper–scissors, a player can also use a mixed strategy
to overcome any psychological advantage that the opponent may possess.

			Paper	Rock	Scissors
			Black Chooses		
			Paper	Rock	Scissors
			1	2	3
White	Paper	1	0	1	−1
Chooses	Rock	2	−1	0	1
	Scissors	3	1	−1	0

Table 33.2. Normal form of the game rock–paper–scissors.

As expected, the double symmetry of the game, namely, that between the two players and among the three possible moves, can be seen in the result:

- in the original game, in which only pure strategies are allowed, the maximin value for white is −1, and the minimax value is 1.

- in the game extended to mixed strategies, the maximin and minimax values are both zero.

- the optimal strategies choose each of the three possible moves equiprobably, with probability 1/3.

All three of these statements can be verified at once with a look at the normal form displayed in Table 33.2.

The game rock–paper–scissors was first investigated formally from this point of view in 1924 by Émile Borel,[2] one of the founders of modern probability theory,[3] who at that time was also a member of the French parliament and indeed, one year later, was briefly the naval minister in the cabinet of his mathematician colleague Paul Painlevé (1863–1933). Borel was the first to discover, in 1921, the advantage of mixed strategies and also the first to recognize the normal forms as a universal description of games.[4] If white and black mix their strategies with probabilities p, q, r and u, v, w, respectively, then white obtains a winning expectation in rock–paper–scissors of

$$(r - q)u + (p - r)v + (q - p)w.$$

This formula shows us that white is protected against a negative expectation only if the three numbers $r - q$, $p - r$, and $q - p$ are equal to zero.

---

[2]Émile Borel, Sur les jeux où interviennent le hasard et l'habileté des jouers, in: *Théorie des Probabilités*, Paris 1924; English translation: On games that involve chance and the skill of the players, *Econometrica* **21**, 1953, pp. 101–127.

[3]See also Chapter 5.

[4]Émile Borel, La théorie du jeu et les équations intégrales à noyau symétrique, *Comptes Rendus de l'Académie des Sciences* **173**, 1921, pp. 1304–1308; English transation: The theory of play and integral equations with skew symmetric kernels, *Econometrica* **21**, 1953, pp. 97–100.

Otherwise, among the three numbers, whose sum is zero, there must be at least one that is negative, so that black can force a negative expectation on white with a well-targeted pure strategy.

Needless to say, Borel's true interest was not the game rock–paper–scissors. Among other things, he was interested in the question whether in baccarat it is advisable to draw another card from the value 5.[5] He was looking for a general method whereby a player in a symmetric two-person zero-sum game can prevent a negative expectation:[6]

> Let us consider a game in which the score depends both on chance and the skill of the players. Let us confine ourselves to the case of two players, A and B, and a game symmetric in the sense that if A and B adopt the same method of play, then their chances are equal. One may propose to investigate whether it is possible to determine a method of play better than all others, i.e., one which gives the player who adopts it a superiority over the player who does not adopt it. Let us first define what we should understand by a method of play. It is a code that determines for every possible circumstance (supposed finite in number) exactly what the person should do.

The restriction to *symmetric games* is thus of great practical utility:

- on the one hand, one knows that neither player can be assured of a positive expectation. If one were to find a mixed strategy that could guarantee a player an expectation of at least 0, then such a strategy must be optimal. If the opponent employs the same strategy, then these two strategies combine to form a saddle point at which both players have expectation 0.

- on the other hand, this restriction causes two possible problems. Most games are asymmetric, even if due only to who moves first. Nonetheless, every game can be seen as part of a symmetric game. Namely,

---

[5] In the second part of the article, "Sur les jeux ...," cited above, Borel refers to Joseph Bertrand, who mentions this question in his textbook on probability theory, *Calcul des probabilités*. Bertrand's book first appeared in 1899. The study of baccarat is described in Chapter II, 33, Problème XIX (second edition, Paris 1907, reprinted New York 1972). Bertrand compares the options for both the player and the bank, assuming that each knows the other's strategy. We will return to Baccarat in Chapter 42.

[6] In his 1921 work mentioned above, Borel restricts his attention to games that can only be simply won or lost (score of $+1$ or -1). It is within this limitation that the winning probabilities are investigated.

one simply plays, as Borel noted,[7] two games with roles interchanged. It is even simpler to play a single game and draw lots to determine who gets which role.

Borel's first success was with symmetric games in which each player possesses three strategies, and then in 1924 with games of five strategies. In particular, he began with an arbitrary 3×3 game, which always has a normal form of the following type:

| | | Black Chooses | | |
		1	2	3
White	1	0	a	$-b$
Chooses	2	$-a$	0	c
	3	b	$-c$	0

Borel constructed a probability distribution depending on the arbitrary values a, b, c with which white can mix her strategies in such a way that she is protected from a negative expectation. That is, if white mixes her three strategies with probabilities p, q, r, then the winning expectations against black must be at least 0. This corresponds to the conditions

$$-aq + br \geq 0,$$
$$ap - cr \geq 0,$$
$$-bp + cq \geq 0.$$

Of course, the conditions $p \geq 0$, $q \geq 0$, $r \geq 0$, and $p + q + r = 1$ must be satisfied as well.

To arrive at the desired probabilities we must investigate a number of cases, depending on the signs of a, b, and c. If none of the numbers a, b, c is negative and at least one of them nonzero, then white can use the following probabilities:

$$p = \frac{c}{a+b+c}, \qquad q = \frac{b}{a+b+c}, \qquad r = \frac{a}{a+b+c}.$$

There does not appear to be a generalization of the procedure designed for the special situation of 3×3 games, and Borel became convinced that

---

[7]In a footnote by Émile Borel, Sur les systèmes de formes linéaires à déterminant symétrique gauche et la théorie générale du jeu, *Comptes Rendus de l'Académie des Sciences* **184**, 1927, pp. 52–53; English translation: On systems of linear forms of skew symmetric determinant and the general theory of play, *Econometrica* **21**, 1953, pp. 116–117. We will discuss in detail the construction of symmetric versions of games in Chapter 36.

such a mixed strategy does not exist for every symmetric game. Yet a game in which no player can be assured of a nonnegative expectation exhibits a rather specific nature. For in such games, in contrast to a game with perfect information or one of the 3 × 3 games that we have looked at like rock–paper–scissors, it is necessary for a successful game to make a psychological evaluation of the opponent. Borel formulated the consequences thus in 1921:

> Since this is the situation, whatever variety is introduced by A into his play, once the variety is defined, it will be enough for B to know it in order that he may vary his play in such a manner as to have an advantage over A. The reciprocal is also true, whence we should conclude that the calculation of probabilities can serve only to facilitate elimination of bad manners of playing ...for the rest, the art of play depends on psychology and not on mathematics.

Has mathematics, then, reached its limit on simple two-person zero-sum games? Or can a saddle point of mixed strategies be found for every two-person zero-sum game? If not, then how far can the difference between the minimax and maximin values be reduced using mixed strategies? We will have more to say about these questions in the next chapter.

34

Minimax Versus Psychology:
Even in Poker?

Two players play two games of poker, where each player gets to open in one of the games. Can a player vary his strategy randomly in such a way as to prevent a loss on average?

With this question we make concrete the suite of problems posed at the end of the previous chapter. We are not going to get into the details of poker, and thus we are not looking for an explicit strategy. Rather, we would like to begin with an investigation of the question whether mixed strategies are worthwhile. Can a negative expectation be prevented in our poker game, independent of psychological factors, just as in two games of chess with alternating colors a player can at least theoretically break even? That is, can one, depending on the state of one's own information, which includes in particular one's own hand and the previous bids, vary one's manner of play randomly in such a way that guarantees an expectation of at least zero?

Poker offers us a typical example, since the property of imperfect information characterizes the game. Each player knows only his own cards and attempts to draw conclusions about the opponents' play: is the opponent's hand really as good as it would appear based on his previous bids? Or is one's own hand good enough to raise the bet, on the assumption that it is the best hand in the game?

Only shortly after Borel's work, and without knowing his pessimistic prognosis, a much younger mathematician set to work on such problems.

It was 1926, and the young Hungarian mathematician John von Neumann had just arrived in Göttingen, at the time one of the centers of world mathematics, after obtaining his degree in mathematics in Budapest and a diploma in chemistry from Zurich. One of von Neumann's interests, though certainly not the main one, was the mathematical aspects of games, and with a serious reason: while Borel, in 1921, had drawn analogies between games and both economics and military tactics, von Neumann saw games as a universal model for decision processes:[1]

> And finally, an event with given external conditions and given actors (assuming absolute free will for the latter) can be seen as a parlor game if one views its effects on the persons involved.

The question he posed about successful play in such games thus is of great importance, since:

> There is scarcely a question of daily life in which this problem does not play a role.

Von Neumann's researches, which include greatly simplified models of poker, led to a lecture on 7 December 1926 before the Göttingen Mathematical Society, whose content was published less than two years later.[2] There he began by restricting the objects under investigation with a formal definition of a game. Like Borel, he stated that games can always be transformed in such a way that every player has a single move on which to decide, and to do so at the same time as all the other players. This form of a game, namely, the normal form, becomes the starting point for his further investigations.

For von Neumann the number of players is irrelevant, though he assumes the game to be of zero sum. He begins by investigating two-person games, and how the winning prospects of the two players, S_1 and S_2, are reflected in the data of the normal form, and he thus recognizes the significance of the maximin and minimax values. The maximin value is, according to von Neumann,

> The best result that S_1 can attain if S_2 sees through him completely ... (Due to the game rules, S_2 was not allowed to know

---

[1] John von Neumann, On the Theory of Games of Strategy, in: Contributions to the Theory of Games IV, *Annals of Mathematics Studies* **40**, Princeton 1959; Werke: Band IV, pp. 1–26.

[2] John von Neumann, Zur Theorie der Gesellschaftsspiele, *Mathematische Annalen* **100**, 1928, pp. 295–320; Werke: Band IV, pp. 1–26, particularly in reference to the work of Borel that von Neumann became aware of only later: J. v. Neumann, *Sur la théorie des jeux, Comptes Rendus de l'Académie des Sciences* **186**, 1928, pp. 1689–1691.

what S_1 will play, and so he must infer this by other means. This is what we mean by "see through.")

Analogously, the minimax value is

The best result that S_2 can attain if he has seen through S_1. If both these numbers are equal, this means that it doesn't matter which of the two players is the better psychologist; the game is so insensitive that the result is always the same ....

The difference in the two values ... means that of two players S_1 and S_2, both of them cannot be simultaneously the cleverer one.

Already in the next sentence von Neumann announces his decisive result:

However, despite the existence of a trick, it is possible to force the equality of the two above-mentioned numbers.

By "trick" von Neumann means that the players choose their rules of behavior in an actual game randomly. Thus von Neumann has the same idea as Borel, except that he carries the approach of mixed strategies through to the end by showing that in every such game, extended as to the possibilities for moving, the maximin and minimax values agree.[3] Thanks to this *minimax theorem*, which holds for every finite two-person zero-sum game, no player need fear that his opponent might see through his strategic plans, on the assumption that the player has mixed his strategies according to the minimax principle, which is always possible. In relation to the poker game that we introduced at the start of this chapter, there thus exists a mixed strategy that—however it might look in detail—prevents a negative expectation.

Von Neumann's work received little attention at first.[4] His results obtained wider recognition only when in 1944, von Neumann published,

---

[3]The significance of Borel's work on the development of game theory was the subject of a 1953 controversy initiated by the the French mathematician Fréchet: Maurice Fréchet, Emile Borel, initiator of the theory of psychological games and its applications, *Econometrica* **21**, 1953, pp. 95–96; Maurice Fréchet, Commentary on the three notes of Emile Borel, *ibid.*, pp. 118–124; J. von Neumann, Communications on the Borel notes, *ibid.*, pp. 124–125.

[4]A notable exception is the short work by René de Possel, *Sur la théorie mathématique des jeux de hasard et de réflexion*, Paris 1936, reprinted in Hevre Moulin, *Fondation de la théorie des jeux*, Paris 1979, where in just under 40 pages the various aspects of games are explained in accessible prose. In reference to Borel, Possel distinguishes games of chance, reasoning, and trickery, where "A game is sensitive to trickery if a player can gain an advantage if he knows his opponent's thoughts." All three types of games and game influences are explored from a mathematical point of view. Von Neumann's minimax theorem is cited with respect to games of trickery.

together with the economist Oskar Morgenstern, who, by the way, was the illegitimate grandson of the German Kaiser Friedrich III, an extensive monograph called *Theory of Games and Economic Behavior*.[5] This publication represented the birth of mathematical game theory, although significant aspects of the theory had by then been known for 18 years.

Von Neumann's original proof of the minimax theorem is a pure existence proof, and thus does not show how to calculate the relevant strategies. Moreover, the proof, though rather elementary in its argumentation, is comparatively lengthy. The proof can be greatly shortened if one makes use of the Brouwer fixed-point theorem.[6] The mathematical content of the theorem, unknown as much to von Neumann as to Borel, had already been discovered and proved several times, the first in 1902 by Julius Farkas (1847–1930) in the form of an abstract theorem on inequalities.[7] This purely algebraic formulation with inequalities makes possible a geometric interpretation, in which convex sets play a role.[8] A set of points is said to be convex if for every pair of points in the set, the line connecting them also belongs to the set (see "The Idea of the Proof of the Minimax Theorem"). Only later, at the end of the 1940s, was the scope of the minimax theorem again extended, this time as a result about solutions of optimization problems. We shall return later to this theme.

The minimax theorem will form the foundation of our further mathematical analysis of two-person zero-sum games, much as Zermelo's theorem did for games with perfect information. We shall investigate games in which explicit results can be calculated as well as those for which only qualitative assertions can be made. We begin with some foundational statements about the nature of two-person zero-sum games that can be formulated directly from the two above-mentioned theorems:

---

[5] For the history of this monograph and the careers of the authors, see H. W. Kuhn, John von Neumann's work in the theory of games and mathematical economics, *Bulletin of the American Mathematical Society* **64**, 1958, pp. 100–122, special edition on the death of John von Neumann; William Poundstone, *Prisoner's Dilemma*, New York 1992; Urs Rellstab, *Ökonomie und Spiele: Die Entstehungsgeschichte der Spieltheorie aus dem Blickwinkel des Ökonomen Oskar Morgenstern*, Chur 1992.

[6] This theorem was mentioned in one of the notes to Chapter 19; see also Tinne Hoff Kjeldsen, John von Neumann's conception of the minimax theorem: A journey through different mathematical contexts, *Archive for History of Exact Sciences* **56**, 2001, pp. 39–68.

[7] See Tinne Hoff Kjeldsen, Different motivations and goals in the historical development of the theory of systems of linear inequalities, *Archive for History of Exact Sciences* **56**, 2002, pp. 469–538.

[8] The first proof of the minimax theorem based on convex sets was found in the mid 1930s by Jean Ville, and was first published in É. Borel, *Traité du calcul des probabilités et de ses applications*, Tome IV, Fascicule II, Applications aux jeux de hasard, Paris 1938, pp. 105–113.

- games without randomness and with perfect information are purely combinatorial. Like chess and go, they are completely determined. Each of the two players can force a result that corresponds to the game value.

- games with elements of chance with perfect information are determinable only up to expectation. Each player can optimize his or her play so that on average, independent of the opponent's strategy, a result is guaranteed that reflects the game value. Otherwise, nothing can be said about the results of individual games.

- without perfect information, players are often compelled to vary their strategies randomly. Even if the rules contain no random elements, the course of the game is not necessarily deterministic. As with games of chance, a player is guaranteed a result corresponding to the game value only up to the expectation: in individual games, the result can be much less favorable. In von Neumann's words:[9]

> In spite of ... chance (via the introduction of expectations...) being eliminated from the games considered, it has reappeared on its own: Even if the rules of the game contain no elements of chance... it is nonetheless absolutely necessary to introduce the elements of chance in specifying how the players are to act. The dependence on chance is so deeply embedded in the nature of the game (if not in the nature of the world) that it is not at all necessary to introduce it artificially in the rules: Even if there is no trace of it in the rules of the game, it establishes itself of its own accord.

The Idea of the Proof of the Minimax Theorem

In the following discussion we shall restrict our attention to the case that white has exactly two strategies, since then the content of the minimax theorem can be presented geometrically in the plane. Nonetheless, the construction that arises is universally applicable and works in the same way when white has more

---

[9]In John von Neumann, On the Theory of Games of Strategy, in: Contributions to the Theory of Games IV, *Annals of Mathematics Studies* **40**, Princeton 1959; Werke: Band IV, pp. 1–26.

than two strategies, though in that case, a higher-dimensional representation is necessary. The geometrically supported arguments that we use can be proved in complete generality using standard techniques of analytic geometry.

We first prove the following: either black can restrict white to a score of at most 0 using a mixed strategy, or white possesses a strategy by means of which she can ensure herself a positive score.

We will indicate the required construction using two examples with the following normal forms:

		Black			
		1	2	3	4
White	1	1	−1	1	2
	2	−2	1	0	1

and

		Black			
		1	2	3	4
White	1	2	−1	1	2
	2	−1	1	0	1

First the scores that white can obtain from black using the various strategies are entered into a rectangular coordinate system: each pure strategy of black yields a point whose first coordinate equals white's score if she chooses her first strategy, and whose second coordinate is the analogue for the case that white chooses her second strategy. Black's mixed strategies can also be represented in this way. They "center" the pure strategies, and indeed, in the same way whether from a geometric point of view or that of the quantitative result. For example, a 1 : 1 mixed strategy lies geometrically at the midpoint between the two pure strategies whose mixture was used. For our two examples, the totals of all mixed strategies are represented by the triangles. As one can see, the region is "anchored" on the points that can be specified by pure strategies. Note also that for every two points in the region, the line connecting them is also contained in the region. Thus beginning with the points that represent the pure strategies, one successively obtains the entire region that represents the mixed strategies:

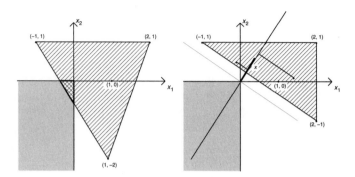

With respect to the negative third quadrant of the coordinate system, which is shaded in the diagrams, we see that there are two possible cases. We shall see that this division into cases corresponds to the two alternatives presented at the beginning of this sidebar:

- in the case represented in the left-hand diagram, black can find a mixed strategy that prevents white from attaining a positive expectation. Here black can use any mixed strategy corresponding to the region of overlap with the third quadrant.

- in the case depicted in the right-hand diagram, in which there is no overlap, one seeks the shortest line connecting the two regions. Such a line always exists, since asymptotic situations, such as one knows from hyperbolas and their axes, are excluded due to the limited size of the triangular strategic region. When a line with minimal distance is found—it need not begin at the origin—then its direction indicates the desired strategy for white:

  - first of all, no coordinate of the direction vector can be negative, since a shorter line could then be found by translating the endpoint in the third quadrant. Furthermore, the vector found has at least one positive coordinate.

  - thus the relationship between the coordinates can be interpreted as a relation by means of which white can mix her strategies. The vector of probabilities, normed to length 1, is shown in the figure as a heavy line and marked with ×.

   – the score that white can achieve with the strategy ×
   has a geometric interpretation: as with white's pure
   strategies, the score is equal to the corresponding coordinate of the geometrically represented black strategy,
   thus in the general case one forms a scalar product.
   And this product is equal to the length that results
   from the projection onto the line that was found, as
   is shown, for example, in the diagram for two points
   that correspond to a black strategy. On one side of the
   dotted line, thus in particular for the entire triangular
   region, the result is always positive.

With this proof the minimax theorem for symmetric games is
clear: since for symmetric games the second alternative can
never arise, for such games the first alternative must always
obtain. That is, black can use a mixed strategy to limit white
to a maximum score of 0. The same holds, of course, for white.

To be able to prove the minimax theorem for asymmetric games,
one introduces a handicap in the form of a fee to be paid whose
value is varied. If one also allows negative fees, then one may always assume that white pays the fee to black. The game is now
investigated for all fee levels from the point of view of the proved
alternative: when does one hold, and when the other? Here the
two alternatives, which basically say nothing but that white is
ahead or that the opposite is true, divide the entire number
line into two halves. The number that divides the two halves
is the minimax value of the game. White can assure herself of
every value smaller than this number as winning expectation.
Likewise, black can limit white's expectation to every amount
that is larger. By passing to the limit, one obtains the minimax
theorem.

Further Literature on the Minimax Theorem and Game Theory

[1] R. Duncan Luce, Howard Raiffa, *Games and Decision*, New York 1957.
[2] Samuel Karlin, *Mathematical Methods and Theory in Games, Programming and Economics*, Reading, MA 1959, 2 volumes.

[3] Melvin Dresher, *The Mathematics of Games and Strategy: Theory and Applications*, Englewood Cliffs, NJ 1961.

Information on the historical development of game theory can be found in:

[4] N. N. Worobjow, *Die Entwicklung der Spieltheorie*, Berlin (East) 1975 (Russian original 1973).

[5] E. Roy Weintraub (ed.), *Toward a History of Game Theory*, Durham 1992.

[6] Norfleet W. Rives, On the history of the mathematical theory of games, *History of Political Economy* **7**, 1975, pp. 549–656.

[7] Robert W. Dimand, Mary Ann Dimand, *The History of Game Theory, Volume 1: From the Beginnings to 1945*, London 1996.

[8] Mary Ann Dimand, Robert W. Dimand, *The Foundations of Game Theory*, Cheltenham 1997, 3 volumes.

35

Bluffing in Poker: Can It Be Done Without Psychology?

The success of a good poker player rests in large measure on the ability to bluff. But on what basis does one decide to bluff? Does successful bluffing depend on an astute psychological estimation of one's opponent? Or is bluffing the expression of an objective mathematical optimality that allows a player to parry an opponent's strategic options according to the laws of the minimax theorem?

Von Neumann's minimax theorem guarantees optimal strategies to every player of a two-person zero-sum game. Thus one should be able—at least in theory—to withstand the cunning tactics of the most hardened poker face. However, the defensive basis of optimal strategies has its drawbacks, since no specific advantage can be drawn from the weaknesses of a player who is clearly not playing up to snuff. Such advantages accrue to the player who can realistically assess his opponent.

What form the minimax optimality takes in a particular case and to what extent it agrees with empirical experience in game playing is an open question. With the question posed at the beginning of this chapter we have particularized the problem to poker and the technique of bluff: should one bluff only when one believes it possible to fool an inexperienced and therefore insufficiently or excessively cautious player? Or is bluff used on occasion even when no extra information about the opponent is available?

Poker is played in a number of variants.[1] What all these variants have in common is that every player places bets that he or she holds the best cards. Only those who remain in the game throughout the entire bidding process can participate in the showdown, where the hands of those remaining players are compared. If all the players except one fold, then that player wins the hand without a showdown. It is clear that good hands can support high bets more than bad ones can. But it would make little sense to use a strict formula for placing bets based on the strength of one's hand, because that would allow the opponent to know your cards. John von Neumann and Oskar Morgenstern have this to say on the subject:[2]

> The point in all this is that a player with a strong hand is likely to make high bids—and numerous overbids—since he has good reason to expect that he will win. Consequently a player who has made a high bid, or overbid, may be assumed by his opponent—a posteriori!—to have a strong hand. This may provide the opponent with a motive for "Passing." However, since in the case of "Passing" the hands are not compared, even a player with a weak hand may occasionally obtain a gain against a stronger opponent by creating the (false) impression of strength by a high bid, or overbid—thus conceivably inducing his opponent to pass.

> This maneuver is known as "Bluffing." It is unquestionably practiced by all experienced players. Whether the above is its real motivation may be doubted; actually a second interpretation is conceivable. That is, if a player is known to bid high only when his hand is strong, his opponent is likely to pass in such cases. The player will, therefore, not be able to collect on high bids, or on numerous overbids, in just those cases where his actual strength gives him the opportunity. Hence it is desirable for him to create uncertainty in his opponent's mind as to this correlation—i.e., to make it known that he does occasionally bid high on a weak hand.

---

[1] Claus D. Group, *Alles über Pokern*, Niedernhausen 1987; Kay Uwe Katira, *Poker und andere Kartenspiele*, Ravensburg 1979; John Scarne, *Complete Guide to Gambling*, New York 1974, pp. 670–701. Historical, strategic, and mathematical aspects of poker are discussed in John McDonald, Poker: an American game, *Fortune* **37**, March 1948, pp. 128–131, 181–187. A supplement to that article can be found in John McDonald, A theory of strategy, *Fortune* **39**, 1949, pp. 100–110; John McDonald, *Strategy in Poker, Business and War*, New York 1950.

[2] John von Neumann, Oskar Morgenstern, *Theory of Games and Economic Behavior*, Princeton 1944.

To sum up: of the two possible motives for Bluffing, the first is the desire to give a (false) impression of strength in (real) weakness; the second is the desire to give a (false) impression of weakness in (real) strength. Both are instances of inverted signaling ... i.e., of misleading the opponent.

Typical properties of poker variants can be translated into simple models that, in contrast to the real variants, can be analyzed mathematically within manageable limits. Like physical models, these offer the advantage that the characteristic properties are highlighted. Here are von Neumann and Morgenstern again:

> However, actual Poker is really a much too complicated subject for an exhausitve discussion and so we shall have to subject it to some simplifying modifications, some of which are, indeed, quite radical. It seems to us, nevertheless, that the basic idea of Poker and its decisive properties will be conserved in our simplified form. Therefore it will be possible to base general conclusions and interpretations on the results which we are going to obtain by the application of the theory previously established.

We begin with a simple game that contains the basic elements of poker:

- each of two players obtains, after putting up an ante of eight units, a high card or a low card, randomly and with equal probability. Moreover, the values of the two cards are independent: they can be thought of as dealt from two different decks.

- the first player can pass or raise his bet from 8 to 12. If he passes, then the showdown takes place at once, where the player with the higher card wins both players' bets. If the cards are the same, each player gets his ante back.

- if the first player has raised, then the second player can decide whether to pass or keep playing. In the first case, he loses his bet. In the second, he also adds an additional four units to the pot to "see" his opponent, that is, to force a showdown. Again, the winner is the player with the higher card.

The rules of the game are shown schematically in Figure 35.1.
Each of the two players can choose from among four pure strategies:

- the strategy of the first player must have a move for the case of a high card and a move for a low card, namely, pass (P) or raise (R). The strategies will be denoted PP, PR, RP, and RR. For example,

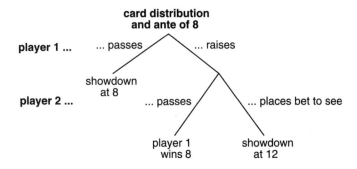

Figure 35.1. The possible decisions in the poker model.

the abbreviation PR means pass with a low card and raise with a high card.

- the strategy of the second player determines his reaction to the opening moves in which the first player has raised the bet. This player must decide for each kind of card whether to pass or to see. Thus he, too, has four pure strategies, namely, PP, PS, SP, and SS.

To obtain the normal form, we must investigate the course of the game for each of the $4 \times 4 = 16$ combinations of strategies. There are four different ways in which the cards can be dealt. Table 35.1 shows the normal form for two selected strategy pairs, where the four equiprobable scores of the first player depend on the hand dealt. Here L stands for a low card, and H for a high card.

In Table 35.1 it is seen clearly that the first player has an advantage. He can protect himself from loss by always passing with the strategy PP. Better is the strategy PR, which passes only on a low card and raises the

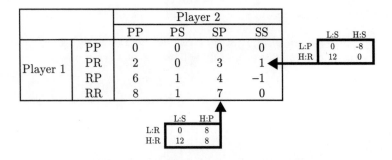

Table 35.1. Normal form of the poker model.

bet on a high card. That is not only intuitively clear, since raising with a high card is completely without risk, but can be seen as well from the normal form: regardless of how the second player decides, the expectation from PR is always at least as high as from PP. Such a situation is characterized by saying that strategy PR *dominates* strategy PP. While dominated strategies can have their place in optimal strategies, their role can always be replaced by a dominating strategy. Thus dominated strategies can be ignored when one seeks only an optimal strategy for each player.

There are other dominated strategies in the normal form presented in Table 35.1 for our primitive poker model. Thus the strategy RR of the first player dominates the strategy RP. A similar situation obtains for the second player, for whom the strategy PP is dominated by PS, and SP by SS. Since only the first player's score is shown in the normal form, the dominance criterion for the minimizing player is obtained by changing the order of the numbers. That is, if the relation exists between two columns that each entry of the first column is at least as big as the corresponding entry of the second column, then the strategy corresponding to the first column is dominated.

In our example, the four dominance properties that we have found are not in the least surprising. They say nothing more than that the first player should always raise with a high card and that the second player should always be prepared to see with a high card. Both decisions are completely safe under the given circumstances and therefore optimal. The dominance relationships thus formally confirm a principle that is obvious from the point of view of the game itself. What is more important is that the normal form can be greatly simplified using these dominances. Namely, if one eliminates the rows and columns associated with the dominated strategies, then one has the 2×2 remainder shown in Table 35.2.

Based on our experience with "even or odd," one at once suspects that both players should randomly mix their strategies in the proportion $1 : 1$. The value of the game is $1/2$. With respect to the individual decisions to be made, the result can be interpreted as follows:

- if a player holds a high card, he can always risk a higher bet; that is, he raises as first player and does likewise as second player.

| | | Player 2 | |
		PS	SS
Player 1	PR	0	1
	RR	1	0

Table 35.2. Poker model: the normal form without dominated strategies.

- with a low card a player passes with probability 1/2. With the same probability he takes a risk; that is, he raises as first player or sees as second player.

Bluffing in the form of raising the bet despite the presence of a low card can be seen to be strategically necessary on the objective mathematical level. The concrete decision whether to bluff is made randomly, without using any psychological impressions. The underlying quantitative framework, that is, the probabilities with which the various strategies are chosen, determine the degree of success that a player can expect on average.

Since the model that we have used is extremely simple, there is much to learn that our model has not been able to reveal. Therefore, we would do well to investigate some additional poker models whose construction has more in common with the real card game. We will require some universal algorithms that allow for the calculation of minimax strategies. We shall look at these in the coming chapters.

36

Symmetric Games: Disadvantages Are Avoidable, but How?

In symmetric two-person zero-sum games, both players are guaranteed the existence of a mixed strategy that can prevent a negative expectation. How can such strategies be found?

Symmetric games were given particular consideration by Borel. As mentioned in Chapter 33, one can restrict attention to symmetric games, since every game can be viewed as part of a symmetric game. A game whose normal form contains n rows and m columns can be "embedded" in a symmetric game in which each of the two players has $m + n + 1$ strategies from which to choose (see Note 1 at the end of the chapter). Moreover, the minimax value of a symmetric game is known a priori; it is zero. Thus a given strategy can be tested relatively easily to see whether it is in fact optimal. This was discussed in Chapter 33: one must check how the given strategy works against every possible strategy of the opponent. That is, the corresponding expectations must be calculated. Then a strategy for white is optimal if none of these expectations is negative, and that can be determined, as Borel showed, by the solution of a system of inequalities.

For example, if a minimax strategy for white is sought for a game with normal form

		Black Chooses			
		1	2	3	4
White	1	0	1	−3	2
Chooses	2	−1	0	1	−4
	3	3	−1	0	3
	4	−2	4	−3	0

then one must solve the system of linear inequalities

$$-x_2 + 3x_3 - 2x_4 \geq 0,$$
$$x_1 - x_3 + 4x_4 \geq 0,$$
$$-3x_1 + x_2 - 3x_4 \geq 0,$$
$$2x_1 - 4x_2 + 3x_3 \geq 0,$$

where the conditions

$$x_1 \geq 0, \quad x_2 \geq 0, \quad x_3 \geq 0, \quad x_4 \geq 0, \quad x_1 + x_2 + x_3 + x_4 = 1,$$

must also be satisfied.

Such systems of inequalities can be seen as a generalization of systems of linear equalities, which are studied in many branches of mathematics. In addition to greater-than-or-equal relations, one might include equalities and less-than-or-equal relations. These, too, can be transformed as necessary into greater-than-or-equal relations: an equation yields two inequalities. In comparison to systems of linear equations, the theory of systems of inequalities has a much briefer tradition.

We should note as well that Borel considered the systems of inequalities associated with symmetric games as unsolvable in general, although problems of this kind had been solved years before in certain special cases. A significant turn of events occurred in 1947 with the development of linear optimization—due in large measure to Dantzig, then a civilian employee of the US Air Force—a discipline founded expressly for applications in the area of military logistics. The objects of study in linear optimization are methods by which, for example, costs can be minimized or yields maximized, to the extent that the controlling parameters, their possible values, and their effect on the quantities to be optimized are completely known, and that the entire system assumes a particular linear form. When Dantzig found that such problems arose repeatedly and frequently took on a typical linear form, he turned for advice to the economist and later Nobel Prize winner, Tjalling Koopmans (1910–1985). However, his hopes of learning about some well-known standard solution methods were soon dashed.[1] Therefore, Dantzig set out on his own to find a practical solution algorithm. Success came in 1947 with the simplex algorithm (see "Linear Optimization" and "The Simplex Algorithm").

Linear Optimization

A typical problem of linear optimization deals with a simple model of a production process and its optimal control. A decision has to be made as to the various quantities in which various products should be produced. One has to take into account the capacities of the various resources, such as labor, machine output, and availability of raw materials, as well as profits that can be achieved above and beyond the production costs. Let us consider a simple example.

Resources A, B, C, and D are used to produce products X and Y in quantities x and y, respectively. The following facts are known:

- the achievable profit in producing products X and Y is 2 monetary units per unit of X, and 3 monetary units per unit of Y. That is, one obtains a net profit P given by

$$P = 2x + 3y.$$

- the use of resources and their limits of capacity are given by a series of inequalities:

  - to produce one unit of product X requires one unit of resource A, while four units of A are required to produce one unit of Y.
  - there are 24 units of resource A available.

---

[1] Only later did it become known that the Russian mathematician Leonid Vital'evich Kantorovich (1912–1986) had studied such optimization problems a decade earlier. However, certain obstacles prevented him from achieving a breakthrough, even though the main ideas are present in his work. In 1975, Kantorovich shared the Nobel Prize in Economics with Koopmans. See the book of Dantzig previously cited, as well as L. V. Kantorovich, My journey in science, *Russian Mathematical Surveys* **42:2**, 1987, pp. 233–270; L. V. Kantorovich, Mathematical methods of organizing and planning production, *Management Science* **6**, 1960, pp. 366–422 (Russian original 1939).

This yields the inequality

$$x + 4y \leq 24.$$

We assume similar conditions for the other resources that yield the following inequalities:

$$
\begin{aligned}
x + 2y &\leq 14 &&\text{from B,} \\
x + y &\leq 10 &&\text{from C,} \\
2x + y &\leq 17 &&\text{from D.}
\end{aligned}
$$

- finally, we recognize real-world limitations not stated explicitly in the model:

$$x \geq 0 \quad \text{and} \quad y \geq 0.$$

Simple situations like that of our example can be displayed graphically. We enter the possible production plans characterized by the numbers x and y into a two-dimensional coordinate system. In our optimization we must consider all pairs (x, y) that satisfy all six inequalities, called *side conditions*. Each of the six inequalities defines a half-plane, that is, a region in the plane lying entirely on one side of a straight line. The equation of this line is found by replacing the inequality sign with an equal sign. If one then forms the set-theoretic intersection of the six half-planes, one obtains a geometric representation of all possible production plans, that is, all those that satisfy the six constraints:

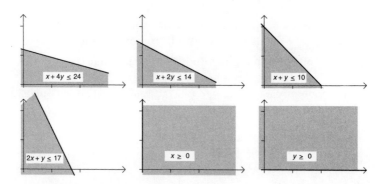

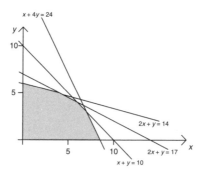

What is the profit distribution within the region of possible plans? In particular, where does it attain its largest value? To answer this question we represent the profit graphically as well. We obtain a set of level lines, each representing a particular value of the profit. In the following figure, such lines are shown for the levels $6, 12, 18, 24, 30$:

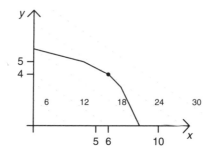

We see at once from the figure that the maximum achievable profit is 24, and that the associated production plan, which corresponds to a vertex of the admissible region, consists in producing $x = 6$ and $y = 4$ units of products X and Y.

Although our example is painfully simple, it reveals some of the typical properties of linear optimization problems, in which—though generally for more than two parameters x and y—optimal values are to be determined:

- one seeks—after a reformulation if necessary—the maximum of a linear function that must satisfy certain side conditions presented in the form of greater-than-or-equal inequalities.

- it is useful to indicate the permissible values of the parameters geometrically. The number of dimensions is equal to the number of parameters, and even with four or more parameters such a representation is useful, and even though the possibility of visualization is lost, the techniques of analytic geometry can be used in these higher-dimensional situations.

- each of the linear side conditions individually limits the admissible region to a half-plane. The set-theoretic intersection of these half-planes yields the range of admissible values for the parameters that underlie the optimization. The following figure represents a simple three-dimensional example:

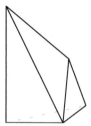

- the admissible region is always convex.

- since the value to be optimized, the target value, depends linearly on the parameters, the achievable values form a collection of parallel hyperplanes—thus the name of a linear subspace whose dimension is one less than that of the space in which it is embedded. In particular, in three-dimensional space a hyperplane is a normal plane. Every hyperplane within the collection corresponds to precisely one possible target value.

- if there exist optimal parametric values, then they can always be found at a vertex of the admissible region.

- however, such optimal vertices do not necessarily exist, nor need they be unique.

  - All the points of a side can represent optimal parameters. Then every vertex of that side is optimal.
  - The admissible region might be empty. Such linear optimization problems have no solution.

    − Optimization problems with an unbounded admissible region may also be unsolvable, since the value to be optimized may grow without bound.

There are a number of algorithms for solving linear optimization problems. The most frequently used is Dantzig's simplex algorithm. For difficult cases, which arise seldom in practice, there is a better algorithm, discovered in 1984 by Karmarkar (see Note 2 at the end of the chapter).

    The connections between linear optimization and game theory were quickly—though not immediately—recognized. As Dantzig later reported,[2] John von Neumann had suspected already in 1947 that the theory of games for two-person zero-sum games was equivalent to linear optimization. And that is exactly what was proved later to be the case. In particular, the equivalent of the minimax theorem was found within linear optimization (see Note 3 at the end of the chapter). Here, however, we do not wish to concern ourselves with the theoretical equivalence, but with how the algorithms of linear optimization can be used to calculate minimax strategies. We will begin by restricting attention to symmetric games and use an old example to demonstrate what is going on. Starting with the system of inequalities already formulated, we start with a small trick: we weaken the side condition

$$x_1 + x_2 + x_3 + x_4 = 1$$

to the inequality

$$x_1 + x_2 + x_3 + x_4 \leq 1$$

and then search in the enlarged region of admissible values x_1, x_2, x_3, x_4 for the maximum of the function

$$x_1 + x_2 + x_3 + x_4.$$

Of course, we already know that the maximum is 1. But we are not really interested in the maximum itself, but in the parametric values x_1, x_2, x_3, x_4 for which the maximum is attained, for these give us the desired strategy. The actual calculation can be done using the simplex algorithm. Beginning with the admissible values

$$x_1 = x_2 = x_3 = x_4,$$

the maximum is increased step by step to the value 1.

---

[2]Interview in Donald J. Albers, Gerald J. Albers, Constance Reid (eds.), *More Mathematical People*, San Diego 1990, pp. 73–77.

The Simplex Algorithm

The idea of Dantzig's simplex algorithm is based on the geometric interpretation of a linear optimization problem. However, the geometric properties, as they play a role in the algorithm, are always characterized purely algebraically: thus we have seen already that one can always find a maximum for a solvable problem by examining the vertices of the admissible region. But what is a vertex? That is, how can a vertex be characterized algebraically, and how can one be determined?

Let us look first at the border of the admissible region. Such a border point is characterized by equality holding for at least one of the greater-than-or-equal side conditions. In the case of vertices and other "special" border points there must be additional identities that are satisfied. The following figure clarifies this idea for two simple examples, where the admissible region is two-dimensional in the left-hand figure and three-dimensional in the one on the right:

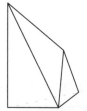

sides: at least one equality
vertices: at least two equalities

faces: at least one equality
edges: at least two equalities
vertices: at least three equalities

One could now attempt to carry out a search of all the vertices of the admissible region. To do this, we determine the vertices by prescribing equality in sufficiently many of the side conditions and then solving the resulting system of equalities. If such a system has a solution and if all the nonnegativity conditions are satisfied, then we have found a vertex. If we have determined all the vertices in this way, then we finally choose the largest one of the possible target values. Since in large optimization problems the number of vertices grows very rapidly, such a process is hardly practicable. A significantly better modus operandi is the following step-by-step method.

Starting at a vertex that has been located, the edges leaving this vertex are then investigated to see how the optimization function changes along the given direction. If no edge goes in a direction in which the function grows, then one has found the maximum. Otherwise, one chooses an edge along which the function increases and follows it to the next vertex. Computationally, this is done by giving up an equality relation in the side conditions and substituting an additional relation. This exchange is called the *pivot step*.

In transforming the geometric idea into computational form, one must note that greater-than-or-equal relationships are algebraically difficult to handle. Therefore, the inequalities are transformed into equalities with a so-called *slack variable*, and these are then solved step by step through a selection of variables. Each such solution corresponds to a vertex, and indeed, the properties of the optimization problem become particularly clear in a neighborhood of this vertex. Let us consider the example that we introduced in the previous sidebar:

$$
\begin{aligned}
T = & & 2x & + 3y, \\
u_1 = 24 & - & x & - 4y, \\
u_2 = 14 & - & x & - 2y, \\
u_3 = 10 & - & x & - y, \\
u_4 = 17 & - & 2x & - y.
\end{aligned}
$$

Here T stands for the target optimal value, that is, the proceeds to be maximized. In addition to the four side conditions, all the variables, including the slack variables u_1, u_2, u_3, u_4, must be nonnegative:

$$x \geq 0, \quad y \geq 0, \quad u_1 \geq 0, \quad u_2 \geq 0, \quad u_3 \geq 0, \quad u_4 \geq 0.$$

This form of the side conditions corresponds to the vertex $(x, y) = (0, 0)$. Clearly, the target T can be improved with respect to the value obtained. To this end, the variables x and y can be increased within certain limits without violating a side condition. In the simplex algorithm, however, only a single variable is selected at each step to be increased starting with the value 0. Since for the same amount of increase the variable y provides a greater increase in the target, we decide to increase the variable y. But how far can we go with this? A look at the four equations shows that we must stop with $y = 6$, since then the variable u_1 reaches the value 0, while the other variables u_2, u_3, u_4 are still positive. In order to obtain information on the growth of the target for the vertex with $x = 0$ and $y = 6$, analogously to what we obtained for the first vertex, the second equation, which provides the bound on the amount of increase, is solved for y and the result substituted into the other four equations:

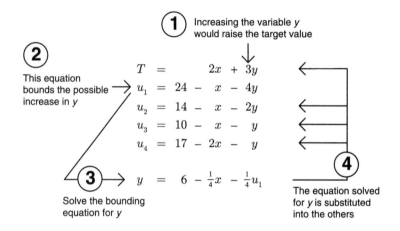

1 Increasing the variable y would raise the target value

2 This equation bounds the possible increase in y

$$T = 2x + 3y$$
$$u_1 = 24 - x - 4y$$
$$u_2 = 14 - x - 2y$$
$$u_3 = 10 - x - y$$
$$u_4 = 17 - 2x - y$$

$$y = 6 - \tfrac{1}{4}x - \tfrac{1}{4}u_1$$

3 Solve the bounding equation for y

4 The equation solved for y is substituted into the others

Geometrically, the form of the system of equalities that arises represents a change to a coordinate system with the point $(x, y) = (0, 6)$ as the origin and axes x and u_1. Purely algebraically, it is simply a transformation of the inequalities to an equivalent system so that the target and side conditions can be considered relative to another base point. The result is

$$T = 18 + \frac{5}{4}x - \frac{3}{4}u_1,$$

$$y = 6 - \frac{1}{4}x - \frac{1}{4}u_1,$$

$$u_2 = 2 - \frac{1}{2}x + \frac{1}{2}u_1,$$

$$u_3 = 4 - \frac{3}{4}x + \frac{1}{4}u_1,$$

$$u_4 = 11 - \frac{7}{4}x + \frac{1}{4}u_1.$$

From the first equation one sees at once that a further increase of the target can be achieved only if the variable x increases above 0. How far this value can grow without making one of the variables negative is given by the third equation, namely, up to the value $x = 4$. Again the equation that gives the limiting value is solved for the variable x and then substituted into the other equations. The result is

$$T = 23 - \frac{5}{2}u_2 + \frac{1}{2}u_1,$$

$$y = 5 + \frac{1}{2}u_2 - \frac{1}{2}u_1,$$

$$x = 4 - 2u_2 + u_1,$$

$$u_3 = 1 + \frac{3}{2}u_2 - \frac{1}{2}u_1,$$

$$u_4 = 4 + \frac{7}{2}u_2 - \frac{3}{2}u_1.$$

To obtain a further increase in the target T, the value of the variable u_1 must be increased above 0. This is possible, if no side condition is violated, up to the limit $u_1 = 2$, for beyond that point the variable u_3 becomes negative. As in the previous steps, the equation giving the limit is solved for the variable to be increased, u_1, and substituted into the other equations:

$$T = 24 - u_2 - u_3,$$

$$y = 4 - u_2 + u_3,$$

$$x = 6 + u_2 - 2u_3,$$

$$u_1 = 2 + 3u_2 - 2u_3,$$

$$u_4 = 1 + -u_2 + 3u_3.$$

As can be seen from the first equation, the value of T cannot exceed 24. This concludes the optimization. We would now like to look graphically at the path covered thus far through the vertices. In the next figure the affected vertices are marked with the letters "a" through "d."

What is significant in this example of the calculational steps is the solution of the equations for a changing number of variables. Step by step, a cleverly chosen equation is solved for a suitable variable, and then this result is substituted into the other equations. Which equation and which variable were to

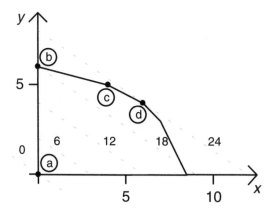

be chosen depended on the current form of the target function and the side conditions.

In the literature, the simplex algorithm is generally described in terms of tables, called simplex tableaus. In this way it is possible to completely describe formulaically the selection of rows and columns in a calculational step and the subsequent transformation of the coefficients. The actual calculations, though, are the same. Simplex tableaus will be described in the next chapter.

If we proceed as in "The Simplex Algorithm," then with a symmetric game we begin our search for a minimax strategy with the system of equations

$$T = x_1 + x_2 + x_3 + x_4,$$
$$u_1 = -x_2 + 3x_3 - 2x_4,$$
$$u_2 = x_1 - x_3 - 2x_4,$$
$$u_3 = -3x_1 + x_2 - 3x_4,$$
$$u_4 = 2x_1 - 4x_2 + 3x_3,$$
$$u_5 = 1 - x_1 - x_2 - x_3 - x_4,$$

where none of the variables $x_1, \ldots, x_4$ and none of the slack variables $u_1, \ldots, u_5$ may be negative. The system is transformed step by step:

$$T = -\frac{1}{3}u_3 + \frac{4}{3}x_2 + x_3,$$

$$u_1 = -x_2 + 3x_3 - 2x_4,$$

$$u_2 = -\frac{1}{3}u_3 + \frac{1}{3}x_2 - x_3 + 3x_4,$$

$$x_1 = -\frac{1}{3}u_3 + \frac{1}{3}x_2 - x_4,$$

$$u_4 = -\frac{2}{3}u_3 - \frac{10}{3}x_2 + 3x_3 + 2x_4,$$

$$u_5 = 1 + \frac{1}{3}u_3 - \frac{4}{3}x_2 - x_3.$$

The result of the first transformation step is interesting in that it yields no improvement. Despite the exchange of the variables x_1 and u_3, no new vertex is reached. Geometrically, one might imagine such a phenomenon as a "multiple" vertex, in which more edges, faces, and so on intersect than the "usual" required minimum.[3] No further vertex is reached in the next steps of the simplex algorithm. Finally, one reaches another vertex with the optimal value $T = 1$. The coordinates are

$$x_1 = 0, \quad x_2 = \frac{3}{8}, \quad x_3 = \frac{1}{2}, \quad x_4 = \frac{3}{8},$$

and so the minimax strategy for the symmetric game has been found.

Of much greater importance than this special result is the fact that the method described always works, even for asymmetric games once they have been symmetrized. The calculation of minimax strategies can therefore fail in practice only as a result of the complexity of the game in question.

Chapter Notes

1. In practice, it is usual either to play two games with roles interchanged or one game with the first player determined by lot. In each case a strategy in the symmetric game encompasses plans for both roles in the original game. From a game-theoretic point of view this is somewhat disquieting, since it

---

[3]One may also demonstrate the multiple character of vertices by slightly altering the data of the optimization problem. Then multiple vertices split. Multiple vertices can lead to problems in the simplex algorithm if a circular chain of variable exchanges results.

makes the normal form much larger. If the players originally have n and m strategies, respectively, then the number of strategies in the symmetric version grows to nm for each player.

Using a different procedure, proposed around 1949 by Brown and Dantzig, the normal form grows much less dramatically. In this variant, the distribution of roles is accomplished with a game like rock–paper–scissors: assume a game in which white has an advantage, that is, a game with positive value. This can be accomplished by raising all the scores by an amount sufficient to make them all positive. This possibly modified game is now symmetrized. To accomplish this, in analogy to rock–paper–scissors, both players are required to choose simultaneously from among three selections: white, black, and one. If both players make the same choice, then the game ends in a draw. With the selection white–black or black–white, a game is played with the roles so assigned. In the other four cases the score is 1, determined as follows: one beats white and loses to black.

Like rock–paper–scissors there is no single best choice, and depending on the value of the original game, all three selections must be chosen with particular nonzero probabilities. The decisions for white and black must be expanded to include strategic decisions for the actual game that might follow. One can always limit oneself to a single role in the original game, namely, the one previously selected. Here both players have $n + m + 1$ strategies.

See D. Gale, H. W. Kuhn, A. W. Tucker, On symmetric games, in: H. W. Kuhn, A. W. Tucker (eds.), Contributions to the theory of games I, *Annals of Mathematics Studies* **24**, 1950, pp. 81–87; R. Duncan Luce, Howard Raiffa, *Games and Decision*, Toronto 1957, pp. 440–442.

2. One can create linear optimization problems for which the usual variants of the simplex algorithm are unsuited for obtaining a solution in polynomially bounded time. The simplex algorithm, which has proved useful in practice, thus exhibits its theoretic limitations. In this sense, a better algorithm is the ellipsoid method, for which the Russian Khachiyan proved in 1979 that it always leads to a solution in polynomial time.

Linear optimization problems to be solved by the ellipsoid method are first transformed into systems of inequalities, for which a solution is then sought; we have done just that implicitly in our symmetrization of games at the beginning of this chapter. Then the solution set of the system is bounded step by step via a series of shrinking ellipsoids. Each step begins with an ellipsoid that contains all solutions of the system of inequalities:

- if the midpoint of the ellipse is a solution, then one is done.

- otherwise, the coordinates of the midpoint violate at least one inequality. This inequality represents geometrically a separating hyperplane. If one transforms this plane linearly, so that the midpoint of the ellipsoid lies on it, then all solutions lie in one of the halves of the ellipsoid separated by this hyperplane. The following figure shows a typical situation, with the solution set shown in cross-hatching, the hyperplane for the violated inequality as a dotted line:

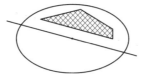

The decisive step of the ellipsoid method now consists in a general procedure for constructing a smaller ellipsoid out of a half-ellipsoid:

- the new ellipsoid completely contains the half-ellipsoid.

- the volume of the new ellipsoid is smaller by a certain factor than that of the previous ellipsoid.

Since one can deform the axes of the ellipsoid, it is sufficient in principle to carry out the construction for a hypersphere. The following figure illustrates the two-dimensional case:

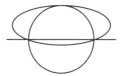

For a system of inequalities whose solution set is bounded but that represents a true volume, this successive shrinking of the ellipsoids must end after a number of steps that can be estimated in advance, where the last midpoint yields a solution. Other systems of inequalities can be transformed as necessary, such as, for example, by allowing oneself some "generosity" with respect to the inequalities and extending slightly the given bounds.

While Khachiyan's method works well in theory, in practice it is generally inferior to the simplex algorithm. So in 1984 Karmarkar presented a decidedly better algorithm. Nonetheless, the simplex algorithm has remained the most popular method for linear optimization.

For more on the methods of Karmarkar and Khachiyan, see Ulrich Derigs, Neuere Ansätze in der Linearen Optimierung, *Operations Research Proceedings* **29**, 1985, pp. 47–58; A. Shrijver, The new linear progamming method of Karmarkar, *Centrum voor Wiskunde en Informatica Newsletter* 8, 1985, pp. 2–14; Neuer Dampf, *Der Spiegel* **49**, 1984, pp. 239–240; Varék Chvátal, *Linear Programming*, New York 1983; Robert G. Bland, Donald Goldfarb, Michael J. Todd, The ellipsoid method, *Operations Research* **29**, 1981, pp. 1039–1091.

3. This is the so-called duality theorem, which relates a given linear optimization problem to another optimization problem, called the dual optimization problem. In the example considered in "Linear Optimization," the
dual problem asks about the values of the individual resources based on a
marginal cost analysis. That is, resources are evaluated on the basis of the
growth in proceeds that accrue from a very small increase in the resources.
In this sense, resources that are never exhausted in the optimal production
plan are worthless. In our example, for the four resources A, B, C, and D
we have the values $a = 0$, $b = 1$, $c = 1$, and $d = 0$.

As with the minimax strategies of a two-person zero-sum game, the solutions of the two dual problems supplement each other, and the solution
of one can be used, even if one has just happened to guess the values by
chance, to confirm the solution of the other. To this end one simply adds
the inequalities with weights that correspond to the solution of the dual.
With regard to the values $a = 0$, $b = 1$, $c = 1$, and $d = 0$, one adds in the
example the second and third inequalities and obtains

$$2x + 3y \leq 24,$$

so that the maximum obtainable income $2x + 3y$ can be *at most* 24. That
it must be *at least* 24 follows from the values $x = 6$, $y = 4$, which satisfy
all the conditions.

What formal requirements must such values a, b, c, d satisfy so that a conjectured optimality of a pair of values x and y can be confirmed? The
procedure can be used analogously when the numbers a, b, c, d are greater
than or equal to zero and the conditions

$$a + b + c + 2d \geq 2,$$
$$4a + 2b + 2c + 2d \geq 3$$

are satisfied. Moreover, the expression

$$24a + 14b + 17c + d$$

can attain at most the value of the maximum to be verified, which in our
case is 24. That is, the values a, b, c, d can be derived from a minimization problem whose starting parameters are transposed with respect to the
original geometric arrangement:

minimize
$$24a + 14b + 17c + d$$
subject to the conditions
$$a + b + c + 2d \geq 2,$$
$$4a + 2b + 2c + d \geq 3,$$

as well as
$$a \geq 0, \quad b \geq 0, \quad c \geq 0, \quad d \geq 0.$$

The duality theorem of linear optimization states in principle that a solution of every solvable optimization problem can be verified by a solution of the dual optimization problem.

In Chapter 37 we will meet some special examples of mutually dual optimization problems and their analyses by two-person zero-sum games from the point of view of the two players.

Further Literature on Linear Optimization

[1] Lothar Collatz, Wolfgang Wetterling, *Optimierungsaufgaben*, Berlin 1971.

[2] Peter Kall, *Mathematische Methoden des Operations Research*, Stuttgart 1976.

[3] David Gale, *The Theory of Linear Economic Models*, Chicago 1960.

[4] Robert Dorfman, Paul A. Samuelson, Robert M. Solow, *Linear Programming and Economic Analysis*, New York 1958.

One can also find overviews in practically any mathematical reference work. Here are some sources written at a fairly elementary level:

[5] *For All Practical Purposes: Introduction to Contemporary Mathematics*, New York 1988.

[6] John Casti, *Five Golden Rules*, New York 1996.

[7] Robert G. Bland, Wirtschaftsfaktor lineare Programmierung, *Spektrum der Wissenschaft* **8**, 1981, pp. 119–130.

37

Minimax and Linear Optimization: As Simple as Can Be

We would like to find the simplest method for calculating the minimax strategies for both players of a two-person zero-sum game given by its normal form.

The method introduced in the previous chapter for calculating minimax strategies by first symmetrizing a game leads for a normal form with n rows and m columns to a linear optimization problem with $m + n + 1$ variables and $m + n + 2$ additional slack variables. Since the size of the optimization problem determines in great measure the amount of work necessary for its solution, one is naturally led to the question of whether the minimax strategies might be found using a less extensive optimization problem. In fact, they can be. The simplest method is that introduced in 1960 by Albert W. Tucker (1905–1995), one of the pioneers of linear optimization and game theory.[1] Tucker's approach manages with only m variables and n additional slack variables. We will demonstrate the method by means of an example. We return to a game discussed in Chapter 34:

---

[1] A. W. Tucker, Solving a matrix game by linear programming, *IBM Journal of Research and Development* **4**, 1960, pp. 507–517.

		Black			
		1	2	3	4
White	1	2	−1	1	2
	2	−1	1	0	1

In searching for minimax strategies for white and black, we start by posing two linear optimization problems whose solutions contain the desired minimax strategies. We begin with the maximizing player white, who mixes her two strategies with probabilities x_1 and x_2 to ensure for herself the greatest possible winning expectation. Formally, we are looking for two numbers x_1 and x_2 for which a maximum value v is obtained under the conditions

$$2x_1 - x_2 \geq v,$$
$$-x_1 + x_2 \geq v,$$
$$x_1 \geq v,$$
$$2x_1 + x_2 \geq v,$$
$$x_1 + x_2 = 1,$$
$$x_1 \geq 0,$$
$$x_2 \geq 0.$$

White's goal:

$$v = \text{max!}$$

Black, on the other hand, seeks four numbers y_1, y_2, y_3, y_4 that yield the smallest possible value of v under the conditions

$$2y_1 - y_2 + y_3 + 2y_4 \leq v,$$
$$-y_1 + y_2 + y_4 \leq v,$$
$$y_1 + y_2 + y_3 + y_4 = 1,$$
$$y_1, \ y_2, \ y_3, \ y_4 \geq 0.$$

Black's goal:

$$v = \text{min!}$$

There are now various ways of transforming the two stated problems into the standard form of an optimization problem. The most direct way consists in treating the value v to be optimized as a variable. However, Tucker's suggested method makes do with one variable and one side condition fewer. One begins with the assumption that the game value, that is, the achievable maximum, respectively minimum, in the first or second optimization problem is positive. This is assured us for the game under consideration based on our investigations in Chapter 34.[2] Otherwise, a sufficiently larger bonus is set for white so that all the entries in the normal form are positive. In what follows, the variables can yield almost optimal values only when v is positive. Therefore, for these values it is permitted to divide individual inequalities by v; we do this to the inequalities for which v is on the right-hand side. Then the variables are replaced as follows:

$$X_1 = \frac{x_1}{v}, \quad X_2 = \frac{x_2}{v}, \quad Y_1 = \frac{y_1}{v}, \quad Y_2 = \frac{y_2}{v}, \quad Y_3 = \frac{y_3}{v}, \quad Y_4 = \frac{y_4}{v}.$$

Together with the supplementary slack variables $X_3, \ldots, X_6, Y_5, Y_6$, which allow us to change the inequalities to equalities, we obtain the following optimization problem:

- under the side conditions

$$X_3 = -1 + 2X_1 - X_2,$$
$$X_4 = -1 - X_1 + X_2,$$
$$X_5 = -1 + X_1,$$
$$X_6 = -1 + 2X_1 + X_2,$$

  and the nonnegativity condition

$$X_1, \ldots, X_6 \geq 0,$$

  white maximizes the value of $-1/v$; that is,

$$-X_1 - X_2 = \text{max}!$$

- analogously, under the side conditions

$$Y_5 = 1 + 2(-Y_1) - (-Y_2) + (-Y_3) + 2(-Y_4),$$
$$Y_6 = 1 - (-Y_1) + (-Y_2 1) + (-Y_4),$$

  together with the nonnegativity condition

$$Y_1, \ldots, Y_6 \geq 0,$$

  black minimizes the value of $-1/v$; that is,

$$(-Y_1) + (-Y_2) + (-Y_3) + (-Y_4) = \text{min}!$$

---

[2]One may also convince oneself of a positive value directly if white mixes her two strategies randomly in the proportion $2 : 3$.

The reason that the second optimization problem is formulated with negative variables will now be made clear. Such a formulation allows both problems, whose side conditions employ the same coefficients, though not in the same order, to be represented using a single array of numbers. Such a *simplex tableau* is to be read both horizontally and vertically (see "Simplex Tableaus and the Rectangle Rule"). The nonnegativity conditions are not considered in the simplex tableau:

	$-T_{\max} =$	$X_3 =$	$X_4 =$	$X_5 =$	$X_6 =$	
$T_{\min} =$	0	1	1	1	1	$\times(-1)$
$Y_5 =$	1	2	-1	1	2	$\times X_1$
$Y_6 =$	1	-1	1	0	1	$\times X_2$
	$\times 1$	$\times(-Y_1)$	$\times(-Y_2)$	$\times(-Y_3)$	$\times(-Y_4)$	

Simplex Tableaus and the Rectangle Rule

It has been shown useful for several reasons to use simplex tableaus in the simplex algorithm: the same factors that save on the amount of writing that has to be done in a manual calculation make possible in a computer program the direct use of data organization and manipulation. Furthermore, two relationships can be represented with the same data structure. Let us look first at a simplex tableau, which has the following general form:

		$\alpha =$	$\cdots$	$\beta =$		
		$\vdots$		$\vdots$		
$A =$	$\cdots$	p	$\cdots$	z	$\cdots$	$\times \gamma$
$\vdots$		$\vdots$		$\vdots$		$\vdots$
$B =$	$\cdots$	s	$\cdots$	r	$\cdots$	$\times \delta$
		$\vdots$		$\vdots$		
		$\times(-C)$	$\cdots$	$\times(-D)$		

Without loss of generality, we may restrict our attention to 2×2 tableaus to spare ourselves the many ellipsis marks $(\cdots)$. A tableau represents the two systems

$$A = p(-C) + z(-D),$$
$$B = s(-C) + r(-D),$$

and

$$\alpha = p\gamma + s\delta,$$
$$\beta = z\gamma + r\delta.$$

With this interpretation it is clear how a simplex tableau can be transformed. Rows can be interchanged with one another, as can columns, provided that the variables at the edges are interchanged along with them. Of greater significance, however, is the interchange of a row with a column. Such an exchange represents a transformation such as occurs in each step of the simplex algorithm. For $p \neq 0$, the first equations of both systems will be solved for C and γ. The result in then substituted into the other equation:

$$C = \frac{1}{p}(-A) + \frac{z}{p}(-D),$$

$$B = \frac{-s}{p}(-A) + \left(r - \frac{sz}{p}\right)(-D),$$

and

$$\gamma = \frac{1}{p}\alpha + \frac{-s}{p}\delta,$$

$$\beta = \frac{z}{p}\alpha + \left(r - \frac{sz}{p}\right)\delta.$$

We see that the transformed system of equations again yields a *common* simplex tableau:

	$\gamma =$	$\beta =$	
$C =$	$\frac{1}{p}$	$\frac{z}{p}$	$\times \alpha$
$B =$	$\frac{-s}{p}$	$r - \frac{sz}{p}$	$\times \delta$
	$\times(-A)$	$\times(-D)$	

Instead of transforming the system of equalities, one could simply calculate with simplex tableaus. Here are the general rules:

- the number at the intersection of the interchanged row and column, the *pivot element* (p), is replaced by its reciprocal.
- the other numbers of the pivot row (z) are divided by the pivot element.
- the other numbers of the pivot column (s) are divided by the negative of the pivot element.
- for the rest of the numbers (r) one follows the *rectangle rule*, whereby one forms a rectangle determined by the pivot element and the current number to be transformed: using the four numbers on the corners of this rectangle, one calculates the new value using the formula $r - sz/p$.
- the variables on the edge are interchanged between the pivot row and pivot column, where in the exchange between the left and bottom edges and conversely the signs are to be changed.

The nonnegativity conditions are not considered in the simplex tableau. They are taken into account indirectly in the simplex algorithm via the starting tableau and the choice of pivot, that is, in the choice of row and column to be exchanged.

Before we begin with the actual simplex algorithm, we would like to interpret the starting tableau: the parametric values corresponding to the tableau arise from all the variables beneath and on the right being equal to zero. Only the Y variables, with $Y_5 = Y_6 = 1$ and $Y_1 = Y_2 = Y_3 = Y_4 = 0$, satisfy the nonnegativity conditions. Regarding the X variables, on the other hand, with $X_3 = X_4 = X_5 = X_6 = -1$ and $X_1 = X_2 = 0$ there are no admissible parametric values for the related optimization problem. Such will be obtained only with the last exchange step.

Since only the Y values are admissible, we must orient the exchange steps to the associated minimization problem, as we established in the previous chapter: except for the left-most column, any column may be chosen for the exchange, provided that a positive number is in the first row. A Y variable associated with such a column reduces the target value $T_{\min}$ when it grows above 0. Once such a pivot column is chosen, one must then find the correct pivot row. The criterion is that this row must offer the sharpest bound for the affected Y variable with respect to the nonnegativity conditions. One seeks, then, among the rows with a positive number within the pivot column one whose quotient from the numbers of the first column and the pivot column is minimal.

If one holds to these two general criteria, namely, positive value in the highest row and minimal quotient with positive denominator, then the rest of the pivot selection is arbitrary. Somewhat arbitrarily, namely in choosing a pivot element of 1 so as to avoid fractions, we begin by exchanging the variables Y_5 and Y_3:

	$-T_{\max} =$	$X_3 =$	$X_4 =$	$X_1 =$	$X_6 =$	
$T_{\min} =$	-1	-1	2	-1	-1	$\times(-1)$
$Y_3 =$	1	2	-1	1	2	$\times X_5$
$Y_6 =$	1	-1	1	0	1	$\times X_2$
	$\times 1$	$\times(-Y_1)$	$\times(-Y_2)$	$\times(-Y_5)$	$\times(-Y_4)$	

In the next step, the variable Y_2 must be exchanged, since only in that way can a further diminution of the target value $T_{\min}$ be achieved. It can be exchanged only with the variable Y_6:

	$-T_{\max} =$	$X_3 =$	$X_2 =$	$X_1 =$	$X_6 =$	
$T_{\min} =$	-3	1	-2	-1	-6	$\times(-1)$
$Y_3 =$	2	2	1	1	3	$\times X_5$
$Y_2 =$	1	-1	1	0	1	$\times X_4$
	$\times 1$	$\times(-Y_1)$	$\times(-Y_6)$	$\times(-Y_5)$	$\times(-Y_4)$	

Now the variable Y_1 must be exchanged, since otherwise, no further reduction in the target value $T_{\min}$ can be achieved. It is exchanged with the variable Y_3:

	$-T_{\max} =$	$X_5 =$	$X_2 =$	$X_1 =$	$X_6 =$	
$T_{\min} =$	-5	-1	-3	-2	-6	$\times(-1)$
$Y_1 =$	2	1	1	1	3	$\times X_3$
$Y_2 =$	3	1	2	1	4	$\times X_4$
	$\times 1$	$\times(-Y_3)$	$\times(-Y_6)$	$\times(-Y_5)$	$\times(-Y_4)$	

In the subrectangle above and to the right, no value is positive. This has two consequences: first, the target value $T_{\min}$ has reached its minimum when the variables beneath, Y_3, Y_6, Y_5, Y_4, are all zero. Second, this is the first tableau for which there are not only admissible Y values but also admissible X values that satisfy all the nonnegativity conditions. And these X values are not only admissible, but even optimal for the target value

$$T_{\max} = -5 - 2X_3 - 3X_4$$

if the variables X_3 and X_4 on the right are both equal to zero. And this is not a special property of this example, but a general property of "dual" optimization problems, since the lower-left subrectangle contains no negative entries.

From the optimal values $T_{\max} = T_{\min} = -5$ one obtains the game value $v = 1/5$. Together with the optimal parametric values

$$X_1 = 2, \quad X_2 = 3, \quad \text{and} \quad Y_1 = 2, \quad Y_2 = 3, \quad Y_3 = 0, \quad Y_4 = 0,$$

we have the probabilities for the desired minimax strategies:

$$x_1 = \frac{2}{5}, \quad x_2 = \frac{3}{5}, \quad \text{and} \quad y_1 = \frac{2}{5}, \quad y_2 = \frac{3}{5}, \quad y_3 = 0, \quad y_4 = 0.$$

And even the optimal values of the slack variables,

$$X_3 = 0, \quad X_4 = 0, \quad X_5 = 1, \quad X_6 = 6, \quad \text{and} \quad Y_5 = 0 \quad Y_6 = 0,$$

give us interesting information. Positive values indicate the pure strategies that the opponent should avoid at all costs against the found minimax strategies. Thus in the case at hand, black should in no case choose the strategies corresponding to the slack variables $X_5 = 1$ and $X_6 = 6$, that is, the strategies "3" and "4." The values of the variables estimate, when multiplied by the value $v = 1/5$, the "costs" of the associated decision: white can then count on an additional winning expectation in the amount of $1/5$ or $6/5$ above the minimax value.

To summarize, if minimax strategies are determined for a game whose normal form is given by the matrix A, then in the case of a positive minimax value v, such a determination can be made in the case of a positive minimax value v from the following tableau:

One must hold to the specified interpretation of the variables in the exchange steps of the subsequent simplex algorithm so that the results above and to the left in the final tableau can be associated with the correct variables. Variables that end up, down, or to the right are all equal to zero. To obtain the probabilities of the minimax strategies and the costs of erroneous decisions, one must finally divide the variable values by v. This can be seen in the final tableau in the upper-left corner: $v = -1/T_{\max} = -1/T_{\min}$.

38

Play It Again, Sam: Does Experience Make Us Wiser?

Does experience alone suffice for discovering all good game strategies? In particular, for every two-person zero-sum game can one arrange a sequence of games that enables one to determine the minimax strategies empirically?

There are probably few players who are able to optimize their strategies with the help of the simplex algorithm. And moreover, most games people play are much too complex for such calculations to actually be carried out. But can good strategies be found without such calculations? Is a game's mature tradition a guarantee that good strategies have been found in an evolutionary trial-and-error manner?

One must say that mixed strategies are as strange to many players as they were to mathematicians for many centuries. Debates among skat players about the correctness of the motto "Always play the ace first" arise not only from the heat of battle, but are also an indication that patterns of play can be considered strictly comparable, in which case mixed strategies would be superfluous. The reason for such beliefs can be found in the fact that minimax strategies do not necessarily represent the highest level of play, since they do not always react sufficiently to recognizably bad play on the part of an opponent. Indeed, there are frequently much more important considerations in play than finding the optimal probabilities for the various moves. Anyone who has played against a wily skat player will acknowledge this: not only does he or she keep track of every point and the trumps that have been played, but the entire course of the game, from the bidding to each individual trick, not to mention such auxiliary aids such as noticing how a beginner always sorts his cards the same way. For the average player there is thus plenty of opportunity to perfect his play by completely evaluating the available information. Before this level has been reached, it makes little sense to conquer one's ignorance with a mixed strategy.

On the other hand, in other games, particularly such strategic games as poker, mixed strategies play a central role in usual play. Thus the concepts of a mixed strategy can be brought into play based on empirical experience. Could one even go so far as to imagine simulated sequences of games by which the minimax strategies could be empirically determined?

Let us again begin with a two-person zero-sum game presented in normal form. We set up a series of games in which both players attempt to improve their strategies progressively. Each game is played according to the normal rules. In particular, in the individual games only pure strategies are used. Mixed strategies enter into the picture only indirectly, namely, as relative frequencies with which the pure strategies were chosen in the previous games. To achieve optimal play, our two players will proceed, according to our assumption, as follows: at the beginning of a game each player evaluates the previous play of his opponent and interprets it as a mixed strategy that the opponent will continue to use and that thus must be countered as well as possible. That is, each player seeks a pure strategy that brings the best result against the opponent's previous average strategy. George W. Brown, who developed this approach of imaginary series of games in 1949, noted the following:[1]

> The iterative method in question can be loosely characterized by the fact that it rests on the traditional statistician's philosophy of basing future decisions on the relevant past history. Visualize two statisticians, perhaps ignorant of min–max theory, playing many plays of the same discrete zero-sum game. One might naturally expect a statistician to keep track of the opponent's past plays and, in the absence of more sophisticated calculation, perhaps to choose at each play the optimum pure strategy against the mixture represented by all the opponent's past plays.

Let us begin with an example, returning to the poker model from Chapter 35:

---

[1]G. W. Brown, Iterative solutions of games by fictitious play, in: T. C. Koopmanns (ed.), *Activity Analysis of Production and Allocation*, Cowles Commission Conference Monograph **13** New York 1951, pp. 374–376.

		Black Chooses			
		1	2	3	4
White	1	0	0	0	0
Chooses	2	2	0	3	1
	3	6	1	4	−1
	4	8	1	7	0

Let us see how the games go:

1. in the first game, neither player possesses any information about the opponent's manner of playing. Since we wish in general to assume that the players will always choose among strategies with equally good prospects the one with the smallest number, each will choose the first strategy for the first game.

2. in the second game, each player starts with the assumption that the opponent will stick to the first strategy. For white, then, it is clear that she should choose her fourth strategy. Black, on the other hand, has the same prospects for all four of his strategies, and so black again decides for his first strategy.

3. in the third game, the picture becomes more interesting. White must again decide on her fourth strategy, since black has thus far chosen only the strategy 1. Black, however, suspects that white is using a mixed strategy in which the pure strategies 1 and 4 are mixed randomly in proportion 1 : 1. White's expectation depends on black's reply: 4, 0.5, 3.5, and 0.

4. at the beginning of the fourth game, the players reason as follows: white assumes that black mixes his first and fourth strategies randomly with proportion 2 : 1, while black assumes that white mixes her strategies 1 and 4 in proportion 1 : 2. On this basis, both players decide on their fourth strategies.

We will spare ourselves the gory details of additional games. If one really wished to carry out these computations, one should write a computer program to do it. For purposes of demonstration in a simple game with not too many rounds, a spreadsheet program will do.[2] Table 38.1 shows the results of the first 50 games. The net results for 100, 1000, 10 000, and 100 000 games can be seen in Table 38.2. The following information is given:

---

[2]Table 38.1 was created with a spreadsheet program. Not shown are the auxiliary columns that reflect the minimization and maximization and the resulting selection of the strategies for the next game.

- the frequencies of the strategies in the previous parties.

- the winning expectations derived from the corresponding mix of strategies.

- the bound on the game value resulting from the two current mixtures of strategies, calculated on the basis of the best possible counterstrategy to be used by each player in the following game.

One can see that the dominated strategies 1 and 3 are chosen by white and black at most in the initial phases. Thereafter, the game is concentrated on those strategies that arise from the two minimax strategies. Altogether, in this game both the relative frequencies and the winning expectations appear to converge. That is, our empirical game simulation yields both the game value and the minimax strategies.

Brown's iterative procedure for improving the strategies of both players simultaneously in a type of learning process works in great generality. In 1951 the mathematician Julia Robinson[3] (1919–1985) showed that the game value can be bounded arbitrarily closely, although the process can be very protracted.[4] However, the average strategies of the two players need not converge. This is not too bad, though, since in every step the strategies determined are also good enough to allow the current bounding of the game value. With the convergence of the bounding values, the two strategies become arbitrarily good.

In practice, the slow convergence has a negative effect, particularly for a game with a large normal form. Thus if the maximal remaining approximation error is to be halved for a game with n and m strategies after a fixed number of games, then the number of games must be increased by a factor of up to 2^{m+n-1}. This is much slower than a Monte Carlo simulation, which we saw in Chapter 15 as an empirical method for determining expectations. There the average error is halved after only four times the number of steps.

There is another point of view for considering a comparison with the Monte Carlo method: in contrast to that method, the one considered here is absolutely deterministic. Of course, in individual steps there can be ambiguity if a player has two or more equally good strategies from which

---

[3] Julia Robinson is best known for her contributions to Hilbert's tenth problem, mentioned in Chapter 28. She did significant work that underlaid the ultimate solution found by Yuri Matiasevich.

[4] Julia Robinson, An iterative method for solving a game, *Annals of Mathematics* **54**, 1951, pp. 296–301; reprinted in Harold W. Kuhn (ed.), *Classics in Game Theory*, Princeton 1997, pp. 27–35; H. M. Shapiro, Note on a computation method in the theory of games, *Communications on Pure and Applied Mathematics* **11**, 1958, pp. 588–593.

Game #	Strategy of White	Black	Previous Strategy Frequencies White 1	2	3	4	Black 1	2	3	4	Winning Expectation for Current Probability Distribution	Game Value Min	Max
0			0	0	0	0	0	0	0	0			
1	1	1	1	0	0	0	1	0	0	0	0.000	0.000	8.000
2	4	1	1	0	0	1	2	0	0	0	4.000	0.000	8.000
3	4	4	1	0	0	2	2	0	0	1	3.556	0.000	5.333
4	4	4	1	0	0	3	2	0	0	2	3.000	0.000	4.000
5	4	4	1	0	0	4	2	0	0	3	2.560	0.000	3.200
6	4	4	1	0	0	5	2	0	0	4	2.222	0.000	2.667
7	4	4	1	0	0	6	2	0	0	5	1.959	0.000	2.286
8	4	4	1	0	0	7	2	0	0	6	1.750	0.000	2.000
9	4	4	1	0	0	8	2	0	0	7	1.580	0.000	1.778
10	4	4	1	0	0	9	2	0	0	8	1.440	0.000	1.600
11	4	4	1	0	10		2	0	0	9	1.322	0.000	1.455
12	4	4	1	0	11		2	0	0	10	1.222	0.000	1.333
13	4	4	1	0	12		2	0	0	11	1.136	0.000	1.231
14	4	4	1	0	13		2	0	0	12	1.061	0.000	1.143
15	2	4	1	1	13		2	0	0	13	1.000	0.067	1.133
16	2	4	1	2	13		2	0	0	14	0.953	0.125	1.125
17	2	4	1	3	13		2	0	0	15	0.917	0.176	1.118
18	2	4	1	4	13		2	0	0	16	0.889	0.222	1.111
19	2	4	1	5	13		2	0	0	17	0.867	0.263	1.105
20	2	4	1	6	13		2	0	0	18	0.850	0.300	1.100
21	2	4	1	7	13		2	0	0	19	0.837	0.333	1.095
22	2	4	1	8	13		2	0	0	20	0.826	0.364	1.091
23	2	4	1	9	13		2	0	0	21	0.819	0.391	1.087
24	2	4	1	10	13		2	0	0	22	0.813	0.417	1.083
25	2	4	1	11	13		2	0	0	23	0.808	0.440	1.080
26	2	4	1	12	13		2	0	0	24	0.805	0.462	1.077
27	2	4	1	13	13		2	0	0	25	0.802	0.481	1.074
28	2	2	1	14	13		2	1	0	25	0.800	0.464	1.036
29	2	2	1	15	13		2	2	0	25	0.795	0.448	1.000
30	2	2	1	16	13		2	3	0	25	0.790	0.433	0.967
31	2	2	1	17	13		2	4	0	25	0.784	0.419	0.935
32	2	2	1	18	13		2	5	0	25	0.776	0.406	0.906
33	2	2	1	19	13		2	6	0	25	0.769	0.394	0.879
34	2	2	1	20	13		2	7	0	25	0.760	0.382	0.853
35	2	2	1	21	13		2	8	0	25	0.752	0.371	0.829
36	2	2	1	22	13		2	9	0	25	0.743	0.361	0.806
37	2	2	1	23	13		2	10	0	25	0.734	0.351	0.784
38	2	2	1	24	13		2	11	0	25	0.725	0.342	0.763
39	2	2	1	25	13		2	12	0	25	0.716	0.333	0.744
40	2	2	1	26	13		2	13	0	25	0.707	0.325	0.725
41	2	2	1	27	13		2	14	0	25	0.698	0.317	0.732
42	4	2	1	27	14		2	15	0	25	0.690	0.333	0.738
43	4	2	1	27	15		2	16	0	25	0.683	0.349	0.744
44	4	2	1	27	16		2	17	0	25	0.677	0.364	0.750
45	4	2	1	27	17		2	18	0	25	0.672	0.378	0.756
46	4	2	1	27	18		2	19	0	25	0.668	0.391	0.761
47	4	2	1	27	19		2	20	0	25	0.664	0.404	0.766
48	4	2	1	27	20		2	21	0	25	0.661	0.417	0.771
49	4	2	1	27	21		2	22	0	25	0.658	0.429	0.776
50	4	2	1	27	22		2	23	0	25	0.656	0.440	0.780

Table 38.1. The first 50 games.

Game	Previous Strategy Frequencies								Winning Expectation for Current Probability Distribution	Game Value	
	White				Black						
Number	1	2	3	4	1	2	3	4		Min	Max
100	1	55	0	44	2	40	0	58	0.5874	0.4400	0.6200
1000	1	514	0	485	2	485	0	513	0.5087	0.4850	0.5170
10 000	1	5026	0	4973	2	4949	0	5049	0.5009	0.4973	0.5053
100 000	1	50 031	0	49 968	2	50 185	0	49 813	0.5001	0.5001	0.5020

Table 38.2. Net results up to 100 000 games.

to choose. But something of the sort can occur equally likely for a step in the simplex algorithm. That is, despite the empirical nature of the basic idea, we are dealing with a purely calculational procedure. In particular, in games with random elements, one must always work with expectations, as we did in our example of the primitive poker model. Individual games are therefore played only in their strategic components, not their random ones according to the actual rules.

39

Le Her: Should I Exchange?

White and black are playing a game to see who draws the higher card. They use a normal 52-card deck, with king high and ace low. If the two cards drawn have the same value, black wins.

The game begins with each player receiving a card, and a third card is laid face down on the table. Then each player has a chance of improving his or her situation. White goes first. She is allowed to exchange cards with black. Unless black is holding a king, he must accept the request for an exchange. Regardless of what transpired on white's turn, black on his turn has the chance to exchange his card for the card lying face down on the table, though if he draws a king, he must return it. Then both players show their cards and score the game.

Which cards should the players exchange, and which not?

The game that we have just described was played in the 18th century under the name "le Her." In contrast to the games with mixed minimax strategies that we have considered thus far, this game exhibits a significantly greater complexity. For example, white has 13 individual decisions that can be combined in any way. Namely, for each card value she has to plan whether an exchange should be made. This gives white 2^{13} pure strategies. Black's cogitations are even more involved, since his decision is based not only on the card he holds, but on the results of white's turn as well.

Despite the apparent complexity of the game le Her, one can obtain minimax strategies fairly easily. They yield white a winning expectation of 0.0251, which corresponds to a probability of winning equal to 0.5125:

- white

  - exchanges all cards up to and including a 6;
  - exchanges a 7 with probability 3/8;
  - retains all cards 8 and above.

- black

  - in the case that white has declined to exchange, black exchanges

    * all cards up to and including a 7;
    * an 8 with probability 5/8;
    * no higher cards.

  - If white has exchanged cards, then black exchanges if and only if his card is worse than the card that he knows white to be holding.

A point of great interest about the game le Her is the fact that the minimax strategies just described were discovered more than two centuries before the systematic investigations of Borel and von Neumann. Little is known about their discoverer, an Englishman named Waldegrave who probably was living in Paris. Waldegrave learned about the game from Pierre Rémond de Montmort (1678–1719), who in 1708 had published a book on games of chance, *Essay d'analyse sur les jeux de hasard*, in which he raised the question of the best strategy for le Her. One can see the difficulties in the solution to this problem from an exchange of letters in the following years between Montmort and Nicholas Bernoulli (1687–1759), a nephew of Jacob Bernoulli, the discoverer of the law of large numbers. In all, 16 letters from the years 1711–1715 contain discussions on le Her.[1] In this exchange are also included suggestions that Waldegrave had made to Montmort. Waldegrave's central idea appears in a letter of Montmort dated 15 November 1713 to a very skeptical Nicholas Bernoulli, which Montmort included as an appendix in the second edition of his book on games of

---

[1]Julian Henny, Niklaus und Johann Bernoullis Forschungen auf dem Gebiet der Wahrscheinlichkeitsrechnung in ihrem Briefwechsel mit Pierre Rémond de Montmort, dissertation, Basel 1973, in: *Die Werke von Jakob Bernoulli*, Band 3, Basel 1975, pp. 457–507; Robert W. Dimand, Mary Ann Dimand, The early history of the theory of strategic games from Waldegrave to Borel, in: E. Roy Weintraub (ed.), *Toward a History of Game Theory*, Durham 1992, pp. 15–28; Robert W. Dimand, Mary Ann Dimand, *The History of Game Theory, From the Beginnings to 1945*, volume 1, London 1996, pp. 120-123; Anders Hald, *A History of Probability and Statistics and Their Applications Before 1750*, New York 1990, Chapter 18, in particular, Section 18.6.

chance.[2] Montmort quotes in this letter a letter from Waldegrave of two days earlier on the subject of le Her.[3]

Bernoulli, Montmort, and Waldegrave were in complete agreement on the optimal strategy for most situations arising in the game. Disagreement was focused on how white was to play a 7, and black, if white had declined to exchange, an 8. These two decisions have properties of the sort that we have come to know from the game "even or odd": There is no one "best" move. Which move is best depends on how the opponent decides in the other situation. Waldegrave therefore suggests that the decision be made randomly. He imagines that a token is drawn from a container containing tokens of two different colors. Waldegrave assumes that white's supply contains a tokens for "exchange" and b tokens for "no exchange," while black's supply contains c for exchange and d for no exchange. After extensive combinatorial considerations, one finally concludes that the winning probability for white is given by

$$\frac{2828ac + 2834bc + 2838ad + 2828bc}{13 \times 17 \times 25(a+b)(c+d)}.$$

This formula, which appears in the first paragraphs of Montmort's letter,[4] serves as Waldegrave's starting point. He recognizes that with a strategic mix by white with the values $a = 3$ and $b = 5$, it doesn't matter whether black exchanges. White always wins with probability

$$\frac{2831}{5525} + \frac{3}{5525 \times 4}.$$

This probability that white wins can also be fixed by black, who, according to Waldegrave, should use the values $c = 5$ and $d = 3$. Then it doesn't matter how white decides to play a 7. Waldegrave considers other proportions to be risky. Regarding player white, whom he calls Paul, Waldegrave remarks:

> It is true that for values of a and b that differ from $a = 3$ and
> $b = 5$, Paul [white] can improve his situation if Peter [black]

[2]Pierre Rémond de Montmort, *Essay d'analyse sur les jeux de hasard*, second edition, Paris 1713, reprint New York 1980, pp. 403–413, and also 321, 334, 338, 348, 361, 376.

[3]Montmort's letter appears in partial English translation in Harold Kuhn, James Waldegrave, Excerpt from a letter, in: William J. Baumol, Stephen M. Goldfeld (eds.), *Precursors in Mathematical Economics: An Anthology, Series of Reprints of Source Works in Political Economics* **19**, London 1968, pp. 3–9, reprinted in Mary Ann Dimand, Robert W. Dimand, *The Foundations of Game Theory*, Cheltenham 1997, volume I, pp. 3–9.

[4]Pierre Rémond de Montmort, *Essay d'analyse sur les jeux de hasard*, second edition, Paris 1713, Reprint New York 1980, p. 404, where the denominator has been transformed into modern notation.

makes the wrong decision. But he will worsen his situation if Peter [black] makes the correct choice.

We can best understand Waldegrave's result by expressing the winning probability for white as a function of the two probabilities $p = a/(a + b)$ and $u = c/(c + d)$:

$$\frac{11327 - (8p - 3)(8u - 5)}{22100}.$$

It is apparent at once that white can eliminate the risk arising from the opponent's decision only with the value $p = 3/8$. The same is possible for black, but only with the probability $u = 5/8$.

Waldegrave's ideas found little positive resonance with his contemporaries, beginning with Bernoulli.[5] But thanks to Montmort's book they were not completely forgotten. Thus in 1865, Todhunter mentioned Waldegrave's explanation of le Her in a historical survey of probability theory. He even mentioned Waldegrave's suggestion that one's strategy should be varied, though not that this should be done in a random manner and which probabilities were the best.[6] Todhunter's book in turn inspired Roland Aylmer Fisher (1890–1962) in 1934 to investigate the game le Her. Fisher, one of the founders of modern statistics, also realized that the optimal strategy is to mix the moves in question in the proportion 3 : 5.[7] That made Fisher, who was apparently unaware of the work of Waldegrave, Borel, and von Neumann, the fourth person to solve a minimax situation with mixed strategies independently of his predecessors. Twenty-five years later, Waldegrave's work was rediscovered.[8]

---

[5]Nevertheless, Bernoulli later adopted Waldegrave's ideas with reservations, and even solved a simple game problem himself using his methods. The game was a variant of "even or odd":

			Black Guesses	
			"Odd"	"Even"
			1	2
White	"Odd"	1	−1	0
Chooses	"Even"	2	0	−4

Without describing how he obtained his solution, Bernoulli suggests the mixed strategy $(4/5, 1/5)$ in a letter of 20 February 1714, which is optimal for both players. See the work of Henny previously cited, p. 502.

[6]I. Todhunter, *A History of the Mathematical Theory of Probability from the Time of Pascal to That of Laplace*, Cambridge 1865, reprint New York 1965, pp. 106–110.

[7]R. A. Fisher, Randomisation, and an old enigma of card play, *Mathematical Gazette* **18**, 1934, pp. 294–297.

[8]G. Th. Guilbaud, Faut-il jouer au plus fin, in: *La Décision*, Éditions du Centre National de la Recherche Scientifique, Paris 1961, pp. 171–182.

We have not gone into the combinatorial thinking that underlies the formula cited above, nor have we discussed game situations in which there is a single best decision. In fact, our interest is not particularly focused on the game le Her. Rather, we would like to find in principle a programmable method whose memory requirements and computation time are not predetermined by the fact that several thousand rows and columns need to be calculated and further processed. How can the complete calculation of a normal form be avoided? The answer to this question is close at hand. Consider the chronological course of the game as a sequence of decisions. On this basis it is relatively easy to counter opposing strategies optimally: assume that a player knows the mixed strategies of his opponent. If he declines to decide among his random move options and the resulting influence of chance in the game, then the game becomes in principle a one-person game of chance, within which the player must attempt to react as optimally as possible. Optimal moves can be calculated by investigating the game situation in reverse order of the course of the game and thus searching for a move that presents the best winning expectation, as we saw in Chapter 17 for the game blackjack. The result is the gradual arrival at a pure strategy that optimally counters the opponent's strategy.

Since in le Her each player draws only once, the chronology of the game is quite simple. If the pure or mixed strategy of the opponent is known, then for each decision point one can determine whether one should exchange or not. To find the total strategy, there are 13 situations to distinguish for each of black and white. For each card value one determines whether an exchange profitably counters the opposing strategy. The decisions that black has to make when white has previously decided to exchange require no detailed investigation. Black, who in this case knows the opponent's card, decides on an exchange precisely when he holds the worse card.

Optimal replies are always made with respect to a fixed strategy of the opponent. Minimax strategies, on the other hand, optimize what is assuredly obtainable without regard to the opponent's actions. But how can such strategies be determined based on optimal replies? Since the procedure introduced in the previous chapter of an imaginary series of games was designed to determine optimal counterstrategies, it makes sense to apply those techniques. In particular, one does not have to compute the normal form. Otherwise, the actual procedure remains unchanged; that is, a continual updating of which strategies are used and how often is carried out during the iteration. On this basis, step by step, in the absence of the normal form using the technique just described, a pure strategy is determined for each player that optimally counters the previous average strategy of the opponent. Since most pure strategies are much too weak ever to be considered the best possible opposing strategy, the number of strategies actually used in these imaginary games remains limited. The result is a simplification at the beginning of the procedure in relation to the usual version. However, due to the additional work for each step, this advantage is soon eaten up.

Optimal replies are not limited to placement within imaginary series of games. They can also be used for computing minimax strategies with the help of the simplex algorithm. Here, as with the imaginary games, one proceeds stepwise, where each step is associated with a small (if possible) number of pure strategies. Then step by step a pair of minimax strategies related to the current selection of pure strategies is calculated with the help of the simplex algorithm. How good these strategies fare in actual play, in which all pure strategies are available to both sides, is seen by determining the optimal counterstrategy to them. The iteration is continued as long as at least one player can improve his prospects with a particular counterstrategy, where the counterstrategy in question is then used in the selection of strategies in the following steps. Thus the minimax optimization is gradually extended to all relevant pure strategies.

We would now like to use le Her to study how quickly such an optimization procedure can lead to the goal. Table 39.1 documents the process, which ends after five steps.

Although both players begin with rather stupid strategies in which no card is exchanged, the tactically interesting strategies appear quickly. Table 39.1 does not contain the mixed strategies that were computed for the current strategy selection with the help of the simplex algorithm. These "relative" minimax strategies serve only in finding optimal counterstrategies. Thus it can be determined how good the minimax strategies found actually are, as can be read from the last two columns of the table, and how the strategic selection is to be expanded, which can be seen from the second and third columns of the next step.

In the last step one obtains the two desired minimax strategies:

	Additional Strategy for		Range of Strategies for		Minimax Value of Selection	Winning Expectation for White If the Minimax Strategy Is Optimally Countered by	
Step	White	Black	White	Black		Black	White
1	never exch.	never exch.	1	1	−0.0588235	−0.2586425	0.1523379
2	exch. to 5	exch. to 6	2	2	0.0104374	0.0063348	0.0406033
3	exch. to 7	exch. to 7	3	3	0.0273303	0.0237104	0.0273303
4	—	exch. to 8	3	4	0.0237104	0.0237104	0.0258824
5	exch. to 6	—	4	4	0.0250679	0.0250679	0.0250679

Table 39.1. Iterative search for a minimax strategy for le Her.

- white exchanges all cards up to 6 with probability 5/8, and all cards up to 7 with probability 3/8.

- black exchanges all cards up to 7 with probability 3/8, and all cards up to 8 with probability 5/8.

These are precisely the strategies found by Waldegrave, though now computed with a concept that can in principle be used for every game that does not present too many decision situations.

40

Deciding at Random: But How?

A player employs a mixed strategy by making a single random decision at the start of a game that determines his play for the rest of the game. Can the random play of a minimax strategy also be organized move by move? That is, can each successive move be determined by a separate random decision?

Such a question is not nearly so academic as it might seem. Let us recall the game le Her, which we investigated in the previous chapter. From a practical point of view, a player has to weigh 13 different situations, determined by the value of the card, and make a yes–no decision for each. In order to avoid being seen through by the opponent, a player does well to vary his strategic plan randomly. Instead of planning a "global" mixed strategy that chooses a probability distribution from among the $2^{13} = 8192$ pure strategies, it is simpler for a player to specify 13 "local" probabilities, namely, one for each decision situation. This means that a player decides whether or not to exchange a particular card randomly according to the specified probability. Such a plan is called a *behavioral strategy*. What is not a priori self-evident is whether the concept of a behavioral strategy is sufficiently encompassing to be used in finding minimax strategies.

Behavioral strategies were first used in 1944 by von Neumann and Morgenstern in their book, *Theory of Games and Economic Behavior*. They used a poker model in which each of two players must decide, based on his or her hand, how to bid. Once again, the strict use of pure strategies has little to recommend it, since they allow the opponent to target his countermeasures effectively. But how can mixed strategies be determined in practice? Can it be done with a behavioral strategy as in le Her; that is, is it possible for a player to decide his bidding behavior for each individual hand? To take a simple example of two possible hands, a weak one and a strong one, together with two possible bids, high and low, von Neumann and Morgenstern explain the typical actions:

> Then there are four possible (pure) strategies, to which we shall give names:
>
> - **"Bold"**: bid "high" on every hand.
> - **"Cautious"**: bid "low" on every hand.
> - **"Normal"**: bid "high" on a "high" hand, "low" on a "low" hand.
> - **"Bluff"**: bid "high" on a "low" hand, "low" on a "high" hand.
>
> Then a 50–50 mixture of "Bold" and "Cautious" is in effect the same thing as a 50–50 mixture of "Normal" and "Bluff": both mean that the player will, according to chance, bid 50–50 "high" or "low" on any hand.
>
> Nevertheless these are, in our present notations, two "different" mixed strategies ....

That is, mixed strategies allow the "luxury" of setting not only the probabilities of the individual decisions, but also the statistical relationships among these individual decisions. We must ask, of course, whether this is ever necessary in practice. It is certainly not so in the case of our poker model, where a player has only a single decision to make in an individual game. Von Neumann and Morgenstern continue:

> This means, of course, that our notations, which are perfectly suited to the general case, are redundant for many particular games. This is a frequent occurrence in mathematical discussions with general aims.
>
> There was no reason to take account of this redundancy as long as we were working out the general theory. But we shall remove it now for the particular game under consideration.

In fact, von Neumann and Morgenstern succeeded in finding optimal behavioral strategies for their poker model. In comparison to the use of mixed strategies, their optimization based on individual decisions offers a significant simplification: for the arbitrary number S of equiprobable hands in their poker model, for each of which there are three possible bids, instead of having to search for 3^S optimal values, the number is reduced to $3S$.[1]

When a player moves, he must do so based on the information currently available to him. Instead of the information that the players possess altogether, here it is a larger or smaller subset, depending on the game. If in a game the states of information of the players do not agree entirely, then one speaks of a game with imperfect information. With respect to individual games, there can be two reasons for this:

- frequently, the results of random influences are known directly to only some of the players. Thus in a card game a player knows only his own cards, but not those of his opponent.

- it is in no way obvious that the behavior of the opponents is clear to a player. For example, which two cards a single player in skat lays face down, how many matchsticks a player takes in his hand in a gambling game, or how a player sets up his pieces at the beginning of a game like Stratego or ghosts, whose type is not apparent because they are laid face down, all these behaviors remain hidden at first from the player's opponents.

These two reasons can be understood together if one imagines random influences in a game as the moves of a fictitious player, where this imaginary player plays according to a fixed behavioral strategy that is known to the players and that corresponds exactly to the random decisions: then all the cases of imperfect information are the result of the fact that a player on his turn knows only in part how the players, both real and fictional, have played in the previous moves. If one imagines that the course of the game is completely determined by the sequence of actions of the players, including the imaginary one, then it becomes clear that there is no other cause of the imperfect information.

The character of a game is largely determined by how comprehensively a player whose turn it is, is informed about what moves have been made thus far in the game. Here there are informational components about which it is actually quite plausible that the player knows: on the one hand, this is information about which the player already had knowledge from the previous moves, and on the other hand, it is about the decisions that the player himself has made. If every player has access to this information, that is, if every player always knows what he has done and known previously,

---

[1]If one also considers that the probabilities of a random decision sum to 1, then in passing from mixed strategies to behavioral strategies the number of parameters is reduced even more, namely, from $3^S - 1$ to $2S$.

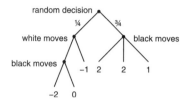

Figure 40.1. The game tree of a game with perfect information.

then we speak of a game with *perfect recall*. That not every game possesses this property lies not so much with the forgetfulness of actual players, which we are not going to consider because of its nonobjective nature, as with that fact that a "player" in the sense that we have been talking about need not be a single person. One can imagine a player as being a team of mutually cooperating individuals who together attempt to maximize their total score. In this case, on the assumption that both partners are not allowed to share their individual knowledge, it is not necessary that some existing information be available at all later moves. We mention again the example of skat, in which the player playing alone must hold out against two cooperating opponents.[2] Since these two do not know each other's cards, their decisions are made on the basis of informational states in which some facts are continually "forgotten," to become known again at later moves: as basis for the decision, sometimes only the cards of one teammate are known, and sometimes only those of the other.

For a player to be able to move, he must know first of all that it is his turn and what moves are available. And he must know the rules of the game, since only then can he weigh the effect of his moves. Regardless of the physical appearance of the game, the rules can always in principle be described graphically, in the form of a game tree. For games with perfect information we have already seen such trees. The simple example displayed in Figure 40.1 contains the elements that are typical of a two-person zero-sum game with random decisions and perfect information. Each node stands for a configuration that in the case of a game with perfect information corresponds to a particular informational state that is openly accessible to all players. Next to each node is indicated whose move it is or whether a random decision is to be made. Each edge symbolizes a move or a random decision, where in the latter case the associated probability is given. The counterpart of a round of playing the game is a path through the game tree that begins at the top node and ends at a node from which no further edge

---

[2]Strictly speaking, this assertion does not hold for the game as such, but only for the part of the game that begins after the bidding phase.

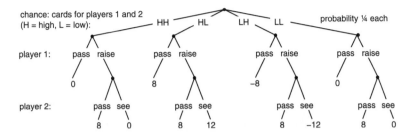

Figure 40.2. Possible moves in the poker model in the form of a game tree.

extends downward. Such nodes representing the end configurations are indicated by the associated end result; in the case of a two-person zero-sum game, it suffices to state the score of one of the players.

Information about which player, depending on the current state of the game, can move and how he can move can also be represented as a game tree for games without perfect information. As an example we consider the poker model that we used in Chapter 35. Its possible moves are shown in Figure 40.2. As one can see from the game tree alone, the game begins with a random move, in which each of the four results is achieved with probability $1/4$. Concretely, each of the two players obtains a card that is either high (H) or low (L). Then player 1 decides whether to raise the ante from 8 to 12. If he passes, then a showdown arises, that is, the players compare their cards, and the player with the better card wins 8 units. If player 1 raises the wager to 12, then player 2 has the choice of passing or placing the additional wager in order to see his opponent.

In contrast to games with perfect information, a game tree such as shown in Figure 40.2 characterizes the underlying game only in part. Of course, one sees all the possible moves that the players have available in the course of the game, but not the informational states on the basis of which the players must make their decisions. Of course, the game tree shown represents a game, but a different one, which can be interpreted as an open version of the poker model, in which each player can see his opponent's hand.

Fortunately, the missing information about the game can easily be added to the game tree. To this end, we consider that a player's partial lack of knowledge about the current game situation can be formally described by saying that for him, at the time of his move, several configurations are subjectively indistinguishable. For example, player 1, if he is holding a high card, cannot tell in deciding between "pass" and "raise" whether the current configuration is "HH" or "HL." Such a distinction could be made

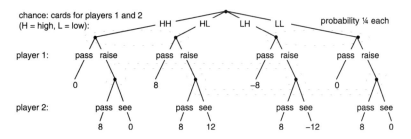

Figure 40.3. Poker model: a game tree augmented with information sets.

only by an observer who possessed all the relevant information. However, from the point of view of player 1, the two configurations form an inseparable unity that relates to his current state of information. Thus formally, one joins these inseparable—from the point of view of the player who is to move—configurations into an *information set*. Then exactly one information set corresponds to each separate informational state. For the poker model, the information sets, of which there are two for each player, are shown in the game tree of Figure 40.3 as dashed lines.

The configurations of an information set can be indistinguishable for the player to move only when its possible moves are analogous to one another: obviously, the number of moves must also agree. However, there need not be a unique correspondence among the possible moves. Only in this way can the decision of a player be implemented independent of the actual position reached in the game. In the game tree of the poker model, the association is in the form of naming the edges representing the moves, namely, "pass" and "see" for the two information sets of player 2.

The description of a game based on its chronological progress is called the *extensive form* of the game. In comparison to the normal form, the extensive form reflects the character of a game in much greater detail. In principle, these two descriptions serve the same purpose, namely, to provide a mathematically precisely definable model for games and above all, for interactive decision processes in economics. Both are thus uniquely representable solely in terms of mathematical objects such as sets and numbers, permitting them to be investigated precisely with the methods of mathematics within the context of game theory. In the case of the properties of perfect information and perfect recall, the analysis begins with the formulation of definitions, which can be done purely formally on the basis of mathematical objects (see Note 1 at the end of the chapter).

Outside of the normal form, which was introduced already in the 1920s by von Neumann and Borel, extensive game models were not investigated until the actual founding of game theory. A first version was described in 1944 by von Neumann and Morgenstern.[3] The version sketched here was given in 1953 by Kuhn.[4] Based on his game model, Kuhn proved that for a game with perfect recall, for every mixed strategy there exists an equivalent behavioral strategy that always yields the same game results; that is, the probabilities of the game results that are possible with a fixed but arbitrary counterstrategy do not change in passing from a mixed strategy to the associated behavioral strategy (see Note 2 at the end of the chapter). In particular, optimal behavioral strategies can always be found for two-person zero-sum games with perfect recall. Each of the two players can thus ensure his minimax result with the help of a behavioral strategy.

For many relatively simple games Kuhn's result ensures that minimax strategies can be described without too great an effort. Let us consider as an example the following symmetric model of two-person poker: each player draws at random one of six cards $1, \ldots, 6$, where the two drawings are independent of each other and equiprobable. Then the two players bid simultaneously, where the six levels $1, 2, 3, 5, 10, 15$ are allowed. The player with the higher bid wins. If they bid the same amount, a showdown follows, where the player with the higher card wins.

How can a player ensure his minimax value of zero as winning expectation? Since there are $6^6 = 46\,656$ pure strategies, it does not seem advisable to base tactical considerations on a mixed strategy, even if many of the pure strategies were to appear with zero probability. On the other hand, an optimal behavioral strategy can be given with little effort, as can be seen from Table 40.1. One sees clearly the presence of planned bluffs. In particular, the highest bid is made not only in the case of the two best cards, but also with the worst, though remarkably, not with the intermediate cards.

This poker model lends itself to experimental study, without the need for a second player. The game can be played with two normal dice and three ten-sided dice of different colors. One first rolls for one's own card, and then makes a bid. Then one rolls the four remaining dice for the opponent's turn. The normal die determines the opponent's card, and the three ten-sided dice the bid,[5] where the selection is made according to the tabulated behavioral strategy.

---

[3] John von Neumann, Oskar Morgenstern, *Theory of Games and Economic Behavior*, Princeton 1944.

[4] H. W. Kuhn, Extensive games and the problem of information, in: H. W. Kuhn, A. W. Tucker (eds.), *Contributions to the Theory of Games II*, Annals of Mathematics Studies **28**, 1953, pp. 193–216, reprinted in Harold W. Kuhn (ed.), *Classics in Game Theory*, Princeton 1997, pp. 46–68.

[5] The interval $[0, 1]$ can be divided into intervals representing the probabilities for the various bids, and the three dice can be used to represent a number between 0.000 and 0.999.

Bid	Card 1	2	3	4	5	6
1	0.35857	0.56071	0.50643	0.46857		
2	0.33786	0.12179	0.41179			
3	0.14143	0.16500		0.51571	0.00429	
5	0.05629	0.12757			0.59286	
10	0.06700	0.02493	0.08179	0.01571	0.14029	
15	0.03886				0.26257	1.00000

Table 40.1. Symmetric poker model: optimal behavioral strategy.

The minimax strategy can be calculated much as in the case of le Her, that is, by an iterative procedure in which a selection of pure strategies is gradually expanded for each player. At each step, a pair of minimax strategies is calculated corresponding to the current selection of strategies. If a player can improve his situation vis-à-vis this pair of strategies with a targeted counterstrategy, then his selection of strategies is expanded to include this counterstrategy. If the targeted counterstrategies bring neither player an improvement, then one is done. In the poker model, this procedure fortunately ends with only a fraction of all pure strategies, for example, with 2×44 pure strategies for an arbitrarily chosen beginning strategy.

The requisite two-fold calculation of optimal counterstrategies for each iteration step is just as simple in this poker model as for le Her. For other games as well that run for several moves, the extensive form is well suited for the move-by-move recursive calculation of optimal counterstrategies. Once mixed minimax strategies have been found, these are transformed into a behavioral strategy based on Kuhn's theorem. To do this, one has only to determine which probability distributions result from a mixed strategy with the various information sets.

Bid	Card 1	2	3	4	5	6
1					-0.18190	-3.33833
2				-0.02524	-0.28524	-3.44167
3			-0.09536			-3.15429
5			-0.07155	-0.23405		-2.66238
10						-2.92262
15		-0.05607	-0.23393	-0.39643		

Table 40.2. Cost of a wrong decision vis-à-vis the minimax strategy.

Table 40.2 shows how costly wrong decisions can be. If, for example, a player believes that holding card 4 he should always bluff with the highest bid, then he loses 0.39643 from his expectation. Since these values are relatively small in comparison to the expected score in a game, bad moves are hardly recognizable in comparison to random selection. Thus in this type of poker, luck plays a larger role than strategically correct play, at least when not too many games are played. This does not invalidate the result, but one's expectations should not be set too high. For in the end, a minimax strategy has a purely defensive nature, and this function is of course fulfilled on average.

Chapter Notes

1. Games with perfect information are characterized by the fact that a player always knows the history of a game, that is, how it is composed of the moves made thus far. In other words, if a player knows something, then all the other players know the same thing. There are thus no configurations that from the point of view of the player to move are indistinguishable. In the formal model one can therefore define perfect information as the property that every information set contains precisely one configuration.

   Perfect recall exists when a player always remembers the decisions about moves made thus far in the game and his previously available information. With respect to the formal model composed of game tree and information sets, within which the player's knowledge, including his memory, is represented by information sets, this corresponds to the following property, which can be used in a formal definition: if a path, corresponding to a game, leads first through position u and later through position v whose information sets U and V belong to the same player, then all game paths to configurations with information set V pass through configurations with information set U in which the same decision was made. Finally, this property says simply that the information set passed through and the decisions made there offer no additional information, since these data can be reconstructed from each individual information set encountered in the further course of the game.

   One should not confuse the concepts of perfect information and perfect recall with the extension of the game-theoretic model offered in 1967 by Harsanyi to such cases in which so-called complete information is not given. With respect to economic applications, one assumes that not all the rules of the game are known to all the players. See Reinhard Selten, Einführung in die Theorie der Spiele mit unvollständiger Information, in: *Information in der Wirtschaft*, Schriften des Vereins für Socialpolitik **126**, 1981, pp. 81–147. Much less detailed and specialized, and thus more comprehensible to a lay audience, is Reinhard Selten, Was ist eigentlich aus der Spieltheorie geworden? *Zeitschrift des Instituts für höhere Studien (IHS-Journal)* **4**, 1980, pp. 147–161, in particular, pp. 151–154.

2. A formal proof of this theorem is relatively complicated. Other than in the original work of Kuhn cited, one can find a proof in Roger B. Myerson, *Game Theory: Analysis of Conflict*, Cambridge 1991, pp. 154–163, 202–204; R. Selten, Reexamination of the perfectness concept for equilibrium points in extensive games, *International Journal of Game Theory* **4**, 1975, pp. 25–55, Chapter 4, reprinted in Harold W. Kuhn (ed.), *Classics in Game Theory*, Princeton 1997, pp. 317–354.

   Kuhn's result appears plausible if one imagines a mixed strategy that corresponds to no behavioral strategy: for such a strategy there is at least one information set for which the probability distribution of the random choice of moves depends on a previous move decision of the same player. Since the player does not forget his move at the previous decision point in the case of perfect information, this move is identical for all configurations of the information set for which we have assumed the given selection of moves. The probability distribution for this move can therefore also be used without changing the expected game result; that is, the altered strategy is equivalent to the original strategy.

41

Optimal Play: Planning Efficiently

In two-person zero-sum games with perfect recall one can easily describe minimax strategies in the form of behavioral strategies at least when the number of possible information sets is not too large. The amount of computation necessary can be quite large, though. To what degree can it be limited?

This quite general question will be our motivation for summarizing the concepts and techniques that we have been exploring in the previous chapters and then pointing to some further results. We will pay particular attention to the amount of computation necessary for the various procedures.

With the simplex algorithm we became acquainted with a method that allows for the computation of minimax strategies. However, the application to such games is limited in practice to games whose normal form is not too complex. Thus even a simple game like le Her, with its 8000 or so pure strategies, can barely be investigated directly.

Simplifications, at first exclusively in the description of strategies, can be achieved if one uses behavioral strategies instead of mixed strategies.[1] With a behavioral strategy, a player makes decisions "locally" about his random behavior, that is, separately for each subjective informational state.

---

[1] However, behavioral strategies are not well suited for *calculations*, since the winning expectation is not linearly dependent on the probabilities that characterize the behavioral strategy. Therefore, from a formal point of view, a behavioral strategy is more difficult to handle than a mixed strategy whose probabilities influence the winning expectation in a linear manner. The realization weights implemented in the course of this chapter have the advantage of possessing both properties, namely, linear effect of the parameters with simultaneous reduction in the number of parameters.

Therefore, a behavioral strategy can usually be characterized with many fewer probabilities than for a mixed strategy.

Although taken together, behavioral strategies are much less diverse than mixed strategies, their concept is quite universal: first, for every mixed strategy there is a behavioral strategy. To obtain it, one evaluates separately, for every information set in which the player in question moves, the probabilities with which the player chooses the various moves. According to Kuhn's theorem, for games with perfect information, the behavioral strategy so constructed is strategically equivalent to the mixed initial strategy; that is, one can exchange the strategies without altering the odds of the game: the various results that are possible for a fixed, but arbitrary, counterstrategy do not change their probabilities. Here the assumed property of perfect recall for individual players, in contrast to teams, is always present in ideal circumstances.

The actual calculation of optimal behavioral strategies was always carried out in the last two chapters via the detour of mixed strategies. The cost in computation and storage was bounded in that the normal form and minimax strategies were calculated on the basis of a manageable selection of pure strategies. The extent to which the winning prospects of a player are worsened by this strategic restriction is at the outset an open question, though it can be determined after the fact if one determines an optimal counterstrategy to such a minimax strategy. Namely, if neither of the two players can improve his prospects with a targeted counterstrategy, then both minimax strategies related to the restricted game are indeed optimal.

The principal component of the criterion just sketched is the calculation of optimal counterstrategies. It is relatively simple based on the extensive form, that is, the description of the chronological course of the game including all possible moves, the reachable configurations, and the information as to what information is available to the player who is to move. To this end, the player, who knows the opponent's mixed strategy and wishes to counter optimally, gradually analyzes every decision situation in reverse order of the game's chronology. Concretely, he looks for a move that brings him the greatest winning expectation, starting with subsequent moves from the results of the optimizations already carried out. Move by move he thus obtains a pure strategy with which the opposing strategy can best be countered.

The recursive method of determining an optimal counterstrategy to an opponent's mixed strategy is comparable to an optimization such as we have seen for one-person games of chance such as blackjack: of course, during an actual game, the optimizing player does not always know the current state of the game, though he always knows the probabilities of all possible game states, which can therefore be collected into a configuration in which the divided random decision is made only later, as is the case for a card that has been dealt but is still face down. To such an extent, a player who knows his opponent's mixed strategy acts within a one-person game with single-element information sets.

Building on the facts that we have again collected here, we see that minimax strategies can be determined iteratively, on the assumption that we are dealing with a two-person zero-sum game with perfect recall:

- at the initial step, each player selects an arbitrary pure strategy.

- an iteration step assumes that each player possesses a selection of pure strategies:

  – first, the normal form is worked out for the selected strategies.

  – then the simplex algorithm is used to calculate a pair of minimax strategies and the associated value.

  – for each of the two minimax strategies an optimal counterstrategy is determined. All decision situations in which a given player moves are optimized in reverse order of the game's chronology. These optimized individual decisions are combined into a pure strategy.

- finally, one checks whether one of the two players can improve his situation with respect to the minimax value with his specific counterstrategy, depending on how he has acted for the current selection of strategies.

  – if neither of the two players found an improvement, then the two minimax strategies that were found are optimal in the complete game.

  – otherwise, the selection of strategies of at least one player are extended by the counterstrategy obtained. This extension takes place precisely when the player in question can thereby improve his result.[2]

According to an observation, probably first stated by Robert Wilson in 1971 in describing a similar iterative procedure,[3] generally only relatively

---

[2]Pure strategies that are not a component of the current minimax strategy can sometimes be removed from the selection in order to simplify the next minimax calculation. However, one must prevent cyclic iteration loops. For example, it is possible to carry out such strategy removal only on steps in which the thus-far narrowest bounding of the minimax value was obtained.

[3]Robert Wilson, Computing equilibria of two-person games from the extensive form, *Management Science* **18**, 1972, pp. 448–459. Wilson supports his claim with the formulation "verified in computational experience on practical problems" (p. 449).

few pure strategies appear in the mix of minimax strategies. One can then hope that the iteration thus described brings about the desired simplification. This conjecture was finally verified at the beginning of the 1990s on the basis of investigations into the question of which properties of a mixed strategy truly influence the results of a game. Using the same methods, a procedure was created by which optimal behavioral strategies could be directly calculated. Thus for each ending configuration, first it is ascertained which decisions are necessary for a game to end in this way. There then results, for each of the two players as well as the random influences, a sequence of individual decisions. And conversely, these three sequences together determine the end nodes. We will now clarify this with an example, involving yet another poker model.

In this model, two players each put in an ante of 4 units. Then each of them draws a card, namely, a high card H or a low card L with equal probability. The two players draw independently of each other. Then the players bid in up to three phases, in which the bet can be raised to 6 or 9 units. Player 1, Jill, begins: either she passes and loses her ante, or she raises her bet to 6 or 9 units. If Jill has raised, then player 2, Jack, must decide whether to pass, see, or raise. The last of these options is available only if Jill raised to 6. If Jack "sees," then Jack must of course raise his bet by the amount that Jill did, and then the winner is determined in a showdown, where the higher card wins. In the case that Jack raises his bet from 6 to 9, then it is Jill's turn again. Her choice is whether to pass or raise the bet.

Let us take a look at the game tree. To keep things simple, Figure 41.1 shows only the decisions to be made and the resulting final score for Jill. The information sets are not shown.

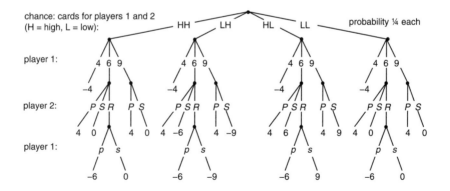

Figure 41.1. Three-stage poker model: bid, raise, see.

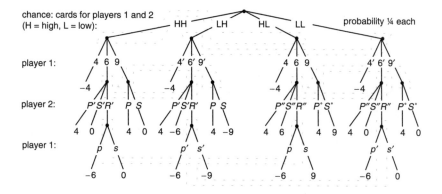

Figure 41.2. Poker model: game tree including information sets.

On his (or her) turn, a player makes a bid based on his own card and any bids that have been made, which of course are common knowledge. The information that is unknown is the opponent's card. If one collects the nodes that are indistinguishable to the player whose turn it is at the time of his decision into information sets, then one obtains the extended game tree depicted in Figure 41.2. Here the naming of the moves has been fitted to the information sets, so that the information sets alone can be determined from the names of the moves.

As can be seen in Figure 41.2, each player must plan his strategic decisions for four different situations. For Jill, our player 1, the four situations are these:

- the decision on a high card to pass or open with a raise to 6 or 9 units. In Figure 41.2 these moves are indicated with 4, 6, and 9.

- the decision on a low card to pass or open with a raise to 6 or 9, indicated with 4', 6', and 9'.

- the decision between passing and seeing on a high card after an opening bid of 6 that the opponent Jack raised to 9, denoted by p and s.

- the decision between passing and seeing on a low card after an opening bid of 6 that the opponent Jack raised to 9, denoted by p' and s'.

Counting up all the possible combinations shows that Jill has 36 pure strategies. However, there are only 16 pure strategies that differ in their effects, since the last two decisions are relevant only after an opening bid of 6.[4] A behavioral strategy for Jill is determined by six parameters.

Jack, our player 2, has the following four decisions to weigh:

- on a high card, against a bid by Jill of 6 he must decide between passing and raising. In Figure 41.2 these moves are denoted by P', S', and R'.

- on a low card and opening bid of 6 he must decide among passing, seeing, and raising. These are denoted by P'', S'', and R''.

- on a high card and opening bid of 9 he must decide between passing and seeing. These are denoted by P and S.

- on a low card and opening bid of 9 he must decide between passing and seeing. These are denoted by $P°$ and $S°$.

Altogether, Jack has 36 different pure strategies. A behavioral strategy for Jack can be characterized with six parameters.

On the basis of this description of the game we can now clarify how one is to understand the concept of the decision sequences "bid–raise–see" in the poker model. Table 41.1 shows the 28 end nodes of the game. In the table are tabulated the three *decision sequences* required by each of the two players so that the game can end at the appropriate nodes. With each combination of three decision sequences is associated at most one end node; that is, if there exists an associated end node, then it is uniquely determined.

Depending on whatever mixed strategy the two players might use, for each end node there is a probability that gives the likelihood that a game will end at that node. In general, each such probability is the product of three separate probabilities corresponding to the decision sequences of the two players and the random elements. One can calculate these three probabilities as they relate to every node, not just the end nodes, as follows:

- for a random influence, one should multiply the probabilities of all random influences that must be reached on the path through the game tree from the start node to the node in question.

- for a player, the probability called the *realization weight* specifies how probable it is that the player makes his decisions in a way required to reach the node in question.[5] There remain the decisions that arise

---

[4]Individual decisions that are irrelevant to the result of a game make it possible to reduce the normal form.

[5]If one assumes a behavioral strategy, then the realization weights are, as with the sequences of random decisions, the products of the individual probabilities.

Node Number Left to Right	Score	Decision Sequences		
		Chance	Player 1	Player 2
1	4	HH	4	
2	4	HH	6	P'
3	0	HH	6	S'
4	-6	HH	$6, p$	R'
5	0	HH	$6, s$	R'
6	4	HH	9	P
7	0	HH	9	S
8	-4	LH	$4'$	
9	4	LH	$6'$	P'
10	-6	LH	$6'$	S'
11	-6	LH	$6', p'$	R'
12	-9	LH	$6', s'$	R'
13	4	LH	$9'$	P
14	-9	LH	$9'$	S
15	-4	HL	4	
16	4	HL	6	P''
17	6	HL	6	S''
18	-6	HL	$6, p$	R''
19	9	HL	$6, s$	R''
20	4	HL	9	$P°$
21	9	HL	9	$S°$
22	-4	LL	$4'$	
23	4	LL	$6'$	P''
24	0	LL	$6'$	S''
25	-6	LL	$6', p'$	R''
26	0	LL	$6', s'$	R''
27	4	LL	$9'$	$P°$
28	0	LL	$9'$	$S°$

Table 41.1. Poker model: end nodes with their decision sequences.

in another way, that is, by chance or from the opponent, even though they, too, influence the end state reached in an actual game.

We now come to the main property of decision sequences and realization weights: if the realization weights for all decision sequences of both players are known, then the winning expectation can be calculated without knowledge of additional details about the two strategies. This is done simply by first determining the probability of each individual end node as the product of the three named probabilities. Altogether, one thereby obtains the probability distribution for the winning score, so that the winning expectation can then be easily calculated. Through this procedure it can also be seen that the probabilities of a mixed strategy affect the player's winning expectation only to the extent that they alter his realization weights.

For our example of the poker model, we can read off the realization weights that occur from Table 41.1. First, to each decision sequence that appears there belongs a corresponding realization weight. Then come the decision sequences associated with the start of one of the given decision sequences, each beginning with the empty decision sequence $\varnothing$. Altogether, the strategies of the two players are characterized by the following realization weights:

- Player 1:

$$x(\varnothing), \ x(4), \ x(6), \ x(6,p), \ x(6,s), \ x(9),$$
$$x(4'), \ x(6'), \ x(6',p'), \ x(6',s'), \ x(9');$$

- Player 2:

$$y(\varnothing), \ y(P'), \ y(S'), \ y(R'), \ y(P), \ y(S),$$
$$y(P''), \ y(S''), \ y(R''), \ y(P^\circ), \ y(S^\circ).$$

As probabilities, the realization weights all lie in the range from 0 to 1. The maximal value of 1 is attained by the realization weights $x(\varnothing)$ and $y(\varnothing)$. The other realization weights are subject to the addition law, which must hold for the results of every decision a player makes. All in all, the following identities must be satisfied:

$$x(\varnothing) = 1,$$
$$x(4) + x(6) + x(9) = x(\varnothing),$$
$$x(6,p) + x(6,s) = x(6),$$
$$x(4') + x(6') + x(9') = x(\varnothing),$$
$$x(6',p') + x(6',s') = x(6'),$$

and

$$y(\varnothing) = 1,$$
$$y(P') + y(S') + y(R') = y(\varnothing),$$
$$y(P) + y(S) = y(\varnothing),$$
$$y(P) + y(S'') + y(R'') = y(\varnothing),$$
$$y(P^\circ) + y(S^\circ) = y(\varnothing).$$

Player 2	Player 1 $\varnothing$	P'	S'	R'	P''	S''	R''	P	S	P°	S°
$\varnothing$											
4	-2										
6		1	0		1	1.5					
$6, p$				-1.5			-1.5				
$6, s$				0			2.25				
9								1	0	1	2.25
$4'$	-2										
$6'$		1	-1.5		1	0					
$6', p'$				-1.5			-1.5				
$6', s'$				-2.25			0				
9								1	-2.25	1	0

Table 41.2. Poker model "bid–raise–see": sequential form.

Conditions derived in this way are in fact sufficient for games with perfect recall. That is, if there are nonnegative values for a player that satisfy these equations, then there is a mixed strategy for them. The corresponding behavioral strategy can be easily constructed, since every information set can be individually investigated.

If a strategy is described using realization weights, one speaks then of a *realization plan*. The effect of realization plans can best be investigated with the help of a table similar to the normal form, the *sequential form* of a game. Such a table appears as Table 41.2 for the poker model that we have been considering. One obtains its entries by analyzing the end nodes, where we may use the data from Table 41.1: for each entry one must determine the end configurations that are reachable with the associated decision sequence. Due to random decisions, this can involve more than one node, as we see for example in the case of the decision sequence pair $4, \varnothing$. In such cases one uses the expectation. Where there is no reachable end node, the gap is indicated by a 0.

As with the generally clearly larger normal form, the sequential form enables one to calculate the winning expectation of the first player as a function of two existing realization plans. Each table entry is reached with a probability that is given as the product of the two associated realization weights.[6]

---

[6]This corresponds to the product $x^t A y$, where x and y are the column vectors of the realization plans of the two players, and A the matrix of the sequential form of the game.

Realization weights were studied by Daphne Koller, Nimrod Megiddo, and Bernhard von Stengel in the early 1990s. They later learned that the Russian Joseph Romanovsky had published some important ideas on this topic in 1962, which were then expanded in 1969 by the mathematician E. B. Yanovskaya. Koller and Megiddo used realization weights to prove that every mixed strategy can be replaced by a completely strategically equivalent strategy that consists of a mix of at most as many pure strategies as there are end nodes in the game tree (see Note 1 at the end of the chapter). This also explains Wilson's observation that optimal strategies generally contain comparatively few pure strategies. Earlier, Koller and Megiddo had already shown that an optimal strategy of a two-person zero-sum game with perfect recall can always be calculated at a cost that is polynomially bounded in the size of the game tree.[7] However, their algorithm is not very practicable.[8] Much simpler is the method that goes back to Romanovsky and von Stengel in which the optimal behavioral strategies are calculated using a linear optimization problem whose size is proportional to the number of end nodes.[9] The variables of the linear optimization problem include, among other things, a player's realization weights and yield as the optimum that player's desired minimax strategy. Additionally, the optimization problem contains additional variables: one associated with the game's beginning and the others with the opponent's information sets. They specify the opponent's best options against the minimax strategy in question. In particular, each of these variables can be interpreted as a proportional winning expectation, where only those games are evaluated that run through the opponent's associated information sets. To obtain an impression of this method, let us look at the optimization problem for our poker model that calculates the second player's minimax strategy:

---

[7] Daphne Koller, Nimrod Megiddo, The complexity of two-person-zero-sum games in extensive form, *Games and Behavior* **4**, 1992, pp. 528–552.

[8] Koller and Megiddo characterize a player's strategy with realization weights. Possible counterstrategies are investigated in the form of pure strategies, where the quality of a counterstrategy is evaluated with a special property of the ellipsoid method that was introduced briefly in a note in Chapter 36 of this book.

[9] J. V. Romanovsky, Reduction of a game with perfect recall to a constrained matrix game (in Russian), *Doklady Akademii Nauk SSSR* **114**, 1962, pp. 62–64, English translation in *Soviet Mathematics* **3**, 1962, pp. 678–681; see also *Mathematical Reviews* **25**, 1963, #1958; Bernhard von Stengel, Efficient computation of behavior strategies, *Games and Behavior* **14**, 1996, pp. 220–246; Daphne Koller, Nimrod Megiddo, Bernhard von Stengel, Fast algorithms for finding randomized strategies in game trees, in: *Proceedings of the 26th ACM Symposium of the Theory of Computing*, New York 1994, pp. 750–759; Bernhard von Stengel, Computing equilibria for two-person games, in: R. J. Aumann, S. Hart (eds.), *Handbook of Game Theory*, volume 3, Amsterdam 2001.

Minimize u_0 under the following conditions:

$$u_0 \geq u(4:6:9) + u(4':6':9'),$$

$$u(4:6:9) \geq -2y(\varnothing),$$

$$u(4:6:9) \geq y(P') + y(P'') + \frac{3}{2}y(S'') + u(p:s),$$

$$u(p:s) \geq -\frac{3}{2}y(R') - \frac{3}{2}y(R''),$$

$$u(p:s) \geq \frac{9}{4}y(R''),$$

$$u(4:6:9) \geq y(P) + y(P^\circ) + \frac{9}{4}y(S^\circ),$$

$$u(4':6':9') \geq -2y(\varnothing),$$

$$u(4':6':9') \geq y(P') + y(P'') - \frac{3}{2}y(S') + u(p':s'),$$

$$u(p':s') \geq -\frac{3}{2}y(R') - \frac{3}{2}y(R''),$$

$$u(p':s') \geq -\frac{9}{4}y(R'),$$

$$u(4':6':9') \geq y(P) - \frac{9}{4}y(S) + y(P^\circ),$$

$$y(\varnothing) = 1,$$

$$y(P') + y(S') + y(R') = y(\varnothing),$$

$$y(P) + y(S) = y(\varnothing)$$

$$y(P'') + y(S'') + y(R'') = y(\varnothing),$$

$$y(P^\circ) + y(S^\circ) = y(\varnothing),$$

and

$$y(\varnothing),\ y(P'),\ y(S'),\ y(R'),\ y(P),\ y(S),$$
$$y(P''),\ y(S''),\ y(R''),\ y(P^\circ),\ y(S^\circ) \geq 0.$$

Where does this optimization problem come from? We will restrict our attention here to a plausibility argument (see Note 2 at the end of the chapter): the first group of side conditions is added to the already known side conditions for the realization weights $y(\varnothing), \ldots, y(S^\circ)$. They characterize the best that player 1 can achieve against an arbitrary realization plan as fixed by the y values. In particular, the variable u_0 is the winning expectation of player 1. Its minimum specifies the most that player 1 can achieve against the strategy of player 2 determined by the realization weights $y(\varnothing), \ldots, y(S^\circ)$. Each of the remaining u variables relates to an information set of player 1 and is denoted by the possible moves, such as, for example, $4 : 6 : 9$ for an opening bid by player 1 holding a high card. The value of such a variable specifies the highest winning expectation that player 1 can achieve against the strategy of his opponent determined by y, where only games that run through the associated information set are considered. This yields for every possible move an inequality whose actual form depends on the further course of the game. For example, player 1 can choose among the moves 4, 6, and 9 in the information set denoted by $4 : 6 : 9$. The possible continuations from that point are reflected in the following inequalities:

4 (pass holding a high card):

$$u(4 : 6 : 9) \geq -2y(\varnothing);$$

6 (raise on 6 holding a high card):

$$u(4 : 6 : 9) \geq y(P') + y(P'') + \frac{3}{2}y(S'') + u(p : s);$$

9 (raise on 9 holding a high card):

$$u(4 : 6 : 9) \geq y(P) + y(P^\circ) + \frac{9}{4}y(S^\circ).$$

We shall see that the right-hand side of each of these inequalities yields the maximal proportional winning expectation that player 1 can achieve with the associated move. Therefore, $u(4 : 6 : 9)$ is equal to the maximum of the three values that can be computationally established by the three inequalities: that $u(4 : 6 : 9)$ is at least as big as the maximum of the three values is obvious. The converse statement, namely, that $u(4 : 6 : 9)$ is at most equal to the maximum, is a result of the minimization process.

To obtain the terms on the right-hand side of such a move inequality, one must take into account the effect of the move in question at each node of its information set. This is done by adding to the winning expectations the contributions that correspond to possible continuations, possibly stretching over a number of moves. Only when an end node is reached or a further move of player 1 arises is the associated portion of the winning expectation not further split up:

- if an end node is reached, the proportional winning expectation is equal to the product made up of:

- the associated scores,

- the associated realization weight of player 2 that is specified by the corresponding y variable,

- the probabilities associated with the previous random moves.

For example, the move 4 leads from the information set denoted by $4:6:9$ to two end nodes. Both possess the score -4 and for player 2 the realization weight $y(\varnothing) = 1$. The probability of the necessary random moves is $1/4$ in each case. Therefore, the proportional winning expectation in the case of the move 4 is equal to $-2y(\varnothing) = -2$.

• if one reaches a node at which player 1 must move again, then, because of the perfect recall, all nodes that belong to the same information set are likewise reached by the relevant continuations. Therefore, the sum of the corresponding portions of the winning expectation is equal to the associated u variable.

For example, the move 6 leads from the information set denoted by $4:6:9$ after an opposing move decision to four end nodes as well as to the two nodes of the information set denoted by $p:s$. The associated paths are indicated with heavy lines in Figure 41.3.

The proportional winning expectation in the case of move 6 thus contains the variable $u(p:s)$ as summand. Altogether, it is equal to

$$y(P') + y(P'') + \frac{3}{2}y(S'') + u(p:s).$$

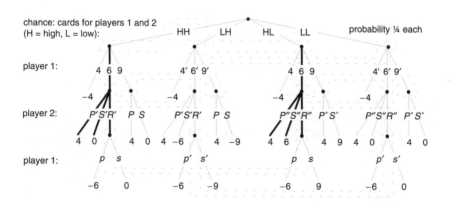

Figure 41.3. Player I makes move 6: $u(4:6:9) \geq y(P')+y(P'')+(3/2)y(S'')+u(p:s)$.

A solution of the linear optimization problem is

$$y(\varnothing) = 1,$$

$$y(P') = 0, \quad y(S') = \frac{2}{9}, \quad y(R') = \frac{7}{9},$$

$$y(P) = 0, \quad y(S) = 1,$$

$$y(P'') = \frac{1}{3}, \quad y(S'') = \frac{5}{18}, \quad y(R'') = \frac{7}{18},$$

$$y(P^\circ) = \frac{1}{2}, \quad y(S^\circ) = \frac{1}{2},$$

$$u(p:s) = \frac{7}{8}, \quad u(p':s') = -\frac{7}{4},$$

$$u(4:6:9) = \frac{13}{8}, \quad u(4':6':9') = -\frac{7}{4},$$

$$u_0 = -\frac{1}{8}.$$

Thus player 2 has the following minimax strategy, which limits the winning expectation of player 1 to $-1/8$:

- if player 1 opens with a bid of 6, then player 2, if holding a high card, sees with probability 2/9 and raises with probability 7/9. With a low card player 2 passes with probability 1/3 in response to the opening bid of 6, sees with probability 5/18, and raises with probability 7/18.

- if player 1 opens with a bid of 9, then player 2 passes holding a low card and sees holding a high card.

Analogously, one can compute a minimax strategy for player 1:

- with a high card, open with a bid of 6 with probability 1/3 and otherwise with a bid of 9.

- with a low card, open with a bid of 6 with probability 1/3, and otherwise with a bid of 9.

- after an opening bid of 6 that is raised by player 2 to 9, always see with a high card and always pass with a low card.

One should keep in mind that behavioral strategies do much more than simplify the description of minimax strategies. Minimax behavioral strategies can also be determined directly from the solution of a linear optimization problem for a two-person zero-sum game with perfect recall. Here the scope of the optimization problem is relatively moderate. In particular, both the number of variables and the number of inequalities can be bounded by the number of parameters that are necessary for characterizing a pair of behavioral strategies. The class of games for which the calculation of minimax strategies is actually realizable is thus significantly enlarged (see Note 3 at the end of the chapter). The most complex minimax calculations to date were accomplished by Daphne Koller and Avi Pfeffer.[10] Using a descriptive language that they invented called GALA and linear optimizations of realization weights, they analyzed poker models with up to 140 000 configurations. In terms of the parameters of the poker model, this corresponds to the following bounds:

- deck of 127 cards, one card per player, 1 round.

- deck of 3 cards, one card per player, 11 rounds.

- deck of 11 cards, five cards per player, 3 rounds.

If perfect memory is lacking for a two-person zero-sum game, then the calculation of minimax strategies can be much more difficult, since the search over the behavioral strategies must be extended. Koller and Megiddo indeed show that in the general case, the calculation of minimax strategies is NP hard with respect to the size of the game tree.[11] Therefore, one may assume that there is no general algorithm with which a minimax strategy can be efficiently determined from the data of the game tree.

Chapter Notes

1. Daphne Koller, Nimrod Megiddo, Finding mixed strategies with small supports in extensive games, *International Journal of Game Theory* **25**, 1996, pp. 73–92, Theorem 2.6.

   In principle, they use a purely dimensional argument, in which the realization weights are used to determine the number of relevant parameters relating to the strategic influence of a player. There can clearly be at most as many of these as the dimension of the set of all realization plans for the player in question, and this dimension is obtained by adding to each of the player's information sets the number of possible moves reduced by 1.

---

[10] Daphne Koller, Avi Pfeffer, Generating and solving imperfect information games, in: *Proceedings of the 14th International Joint Conference on Artificial Intelligence*, Montreal 1995, volume 2, pp. 1185–1192; Daphne Koller, Avi Pfeffer, Representations and solutions for game-theoretic problems, *Artificial Intelligence* **94**, 1997, pp. 167–215.

[11] Daphne Koller, Nimrod Megiddo, The complexity of two-person-zero-sum games in extensive form, *Games and Behavior* **4**, 1992, pp. 528–552. In this paper, Koller and Megiddo construct special two-person zero-sum games in which one of the two players does not possess perfect recall. In this class of games, the question whether a player can be certain of achieving a specified score via a mixed strategy is NP hard.

In many games, the dimension of the realization plans is significantly smaller than the universal bound that is oriented to the number of end nodes. Thus in le Her one can describe white's realization weights with 13 parameters. The same holds for black, if one excludes obviously dominated moves that relate to the situation in which white has previously exchanged. In the sequel each le Her player possesses a minimax strategy that contains at most 13 pure strategies.

That mixed strategies are strategically equivalent when they exhibit equal realization plans was proven in 1969 by Yanovskaya: Ye. B. Yanovskaya, Quasistrategies in positional games (in Russian), *Izvestiya Akademii Nauk SSSR Tehnicheskaya, Kibernetika* **1**, 1970, pp. 14–23; English translation in *Engineering Cybernetics* **1**, 1970, pp. 11–19; see also *Mathematical Reviews* **43**, 1972, #2995.

2. A precise derivation is based on transforming the requirements of an optimal counterstrategy into a linear optimization problem and then investigating the dual optimization problem to the one so derived. (Sufficiently detailed descriptions of this duality can be found in many books on linear optimization.)

One begins with the sequential form of a game whose data form a matrix A. Using this matrix, one can calculate the winning expectation $x^t A y$ from realization plans that appear as column vectors x and y. The requirements on the realization plans have the form

$$Ex = e, \quad x \geq 0;$$
$$Fy = f, \quad y \geq 0.$$

Here e and f are column vectors of suitable dimension whose coordinates, except for the first, which is equal to 1, are equal to zero.

One may now characterize for player 1 the optimal counterstrategy with respect to an arbitrary but fixed realization plan y for player 2 by means of a realization plan x that solves the following linear optimization problem:

$$x^t (Ay) = \max!,$$
$$Ex = e,$$
$$x \geq 0.$$

The dual optimization problem to this one whose minimum is equal to the maximum of the original optimization problem is

$$u^t e = \min!,$$
$$E^t u \geq Ay.$$

In comparison to the original optimization problem, this dual problem has the advantage that it can be studied relatively easily for all possible realization plans y. Therefore, it can be used to determine player 2's minimax strategy:

$$u^t e = \min!,$$
$$E^t u \geq Ay,$$
$$Fy = f,$$
$$y \geq 0.$$

As a special case this yields the optimization problem as it was posed for the example of the poker model "bid–raise–see."

3. The construction of linear optimization problems can be accomplished completely formally, as we saw previously. The starting point for the examples introduced in Chapters 39 through 41 was an object-oriented implementation of the corresponding configurations with methods such as

- **NumberofMoves**: the number of moves, where for end configurations the value is 0.

- **ToMove**: returns the player whose turn it is to move (player 1, player 2, or chance).

- **Move (MoveNumber, Prob)**: generates a new instance of a configuration corresponding to the configuration reached on the move. For random moves, the move probability **Prob** is returned.

- **InformationSet**: returns—except for end configurations and before random moves—the current information set in the form of a unique name (**String**).

- **Score**: for end configurations returns the score of player 1.

This already suffices for an implementation of the algorithms described for calculating minimax strategies.

42

Baccarat: Draw from a Five?

Should a baccarat player whose first two cards total five request another card?

The game of baccarat—also known as chemin-de-fer—with its over 500 year history, is the most widespread casino card game after blackjack.[1] As with blackjack, the game is usually played with several decks of cards. To win, a player must draw a higher hand than that of the bank; equal hands result in a draw. The values of the cards are as follows: the ace has value 1, cards 2 through 9 have their face values, and face cards count zero. The card values are summed modulo ten. Thus an 8 and a 6 total 4, while a jack and an ace total 1.

A game of baccarat begins with the player and banker each receiving two cards face down, which are not revealed to the other player. If either player has a hand worth 8 or 9, then both players show their hands and the game is scored. Otherwise, the player decides whether he wishes to have an additional card dealt. If he chooses to receive a card, it is dealt face up. Then it is the banker's turn. He, too, is allowed to receive a third card, where in making his decision he can take into account his own hand, the decision of the player, and the exposed card if such exists. At this point the game is over; banker and player reveal their hands, and the game is scored.

Let us consider the odds of this game at first from a purely intuitive point of view. The player and the bank have a choice about an additional card only when both original two-card hands have values in the range 0

---

[1]John Scarne, *Complete Guide to Gambling*, New York 1974, pp. 459–479.

Banker's Hand	Player's Third Card: Banker...	
	...Draws	...Does Not Draw
0–2	N, 0–9	—
3	N, 0–7, 9	8
4	N, 2–7	0, 1, 8, 9
5	N, 4–7	0–3, 8, 9
6	6–7	N, 0–5, 8, 9
7	—	N, 0–9
8–9	—	Banker shows hand!

Table 42.1. Banker's strategy as usually prescribed in casino play. Here "N" represents the case that the player has not elected to receive a third card.

to 7. To obtain the best possible hand, player and banker should draw a third card from low hands, while with a hand of 7 or perhaps a bit below, they should refrain from drawing. But the player in particular should consider that with his decision he gives the banker an idea of the value of his original hand. Since the third card is dealt open, these indications, even if only within certain limits, can be extrapolated to the value of the entire hand. Thus the banker can adjust his strategy to the extent that the player's actions provide information about his original hand.

The way baccarat is played in casinos, a number of wagerers can bet on the outcome of the game. Therefore, the player and banker are quite limited in the decisions that they are permitted to make, with this limitation of course in the respective party's favor. Thus the player must take a third card if his hand totals 4 or less, while he may not draw if the hand is worth 6 or 7. Only for a hand of 5 is he allowed to choose. The banker's strategy is more complex, since in addition to his own hand, the player's decision and the value of a possible exposed card must be taken into account. The banker's permissible choices are shown in Table 42.1.

Like blackjack, baccarat is almost symmetric, so that the advantage held by the banker is not immediately apparent. In blackjack, the bank's advantage lies in the fact that the player must draw first and risks going bust with a hand over 21, while in baccarat the advantage lies solely in the banker's increased amount of information: the banker knows the player's decision, which allows an indirect conclusion about the hand; he also knows the value of the exposed third card. Therefore, the banker has available a more nuanced reply to the game situation.

In contrast to blackjack, the role of the banker in baccarat is generally not taken by a casino employee. Instead, the players rotate this function among themselves, whence the name chemin-de-fer. For managing the game the casino takes five percent of the winnings that the banker achieves.

There are several variants to the rules of baccarat:

- there can be two players, each with his own hand, who play against the bank. The banker must attempt to maximize its decision about a third card based on the decisions, open cards, and wagers of both players.[2]

- in principle, the player, and above all, the banker, can be allowed to make their choices as to a third card freely, at least when each is playing only for himself. Banker and player are then engaged in a two-person zero-sum game without perfect information in which the optimal strategies may perhaps be mixed.

An early investigation of baccarat is contained in a book by Bertrand that appeared in 1889. Although this study left rather much to be desired, it nevertheless served Borel as a useful introductory example.[3] The game-theoretic problem underlying the second of our baccarat variants was solved in 1957 by John Kemeny and Laurie Snell.[4] They begin with two strategies for the player, determined by his behavior with a hand of 5, and a total of 2^{88} strategies for the bank, characterized by its 8×11 information sets. The large number of information sets results from the combination of eight possible values that the bank can obtain with its first two cards—other than the case of a game-ending 8 or 9—and the ten possible card values that the player can draw as third card; the 11th possibility represents the case that the player declines a third card. For the sake of simplicity, they assume an infinite deck of cards, so that cards already drawn do not influence the probabilities for further cards to be drawn.[5]

With the method of iteratively extended strategy selection, which we have discussed in the previous two chapters, the astronomical number of

[2]This variant, called baccarat-en-banque or baccarat à deux tableaux, is investigated in Sherry Judah, William T. Ziemba, Three person Baccarat, *Operations Research Letters* **2** 1983, pp. 187–192. For more on the game, see the cited work of John Scarne, pp. 478–479.

[3]See Chapter 33.

[4]John G. Kemeny, J. Laurie Snell, Game-theoretic solution of Baccarat, *American Mathematical Monthly* **64**, 1957, pp. 465–469. See also Richard A. Epstein, *The Theory of Gambling and Statistical Logic*, New York 1977, pp. 193–196.

[5]This assumption is not made in F. G. Foster, A computer technique for game-theoretic problems I: Chemin-de-fer analysed, *The Computer Journal* **7**, 1964, pp. 124–130; reprinted in David N. L. Levy, *Computer Games II*, New York 1988, pp. 39–52. See also M. G. Kendall, J. D. Murchland, Statistical aspects of the legality of gambling, *Journal of the Royal Statistical Society, Ser. A* **127**, 1964, pp. 359–391.

bank strategies does not come into play. After only a few steps, the procedure terminates with, depending on the initial strategies, for example, 2×5 strategies, and returns the following minimax strategies:

- the player takes a third card on 5 with probability 9/11 and declines with probability 2/11.

- the bank's strategy corresponds almost completely with the fixed strategy tabulated above. The only exception is the situation in which the player does not choose a third card and the bank has a hand worth 6. Then the bank draws a third card with probability 0.3754.

The game value reflects the bank's slight advantage: it is -0.0128.

Further Mathematical Publications on Baccarat

[1] Edward O. Thorp, William E. Walden, A favorable bet in Nevada Baccarat, *Journal of the American Statistical Association* **61**, 1966, pp. 313–328.

[2] Edward Thorp, William Walden, The fundamental theorem of card counting with applications to trente-et-quarante and baccarat, *International Journal of Game Theory* **2**, 1973, pp. 109–119.

[3] Edward Thorp, *The Mathematics of Gambling*, Hollywood 1984, pp. 29–39.

43

Three-Person Poker: Is It a Matter of Trust?

Three players are playing poker. Can two players regulate their play so as to disadvantage the third player without cheating?

All the games without perfect information that we have thus far considered have been two-person zero-sum games. Even the games with perfect information that we looked at in Part II were almost exclusively two-person zero-sum games. The only exception was a three-person nim variant that was mentioned in Chapter 20 to point out the main differences between two-person and many-person games. Starting with any configuration of this nim variant, a strategy could be found for each of the three players that together formed an equilibrium.

Such an equilibrium is generally characterized by the fact that no single player can improve his situation by changing his strategy unilaterally. The strategy of each player is thus optimal to the extent that it represents his best reply to the announced strategies of his opponents. If one assumes that every player is aware of the fact that each player attempts to maximize his own advantage, then a certain stability can be assumed to exist for this equilibrium. Conversely, situations in which the opposing strategies do not form an equilibrium are characterized by the fact that at least one player has cause to be dissatisfied with his strategy. This yields the following consequence: if there is a common method of play among experienced players, then this corresponds to an equilibrium.

If one of the players makes an "error," then the supposed stability of an equilibrium disappears. Here the player who has "erred" is not necessarily the only player to suffer a loss vis-à-vis the result of the game associated with the equilibrium, and so it is, in fact, possible that the player whose play has deviated from the equilibrium has not actually made an error. That is, the player may not have accidentally acted contrary to his own interests. It is conceivable that he has acted quite consciously against his own interests in order to achieve a larger total win in concert with another player, so that the pair together thereby achieve a more attractive net result.

In contrast to two-person zero-sum games, the score corresponding to an equilibrium does not represent a guaranteed achievable expectation. Furthermore, the certainty that exists in two-person zero-sum game experiences here yet a further restriction: a game can in general possess several equilibria with varying game results. Thus it is not obvious that the result of a game will be that corresponding to a particular equilibrium.

However, if one is prepared to accept these two limitations, then Zermelo's theorem and von Neumann's minimax theorem can be generalized in a weaker form to arbitrary finite many-person games, including those without zero-sum character. Both theorems were discovered in 1950:

- in a game with perfect information, each player possesses a pure strategy, and these together form an equilibrium.

- in a game without perfect information, there exists for each player a mixed strategy, and these together form an equilibrium.

The first theorem is that mentioned in Chapter 20, which goes back to Kuhn. Starting with the extensive form of a game, such strategies can be constructed recursively move by move in the reverse order of the game's chronology.

The second theorem is the Nash equilibrium theorem, which the 21-year-old John Nash proved in his dissertation and for which he received the Nobel Prize in economics over 40 years later.[1] John Harsanyi (1920–2000) and Reinhard Selten (1930–), who further developed the concept of a

---

[1] John Nash, Equilibrium points in N-person games, *Proceedings of the National Academy of Sciences of the USA* **36**, 1950, pp. 48–49; John Nash, Non-cooperative games, *Annals of Mathematics* **54**, 1951, pp. 286–295; both articles are reprinted in Harold W. Kuhn (ed.), *Classics in Game Theory*, Princeton 1997, pp. 3–4, 14–26. The dissertation is essentially the same as the second-named publication. A facsimile of the dissertation can be found in Harold W. Kuhn, Sylvia Nasar (eds.), *The Essential Nash*, Princeton 2002, pp. 53–84. Background on the dissertation appears in Harold W. Kuhn et al., The work of John F. Nash Jr. in game theory, Nobel Seminar 8 December 1994, *Duke Mathematical Journal* **81**, 1995/96, pp. i–v, 1–29; Sylvia Nasar, *A Beautiful Mind: A Biography of John Forbes Nash, Jr.*, New York 1998, Chapter 10. This book was the

strategic equilibrium, were honored together with Nash.[2] Nash's theorem is a pure existence theorem, offering no method of actually obtaining such a *Nash equilibrium*, as it is usually called, nor of determining whether there is more than one (see Note 1 at the end of the chapter). Nash's theorem and his concept of equilibrium form the basis of *noncooperative game theory*, in which the rational behavior of players is studied under the assumption that the players cannot make binding agreements about their behavior and division of winnings. In economics, such models are useful in the theoretical study of markets and prices, but how can they help us in real games that people play?

Nash himself gave a simple example of an application to a game as an appendix to his proof.[3] Since he sees the application of his concept primarily to games in which the generally accepted customs of a fair game contain a noncooperative manner of playing, Nash investigates a poker game for three players. Because of its great complexity, Nash is forced to restrict his attention to a very simple model:

- each of three players receives a card at random and with equal probability, either a high card or a low card. Moreover, each of the three cards is drawn independently of the others. Each player knows only his own card.

- for the bidding, which is open, there are two levels; whoever wishes to raise the ante of 1 to 2 may do so.

- in the first round of bidding the three players have the chance in turn to open with a bid of 2. If no one does so, the players receive their antes of 1.

- once a player has opened, the two other players have the opportunity to match the bet or pass. Then the game is over and the scoring begins: the entire stake is split evenly among those players who bet 2 units and hold the comparatively highest cards.

---

basis of the film *A Beautiful Mind*, which received four Oscars in 2002, including best film of 2001.

[2]Eric van Damme, On the contributions of John C. Harsanyi, John F. Nash and Reinhard Selten, *International Journal of Game Theory* **24**, 1995, pp. 3–11; Joachim Rosenmüller, Nobelpreis für Wirtschaftswissenschaften: die Spieltheorie wird hoffähig, *Spektrum der Wissenschaft* **12**, 1994, pp. 25–33; Bluffen und drohen, *Der Spiegel* **42**, 1994, pp. 134–136.

[3]See the 1951 work of Nash cited above. A more extensive and generalized (with respect to the bidding levels) version of the poker model is described in J. F. Nash, L. S. Shapley, A simple three-person poker game, in: H. W. Kuhn, A. W. Tucker (eds.), *Contributions to the Theory of Games I*, *Annals of Mathematics Studies* **24**, 1950, pp. 105–116. For a particular bidding level this model also appears in Ken Binmore, *Fun and Games*, Lexington 1992, pp. 593–601.

Aside from the original random move, a game can last for five moves, in which each of the first two players plays up to two times. Clearly, there are moves that are dominated by others and therefore should not be used. Thus a player with a high card would suffer an inevitable loss against the maximum achievable were he to pass after another player had opened. Aside from such situations, in the two rounds there remain the following decisions to be investigated strategically. We give the yes–no decision to be made as well as the information that the player possesses with respect to the previous history of the game:

First Round:

Player 1:

- Open on a low card?
- Open on a high card?

Player 2:

- Open on a low card if player 1 didn't open?
- Open on a high card if player 1 didn't open?
- Match the bet on a low card if player 1 opened?

Player 3:

- Open on a low card if players 1 and 2 didn't open?
- Match the bet on a low card if player 2 opened?
- Match the bet on a low card if player 1 opened and player 2 passed?
- Match the bet on a low card if player 1 opened and player 2 matched the bet?

Second Round:

Player 1:

- Match on a low card if player 3 opened?
- Match on a low card if player 2 opened and player 3 passed?
- Match on a low card if player 2 opened and player 3 matched?

Player 2:

- Match on a low card if player 3 opened and player 1 passed?

- Match on a low card if player 3 opened and player 1 matched?

How should the players behave? In particular, what does a Nash equilibrium look like? Since the poker model is a game with perfect recall and Kuhn's theorem on behavioral strategies holds for multiperson games, for each equilibrium strategy there exists a strategically equivalent behavioral stratgegy. In describing it, equivalent behaviors in different information sets are treated together as much as possible:

Player 1:

- Does not open on a low card.
- Opens on a high card with probability 0.3084.
- Passes in the second round on a low card.

Player 2:

- Opens on a low card if player 1 did not open with probability 0.0441.
- Opens on a high card if player 1 did not open with probability 0.8257.
- Passes on a low card if player 1 opened.
- Passes in the second round on a low card.

Player 3:

- Opens on a low card if no other player opened with probability 0.6354.
- Passes on a low card if another player has opened.

These three behavioral strategies turn out to constitute the only Nash equilibrium. The associated winning expectations of the three players are −0.0735, −0.0479, and 0.1214.[4] Of particular interest is the fact that players 1 and 2 do not always open on a high card, even though they are at no risk of losing anything. This amounts to a reverse bluff, where the

---

[4]These values are in fact irrational numbers, so that in contrast to a two-person zero-sum game, they cannot be calculated by means of a linear system of equations from the game parameters. For example, the value for the second player is equal to

$$\frac{16 - \sqrt{321}}{40}.$$

player acts despite his high hand as though it were a bad one. Like a bluff, this has the effect of revealing as little valuable information as possible about his hand.

We would now like to consider how such Nash equilibria can be determined. For practical play they possess, for reasons that we have mentioned, only limited value. But what conclusions can we draw from the three winning expectations? Since Nash equilibria assume rational behavior and no prior agreements between players, the three values, particularly since those in the poker model are uniquely determined, can be seen as typical characteristics of the game that express the winning prospects of the players. In particular, one sees that the third player has an advantage. Whether he can actually realize his positive expectation depends on how rationally his opponents play. Do they play rationally, or does one of his two opponents perhaps commit a blunder?

Less speculative, namely, without assuming the rational behavior of the opponents, is the worst-case analysis. Here one attempts to maximize the winning expectation of the player that is guaranteed regardless of the strategies of the other two players. That corresponds to an analysis such as we have seen for two-person zero-sum games, but now in competition with a two-person coalition.

We begin with the third player, for whom we can suppose an advantage due to the Nash equilibrium. In fact, player 3 can ensure himself an average score of at least $1/64$ against a coalition of players 1 and 2, who set their strategies in common:

Player 3:

- Opens on a low card, when no other player has opened, with probability $3/16$.
- Matches bets on a low card when player 2 has opened with probability $1/4$.
- Passes on a low card if player 1 has opened.

The minimax behavioral strategy of the two cooperating players is the following:

Player 1:

- Does not open on a low card.
- Opens on a high card with probability $3/4$.
- Passes in the second round on a low card.

Player 2:

- Opens on any card if player 1 has not opened.
- Does not match on a low card if player 1 has opened.
- Passes in the second round on a low card.

With such cooperation, in which the two players agree on their behavioral plans for the game, players 1 and 2 can greatly reduce their net loss in comparison to that indicated by the Nash equilibrium. Are they cheating? As Nash noted, such cooperation violates what is considered good manners in poker and is unusual. Strictly speaking, though, there is no rule against such coordinated plans. Every player can act in this coalition in a manner in which he could act if playing alone. In particular, there is no illegal exchange of information.

Player 1 acting alone has a worse minimax value than player 3 had. It is $-1/12$. This expectation can be ensured with the following strategy:

Player 1:

- Does not open on a low card.
- Opens on a high card with probability $19/27$.
- Matches in the second round on a low card if player 3 has opened with probability $2/27$.
- Passes in the second round on a low card if player 2 has opened.

The minimax strategy of the two cooperating players is as follows:

Player 2:

- Acts defensively on a low card; that is, he passes or opens according to whether player 1 opened or not.
- Opens on a high card if player 1 did not open.
- Passes in the second round on a low card.

Player 3:

- Opens on a low card if no other player has opened with probability $2/3$.
- Passes on a low card if another player has opened.

Player 2 on his own has somewhat better prospects than player 1. The winning expectation that the second player can attain is $-5/88$. The minimax strategy is as follows:

Player 2:

- Opens on a low card if player 1 did not open with probability $3/11$.

- Opens on a high card if player 1 did not open with probability $7/11$.

- Passes in the second round on a low card.

As we did in our investigation of the other two coalition constellations, we would like to see how the cooperating players can limit the expectation of player 2 to his minimax value. It can be seen that there is no behavioral strategy that accomplishes this. The two partners must agree on a suitable mixture of their total strategy. That is, to ensure their minimax value, the partners require a cooperation that goes beyond agreeing on their decision probabilities but making their individual moves independently. In the case before us, what is required is rather a random decision made before the start of the game that sets the *common* plan of the two coalition partners. There are no strategically equivalent behavioral strategies; Kuhn's theorem is not applicable, since it assumes perfect recall for the two cooperating partners, and that is violated here.

A mixed minimax strategy for the cooperating players 1 and 3 looks like this:[5]

Player 1:

- Does not open on a low card.

- Opens on a high card with probability $3/11$.

- Passes in the second round on a low card.

Player 3:

- Opens on a low card if no other player has opened with probability $13/22$, and in fact with conditional probability $13/16$ in

---

[5]Since a calculation based on imperfect recall exceeds the range of the behavioral strategies, the simplified algorithms that we have used in the previous chapters lead only partially to the goal for coalitions in the poker model under consideration here. With correspondingly high expenditure of effort, one can start directly with the normal form; here dominated strategies can be omitted.

the case that the strategy currently determined by chance for player 1 does not call for an opening on a high card (see Note 2 at the end of the chapter).

- Passes on a low card if another player has opened.

The strategy described for an information set goes beyond what a behavioral strategy can offer: if players 1 and 2 did not open, then player 3 on a low card should not align his reaction to the strategy of his partner, which has not become public knowledge. To be sure, player 3 knows that his partner player 1 has not opened. But he does not know whether player 1 has made his decision based on his low card or on the 3 : 8 random decision on a high card. To that extent, player 3, without additional information, cannot mix his strategy as would actually be required in coordination with the mixed strategy of his partner. It is necessary that both players be able to use a random number in common to set their strategy.

The mixed minimax strategy of the coalition of players 1 and 3 makes clear once again the central importance of information in a game and the strategic losses that can arise from a restriction on the flow of information. Since it is generally not allowed for a player to make statements about his informational state, he can transmit his information at best indirectly: in *signaling*, the information to be transmitted is passed in accordance with the rules of the game, proceeding in the course of a move in such a way that the informational state of the receiving player is influenced. Put thus abstractly, the matter sounds much more complicated than it really is in practice. Thus in our poker model, player 1 cannot directly inform his partner what card he is holding in his hand, but he can indirectly hint at this information by using the decision, which will become known to his fellow players, of opening or not. Of course, such a transfer of information has its limitations:

- with each transmitted piece of information the course of the game is influenced. Thus move alternatives are unsuitable for signaling if they too greatly damage the winning prospects of the player.

- the information cannot necessarily be targeted. For example, a bid made openly in a poker game is known to all players. Therefore, moves must be weighed so that an opponent can derive the least possible information from them. Thus a move that gives an opponent information can turn out to be disadvantageous. Both moves that could signal a partner and those that would positively influence the prospects of the player whose move it is can be thus affected.

Von Neumann and Morgenstern mentioned the phenomenon of signaling in their monograph, referring to the familiar bidding conventions in bridge,[6] for which there are various systems. Von Neumann and Morgenstern explain their very formal definition of signaling as a trick that passes information indirectly. They distinguish two cases. In one case, information is passed between one and the same "player," that is, between members of a coalition, while in the other, it is between two different players:[7]

> In the first case—which, as we saw, occurs in Bridge—the interest of the player ... lies in promoting the "signaling," i.e., the spreading of information "within his own organization." The desire finds its realization in the elaborate system of "conventional signals" in Bridge. These are parts of the strategy, and not of the rules of the game ... and consequently they may vary, while the game of Bridge remains the same.

> In the second case—which, as we saw, occurs in Poker—the interest of the player... lies in preventing this "signaling," i.e., the spreading of information to the opponent... This is usually achieved by irregular and seemingly illogical behavior... This makes it harder for the opponent to draw inferences.

What forms of signaling are compatible with the spirit of the rules of the game? Although this question cannot be answered with the methods of mathematical reasoning, mathematical methods nonetheless give us some indications of what might follow from particular sets of rules.

Let us summarize our results about the poker model. In the case that players are allowed to cooperate and that they can even mix their strategies in a common process, the poker model offers the following winning prospects: the expectatations that each player can achieve on his own are $-1/12 = -0.0833$, $-5/88 = -0.0568$, and $1/64 = 0.0156$. This leaves a remainder of $263/2112 = 0.1245$ available as a "bonus" for a coalition of two players, should it come into being, to capture and divide between them. If no coalition is formed, and each player gives his "best" in the sense of an all-around reasonable method of play, then, as implied by the Nash equilibrium, each of the three players can increase his expectation.

---

[6]Bridge is a card game played by four persons. The players sitting opposite play as a team. In the first phase of the game there is a round of bidding for the number of tricks that can be made. The bids allow for certain conclusions to be drawn about the quality of the hands, and this is necessary so that the partner can decide what to bid after evaluating his own hand.

[7]John von Neumann, Oskar Morgenstern, *Theory of Games and Economic Behavior*, Princeton 1944.

The greatest increase is achieved by player 3, who can take home a full 85% of the bonus.

With our discussion of the three-person poker model we came into contact with certain concepts and approaches that arise from the game theory of multiperson games. We were able to see that the results obtained do not exhibit the uniqueness that we had come to expect from two-person zero-sum games. For the way in which usual types of games are played, one thus obtains results that are of limited value. Perhaps, then, we should not go further into the underlying concepts of game theory in the limited scope of this book. We give a brief view of the situation in "The Theory of Multiperson Games."

The Theory of Multiperson Games

The effect of coalitions on the course of a game was first investigated in 1928 by John von Neumann. Von Neumann found in the case of coalitions that the idea of stability that existed in two-person zero-sum games could be carried over to multiperson games at least in limited form. Thus for three-person zero-sum games he analyzed the three possible coalitions that two players could form against the third. If the two partners act as a single player, then the minimax theorem holds. There thus arise three minimax values, just as we have seen for our poker model. Von Neumann writes:[8]

> The three-person game is significantly different from that for two persons. The methods of the individual players fade in importance: They offer nothing new, since the (necessarily occurring) formation of coalitions makes the game into a two-person game. But the value of the game for a player depends not only on the rules of the game, but much more on which of the three equiprobable coalitions comes into being. It makes itself felt in what is completely foreign to the stereotypical and equalized two-person game: a battle.

---

[8] John von Neumann, On the Theory of Games of Strategy, in: Contributions to the Theory of Games IV, *Annals of Mathematics Studies* **40**, Princeton 1959; Werke: Band IV, pp. 1–26.

From von Neumann's idea of studying multiperson zero-sum games from the point of view of coalition scenarios there later arose an independent branch of game theory called *cooperative game theory*, in which every game is reduced to the data that encompass every minimax value that an arbitrary coalition of players can achieve against the remainder of the players.[9] Then, exclusively on the basis of these data, which reflect the "power" of the individual players, the possible coalitions are investigated to establish conditions under which each participating player is prepared to join a particular coalition rather than "seek his fortune" in another union. In detail, this takes place using various concepts based on the goals of the players given more or less plausible fundamental assumptions. The first such concepts were introduced in 1944 in von Neumann and Morgenstern's monograph.

So that the members of a particular coalition can decide how their commonly earned winnings are to be divided, the game and its environment must possess certain properties. Thus the players must be able to communicate with one another, and their total winnings must be both divisible and transferable. Moreover, it must be ensured that agreements made are kept. It is not only in games, but in economic applications as well, that such assumptions about exogenous decision processes, that is, those not covered by the rules, are often problematic. Depending on what assumptions are made, one unsurprisingly obtains differing results.[10]

In contrast to cooperative game theory, the approaches to non-cooperative game theory do not have to contend with game processes that go outside the bounds of the rules of the game. Although the name does not suggest it, coalitions are not excluded in noncooperative game theory. However, a coalition can be formed only if the negotiations take place within the confines of permitted moves.

---

[9]Formally, this approach leads to the so-called characteristic function that associates with every coalition C a minimax value $v(C)$ that this coalition can achieve against the rest of the players. That the strategic possibilities grow with larger coalitions finds expression in the *superadditivity* of the characteristic function: if two subsets C and D of players contain a common element, then $v(C \cup D) \geq v(C) + v(D)$.

[10]An overview of the various approaches to cooperative game theory can be found, for example, in Chapter 6 of Manfred J. Holler, Gerhard Illing, *Einführung in die Spieltheorie*, Berlin 1991.

Nash equilibria in noncooperative game theory possess a fundamental significance. What is problematic is that there can exist
a number of inequivalent Nash equilibria. Although this ambiguity can never be completely eliminated, significant reductions
are possible by focusing on special equilibria as particularly
plausible. Criteria for these arise from a near-stability with
respect to small variations of the opposing strategies as well
as limits on scoring established in the rules of the game. For
games in extensive form, one can investigate all the decisions
to be made in the game on the basis of whether an equilibrium
strategy actually provides a good move for the player in question: for configurations that cannot actually be encountered
using the strategies of the equilibrium this is not at all obvious.
The leading concepts of such a refinement of Nash equilibria
come from John Harsanyi and Reinhard Selten, who, as we
have already mentioned, shared the Nobel Prize in economics
with Nash. Despite all the refinements, the problem remains
that the behavior derived from a Nash equilibrium is optimal
only if the other players behave reasonably and competently.[11]

Chapter Notes

1. Nash's proof rests on the idea that the players can modify their strategies
   simultaneously in a way reminiscent of the fictitious series of games presented in Chapter 38. Here one investigates, from the perspective of the
   individual players, a given combination that contains a mixed strategy for
   each player: with what pure strategies could a player counter the opposing
   strategies better than with the strategy mix that he is actually using? And
   by how much could he thereby increase his winning expectation? Building
   on this idea, Nash defines a transformation that calculates a new strategy
   for all players simultaneously in which the probabilities are raised for the
   pure strategies that for the affected player represent a better reply than the
   strategy mix being used, while those of others are diminished. Using the

---

[11]Refinements of the Nash equilibrium are discussed in Sections 3.7 and 4.1 of the
book by Holler and Illing cited above. More information can be found in Christian
Rieck, *Spieltheorie, Einführung für Wirtschafts- und Sozialwissenschaftler*, Wiesbaden
1993, Chapter 5; Roger B. Myerson, *Game Theory*, Cambridge 1991, Chapters 4 and 5.

Brouwer fixed-point theorem, which had been used by John von Neumann in his proof of the minimax theorem, one proves that this transformation has a fixed point, which must represent an equilibrium of strategies because of Nash's concretely constructed formula.

Nash's construction in which the players simultaneously transform their strategies is not particularly complicated. Let us consider one of the players—the transformation is analogous for the other players—who has mixed his pure strategies $s_1, s_2, \ldots, s_m$ in proportions given by the probabilities $x_1, x_2, \ldots, x_m$, where $x_i \geq 0, i = 1, \ldots, m$, and $x_1 + x_2 + \cdots + x_m = 1$. The player now compares his mixed strategy with the pure strategies to see which pure strategies could have better countered the opponents' strategies. For a comparison, the score improvements $a_1, a_2, \ldots, a_m$ are calculated that the player could have achieved with the pure strategies instead of his mixed strategy. For a pure strategy s_i that worsens the state of affairs with respect to the mixed strategy, the value $a_i = 0$ is assigned. Nash's transformation now provides for the player's mixing his strategies with probabilties

$$\frac{x_1 + a_1}{1 + a_1 + a_2 + \cdots + a_m},$$

$$\frac{x_2 + a_2}{1 + a_1 + a_2 + \cdots + a_m}, \quad \ldots,$$

$$\frac{x_m + a_m}{1 + a_1 + a_2 + \cdots + a_m}.$$

The other players proceed analogously and simultaneously. Nash's transformation is clearly continuous. Since the set of all strategic combinations corresponds to a higher-dimensional hypercube with boundary, the Brouwer fixed-point theorem can be applied, which guarantees the existence of a fixed point. Such a point is a Nash equilibrium: from among the strategies $s_1, s_2, \ldots, s_m$ that are contained with positive probability in the current mix, the "worst" is selected, that is, the strategy that brings the player the least winning expectation against the current opposing strategies and therefore brings no improvement over his mix. For such a strategy s_i one has $x_i > 0$ and $a_i = 0$, so that the fixed-point property of the affected coordinate immediately yields $a_1 + a_2 + \cdots + a_m = 0$ and therefore $a_1 = a_2 = \cdots = a_m = 0$. That is, the player cannot increase his winning expectation against the current opposing strategies.

Finally, we mention that the transformation need not be applicable for computing an equilibrium using an iterative procedure, since the transformation can consistently increase the distance to an equilibrium.

An overview of computational methods for Nash equilibria can be found in Richard D. McKelvey, Andrew McLennan, Computation of equilibria in finite games, in: Hans M. Amman (ed.), *Handbook of Computational Economics*, Amsterdam 1996, pp. 87—142.

2. In general, it suffices to determine "globally" by chance only a portion of the decisions contained in a mixed strategy. The random selection of the other moves can follow "locally" conditioned on the globally made decisions:

- under some circumstances, such decisions must be made globally, that is, with a common random choice, whose information sets display the character of a signal, by which is meant that the player whose turn it is later apparently "forgets" his available knowledge or the move made, for example, because the partner whose move comes later simply does not obtain this information.

- if the decisions of the part of the strategy to be set globally are set, then the moves in the other information sets can be chosen locally, that is, on the basis of a suitable number of behavioral strategies.

See G. L. Thompson, Signaling strategies in n-person games, in: H. W. Kuhn, A. W. Tucker (eds.), *Contributions to the Theory of Games II*, *Annals of Mathematics Studies* **28**, 1953, pp. 267–277; G. L. Thompson, Bridge and signaling, *ibid.*, pp. 279–289.

44

QUAAK! Child's Play?

Two players gamble according to the following rules: at the start of the game, each player receives 15 chips, with which they play over several rounds. In each round, each of the players takes a certain number of his remaining chips in his closed hand; permitted are the numbers 0, 1, 2, and 3. After the two players have made their secret choices, they open their hands and compare. If one player has more chips in his hand than his opponent, he receives a point. After the round, the chips that were wagered are set aside. A player wins when he succeeds in accumulating three points more than his opponent. If that does not occur, then the game is a draw. What is the best strategy?

This game appeared in 1994 as a children's game with the name QUAAK![1] The current game state is indicated by a playing piece in the form of a frog that is moved back and forth over seven squares, representing the possible point differences. Even those who have passed beyond the recommended top age of twelve for this game might find some amusement in trying their strategic talents in a few rounds of the game. And those who find the principle of the game entertaining but would like a game with more variety might like to try the game "Beat the Buzzard" by Alex Randolph.[2]

---

[1] Published by Otto Meier Verlag, Ravensburg. The author is Dirk Hanneforth.

[2] A game of Beat the Buzzard always last 15 rounds. Instead of chips, each player obtains a pack of 15 cards with values from 1 to 15. Moreover, the number of points that can be won in a round is not fixed, but is determined at the start of each round by drawing a card. There are 15 such cards altogether, so that the numbers of points played for are always the same, but in a random order for each game. Additional special

From the game-theoretic point of view, Quaak! is a two-person, nonrandom, zero-sum game with perfect recall, but without perfect information: perfect information is lacking because the players move simultaneously. Therefore, a player cannot evaluate fully the effect of his move at the time the decision for that move is made. Since the opponent is faced with the same uncertainty, each player has an interest in seeing to it that his decision cannot be foreseen by the opponent. Thus a player would do well to make his decision randomly. In particular, let us look for a minimax strategy, which will allow a player to protect himself from a negative expectation, which is possible due to the symmetry of the game. As a consequence of the perfect recall, such a minimax strategy can be constructed in the form of a behavioral strategy. This encompasses a probability distribution for the possible moves for every informational state that can arise for a player in the course of a game. Clearly, those informational states that are equivalent in view of the further move possibilities despite various prehistories can be combined. That is, it is simply the following three values that constitute the determining parameters of an informational state:

- the number of chips that remain to player 1: $0, 1, \ldots, 15$.

- the number of chips that remain to player 2: $0, 1, \ldots, 15$.

- the current score difference from the viewpoint of the first player: -2, $-1, 0, 1, 2$.

The fact of the existence of such an intermediate state is known to both players; a player is uninformed only about the decision of his opponent that is being made in parallel to his own. Thus an intermediate state such as shown in Figure 44.1 can be seen as a closed partial game for which there exists a uniquely determined minimax value. Using this approach, we can determine the associated minimax values and the necessary strategies move by move, where we do so as always in the reverse game chronology. In each case, a minimax value for a given configuration must be determined only for one double move. That is, the desired minimax value of a given configuration is calculated as the solution of a linear optimization problem for the minimax values of those configurations that can arise in a double move. This is necessary for all configurations that can arise from the starting configuration of the game.

One small peculiarity is the fact that in exceptional cases the game never ends, namely, when both players continue to play zero chips even though

---

rules make the game even more complex. The two-person version can be analyzed by the approach we take here for QUAAK! However, the intermediate states are so numerous that one can hardly hope to undertake a complete analysis.

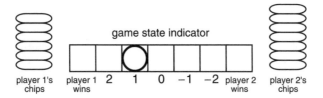

Figure 44.1. QUAAK! configuration 6-7-1.

one of the players could win with a different decision. By evaluating a certain number of repeated moves as a draw, the player who is ahead is forced to abandon such defensive and completely unprofitable behavior. We will go into this problem a bit further in our description of the calculations to be made.

Let us begin with a simple situation in which each player has two chips and the first player has almost won with a point advantage of 2:

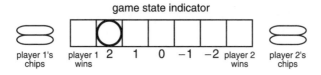

Each player has three possible moves: 0, 1, or 2 chips. If the first player chooses more chips than the second, he wins at once. If both players choose two chips, then they have each "shot their wad," and the game ends in a draw. With one chip each the game also comes to a draw, since the second player can avoid a loss by playing one chip on the next move. If the first player plays no chips, but his opponent two, then the first player is assured of a win in the next two moves. But if the second player plays exactly one chip more than the first player, then the game ends in a draw. Altogether, these not entirely obvious, but nonetheless not complicated, considerations yield the normal form tabulated in Table 44.1.

		Player 2 chooses		
		0	1	2
Player 1	0		0	1
chooses	1	1	0	0
	2	1	1	0

Table 44.1. Normal form of a double move: each player has two chips left. The point difference is 2.

The entry missing from Table 44.1 represents the event that neither of the two players plays a chip and the configuration remains the same. In order to avoid an unending repetition of moves, one could declare the game a draw if no chip is played a certain number of times. For the move immediately before such a draw, one has the normal form

		Player 2 chooses		
		0	1	2
Player 1	0	0	0	1
chooses	1	1	0	0
	2	1	1	0

which has a minimax value of $1/2$. The first player chooses zero or one chip with equal probability, while the second player decides equiprobably between one or two chips. If the state of the game is two moves until the game is called a draw, then the normal form that takes account of the value $1/2$ just determined is given by

		Player 2 chooses		
		0	1	2
Player 1	0	1/2	0	1
chooses	1	1	0	0
	2	1	1	0

Again the minimax value is $1/2$, and the players can use the same mixed strategy as in the move previously considered. Because of the stability just achieved with the value $1/2$, we do not consider further moves at a greater distance from the "horizon" where a draw is called, and thus $1/2$ is the minimax value of the game with unchanged rules: thus in practice, this kind of draw can be avoided in that the player who is ahead, player 1, has no interest in deciding on a zero-chip move with probability 1.[3]

---

[3] Such recursive games have been investigated in general form by Everett: H. Everett, Recursive games, in: H. W. Kuhn, A. W. Tucker (eds.), *Contributions to the Theory of Games III*, *Annals of Mathematics Studies* **39**, 1957, pp. 47–78, reprinted in Harold W. Kuhn (ed.), *Classics in Game Theory*, Princeton 1997, pp. 87—118; R. Duncan Luce, Howard Raiffa, *Games and Decision*, New York 1957, pp. 461–467.

As our second example let us consider the configuration in which each player has three chips left, and the first player is ahead by two points:

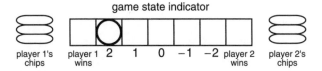

game state indicator

player 1's chips | player 1 wins | 2 | 1 | 0 | −1 | −2 player 2 wins | player 2's chips

Again, the first player wins at once if he plays more chips than his opponent. The second player can avoid a loss only if he plays as many chips as his opponent or one chip more. In the case that each player plays one chip, we obtain the configuration that we just investigated, so that we can take the just calculated minimax value of $1/2$ as the winning expectation. We thus obtain the following table:

		Player 2 chooses			
		0	1	2	3
Player 1	0		0	1	1
chooses	1	1	1/2	0	1
	2	1	1	0	0
	3	1	1	1	0

The omitted special case in which neither player plays a chip can be solved as in the first example. That is, one begins with the normal form entry in the upper left corner with the number 0, and then calculates the minimax value and enters this value in the upper left corner of the normal form, and so on. Again this iteration becomes stable from the second minimax calculation, this time with a value of $3/5$. Thus the first player mixes his moves randomly in the proportion $1:2:0:2$, and the second in the proportion $0:2:1:2$. The normal form reflecting this move is then the following:

		Player 2 chooses			
		0	1	2	3
Player 1	0	3/5	0	1	1
chooses	1	1	1/2	0	1
	2	1	1	0	0
	3	1	1	1	0

```
Player │ Player 1's Advantage (Five Times Chip Advantage Plus Six Times Point Difference)
 1's   │ ..-40..-35..-30..-25..-20..-15..-10...-5....0....5...10...15...20...25...30...35...40....
Chips  │ ...|....|....|....|....|....|....|....|....|....|....|....|....|....|....|....|....|....|....
   1   │ 1111111111111111111111111111111111111111.1111..111...11....1................................
   2   │ 2222222222222222222222222222222mm111222222.2222..222...2m....2..............................
   3   │ 3333333333333333333333333333m3333mmm113222232333.233m..23m...13....2........................
   4   │ 33333333333333333333333m3333m333mmmmm13m22233333m2333m.23mm..113...23....2..................
   5   │ 3.3333333333333333333m3333m333mmmmmm3m22233333m3333m233mm.1113..223...23....2...............
   6   │ ..3333.333333333m3333m333mm33mmmmmmmm3mm22mm333m333mm33mmm11113.2223..233...23....2.........
   7   │ ..333..3333.3333m333mm333mm33mmmmmmmmmmmm2mm3mm333mm33mmmmm111322223.2333..233...23....2..
   8   │ ..33..333..333m.333m.333mm33mmmmmmmmmmmmmmmmm3mmm3mmmmmm1132222323333.2333..233...23..
   9   │ ..3....33...333..333m.33mmm3mmmmmmmmmmmmmmmmmmm3mmm3mmmmmmm13m22233333323333.233..
  10   │ .......3....33...33m..33mm.3mmmmmmmmmmmmmmmmmmmmm3mmmm3m222333333333332333.2333..
  11   │ ...........3....33...3mm..3mmm.mmmmmmmmmmmmmmmmmmmmmmmmm3mm2m3m333333333333323333.2
  12   │ ...........3...3m...3mm..3mm.mmmmmmmmmmmmmmmmmmmmmmmmmmm3mmm3mm333333333333333323
  13   │ ...............3...3m..mmm..mmmm.mmmmmmmmmmmmmmmmmmmmmmmm3mmm3mmm333333333333333333
  14   │ ...................3....mm...3m..mmm..mmmm.mmmmmmmmmmmmmmmm3mmm3mm33m3333333333333333
  15   │ ......................m...mm...mmm..mmmm.mmmmmmmmm3mmmm3mmm3mm333333333333333
```

Table 44.2. Minimax strategy: moves for player 1 without mixing; an "m" indicates a mixed strategy.

On the whole, there is really nothing more to say, since the calculations for additional configurations run completely analogously. The results are so extensive that they cannot be reproduced here in their entirety. We will restrict our attention to two aspects for which we give the minimax behavior of the first player.[4]

Table 44.2 shows all configurations in which the first player does not need to mix his moves for a minimax strategy. To be able to assemble the large quantity of data into a table, the configurations are characterized in a special way. In addition to the number of chips that player 1 still has, a value is given that can be interpreted as a rough measure of the first player's advantage. For this value, player 1's chip advantage is counted five-fold, and the point difference six-fold, so that each configuration can be uniquely characterized by the two parameters. The entries in the table correspond to the configurations in which the first player can select a fixed move as his minimax behavior. Where this is not possible, the letter "m" appears, meaning that the first player must mix his possible moves to obtain a minimax strategy. The empty entries are for configurations that cannot arise from the starting configuration of 2×15 chips via minimax behavior of the first player.

As Table 44.2 shows, the first player can avoid a random mixture of his moves above all when a game is almost decided. If we take, for example, the configuration in which with a point difference of -2 the first player has six chips and his opponent a single chip. For the values 6 and $5 \times (6 - 1)$

---

[4]Of course, from our results one could derive optimal moves for the second player as well. To do this, using the symmetry of the game one has only to interchange the chip numbers of the two players and change the sign of the point difference.

$+ 6 \times (-2) = 13$ for the measure of advantage, one sees in the table the entry 1, and in fact, the choice of one chip is the optimal move for the first player. If he plays no chip, then he can be threatened with an immediate loss. On the other hand, every choice of more than one chip is pure waste, which costs an otherwise certain victory.

Table 44.3 encompasses a portion of the configurations of Table 44.2 that are denoted by "m," that is, configurations in which the first player must mix his strategies to obtain a minimax strategy. Table 44.3 tabulates all configurations in which the two players have up to a total of 13 chips that can arise from the starting configuration of 2×15 chips via a strategy of the first player. For each of these configurations are shown the minimax value, probabilities of the moves, and the loss for the opponent that he can expect on average from a wrong decision at the given move.

As can be seen from Table 44.3, the expected losses from incorrect play allow for a sufficient number of "'traps" in the game. A comparatively high loss of 0.3690 with respect to the minimax value of 0.2074 is suffered by the second player on average if he chooses three chips from the configuration of 7 and 5 chips and zero point gap, as shown in the following figure:

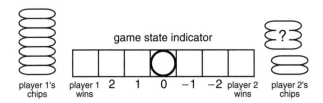

By looking at the normal form, we can see why the move indicated in the figure is so bad for player 2. It is necessary to iterate to determine the top left value, and the iteration must be done several times:

		Player 2 chooses			
		0	1	2	3
Player 1	0	0.2074	0.0000	1.0000	1.0000
chooses	1	0.4455	0.1317	0.0000	1.0000
	2	0.0000	0.3750	0.0000	0.0000
	3	0.0000	0.0000	0.2338	0.0000

The normal form shows clearly that the minimax values achievable in one move run between a draw and a win for the first player. The two extreme values are achieved precisely when one player chooses at least two chips more than his opponent and therefore pays too much, so to speak, for

Chips of Player		Point Gap	Minimax Value	Player 1 Probabilities				Player 2 Expected Loss			
1	2			0	1	2	3	0	1	2	3
2	2	−2	−0.5000		0.5000	0.5000			0.5000		
2	2	2	0.5000	0.5000		0.5000		0.2500			
3	3	−2	−0.6000		0.4000	0.2000	0.4000			0.2000	
3	3	2	0.6000	0.2000	0.4000		0.4000	0.3200			
3	4	−1	−0.2338	0.3897	0.4676	0.1427					0.2338
3	4	2	0.5000	0.5000			0.5000	0.2500			
4	3	−2	−0.5000		0.5000	0.5000			0.5000		
4	3	1	0.2338	0.2338	0.2986	0.4676					0.2986
4	4	−2	−0.6407		0.4690	0.1717	0.3593			0.0859	
4	4	2	0.6407	0.2814	0.3593		0.3593	0.2582			
2	7	2	−0.3333	0.6667		0.3333		0.4444			
4	5	−1	−0.3750		0.6250	0.3750			0.2289		0.1875
4	5	2	0.5000	0.5000		0.1169	0.3831	0.2500			
5	4	−2	−0.5000		0.5000	0.5000			0.2500		
5	4	1	0.3750	0.3750		0.6250		0.0781			
7	2	−2	0.3333		0.3333	0.6667			0.6667		
4	6	0	−0.1317	0.3512	0.5633	0.0855					0.1317
4	6	2	0.1895	0.8105			0.1895	0.1536			
5	5	−2	−0.7593		0.3851	0.1335	0.4814			0.2941	
5	5	2	0.7593	0.2045	0.3141		0.4814	0.1915			
6	4	−2	−0.1895		0.1895	0.8105			0.1895		
6	4	0	0.1317	0.1317	0.3050	0.5633					0.3050
3	8	2	−0.2308	0.3077	0.4615		0.2308	0.8521			
5	6	−1	−0.4455		0.6386	0.1554	0.2060				0.0330
5	6	2	0.5000	0.5000		0.3750	0.1250	0.2500			
6	5	−2	−0.5000		0.5000	0.5000		0.1331			
6	5	1	0.4455	0.3158	0.1297	0.5545		0.0709			
8	3	−2	0.2308		0.2308	0.1538	0.6154			0.3077	
3	9	2	−0.3333	0.6667			0.3333	0.4444			
4	8	1	−0.7165	0.3265	0.2482	0.4253					0.2912
5	7	0	−0.2074	0.3021	0.5531	0.1448					0.1736
5	7	2	0.2727	0.7273			0.2727	0.1983			
6	6	−2	−0.8080		0.3462	0.1738	0.4799			0.2382	
6	6	−1	−0.1778	0.3557	0.2209	0.3557	0.0677				
6	6	1	0.1778	0.1778	0.1576	0.3557	0.3088				
6	6	2	0.8080	0.2129	0.3072		0.4799	0.1511			
7	5	−2	−0.2727		0.2727		0.7273	0.2210			
7	5	0	0.2074	0.2074	0.3690	0.4235					0.3690
8	4	−1	0.7165	0.3686	0.4253	0.2062					0.2835
9	3	−2	0.3333		0.3333		0.6667	0.6667			
4	9	2	−0.1943	0.4172	0.3886		0.1943	0.6961			
5	8	1	−0.0650	0.3133	0.4934	0.1934					0.0650
6	7	−1	−0.5210		0.6043	0.1657	0.2300				0.0589
6	7	2	0.6891	0.3782			0.6218	0.1933		0.1012	
7	6	−2	−0.6891		0.3782		0.6218	0.2668	0.3109		
7	6	1	0.5210	0.3142	0.2069	0.4790		0.0493			
8	5	−1	0.0650	0.2816	0.2251	0.4934					0.2916
9	4	−2	0.1943		0.2712	0.1317	0.5971			0.1756	

Table 44.3. Minimax strategy for player 1 for which moves with up to 13 chips must be mixed.

his point. This is also plausible from the point of view that a choice of three chips for either player is quite daring and is an extremely disadvantageous decision against a minimax strategy of the opponent: the expected loss is 0.1736, respectively 0.3690, and the probabilities for the optimal move mix are

- for player 1: 0.2074; 0.3690; 0.4235; 0.000;

- for player 2: 0.3021; 0.5531; 0.1448; 0.000.

One can make a significant error even in the first move, when each player still has 15 chips. The normal form for the first move looks like this:

		\multicolumn{4}{c}{Player 2 chooses}
		0	1	2	3
Player 1	0	0.0000	−0.2474	−0.0570	0.0863
chooses	1	0.2474	0.0000	−0.2340	−0.0460
	2	0.0570	0.2340	0.0000	−0.2230
	3	−0.0863	0.0460	0.2230	0.0000

Therefore, the players should begin by randomly mixing their possible moves with the probabilities

$$0.1212, \quad 0.2272, \quad 0.0000, \quad 0.6515.$$

Note that the choice of two chips already represents an error that lessens the winning expectation by 0.0852 against the given minimax strategy, with the winning expectation falling from 0.5 to 0.4574.

45

Mastermind:
Color Codes and Minimax

To what extent can the "encoder" who chooses the color code in a game of Mastermind influence the winning expectations of the game?

In Chapter 32, we optimized the search strategies for the game Mastermind from two different points of view. We investigated the minimum number of moves necessary to ensure that the code could be broken: in 6^4 Mastermind, that is, for codes of length 4 with a choice of 6 colors, the number of moves is five. We also minimized the expected number of moves under the assumption that the code had been selected at random and equiprobably: in 6^4 Mastermind, this minimum is 4.340 moves.

In both of these approaches, the character of Mastermind as a two-person game comes into play little or not at all, due to the relatively passive role of the encoder. Thus the question posed at the beginning of this chapter opens up another point of view for our consideration. We begin by describing Mastermind in the sense of a game-theoretic model.

Mastermind is a nonrandom two-person zero-sum game without perfect information, but with perfect recall. The encoder has a single decision to make, at the start of the game, associated with a single one-element information set. Much more complex in their structure are the decision situations of the decoder. Each of his information sets reflects the previous course of the game of which he is aware, namely, the questions posed and their answers. Of greatest importance are not the details of the questions and answers, but only the possible conclusions that can be drawn from them. In particular, for the decoder, the informational states are characterized by the sets of codes that are compatible with the previous questions and answers and are therefore still possible.

Due to the perfect recall, the mixed minimax strategies can be obtained in the form of behavioral strategies. A further simplification of the minimax analysis can be derived from the symmetries that arise in Mastermind due to the possible color and code positional permutations: if one were first to find a minimax strategy that failed to respect all these symmetries, one could then derive from it a symmetric minimax strategy by mixing the asymmetric strategy with all the symmetric associated strategies that can arise from it by permuting the colors and positions. Therefore, one can restrict one of the two players to such symmetric strategies without harm. This restriction leads to great simplifications for the opponent: if the encoder has selected his code with symmetrically distributed probabilities, then the number of moves for decoding does not change if the opponent modifies his search strategy with color and position permutations. That is, such mutually transformable decoding strategies mutually dominate one another and can thus be rejected except for one, which again greatly simplifies the game.

If one has found minimax strategies in the twice simplified game, then one can construct minimax strategies for the original game at once: here the encoder's symmetric strategy can be carried over without change. On the other hand, the decoding strategy must be symmetrized; that is, it is mixed with all of its "mirror images."[1] That all of this sounds much more complicated than it really is can be shown by examining a simple Mastermind variant.

The game of 3^2 Mastermind, with its nine codes

$$11, \quad 12, \quad 13, \quad 21, \quad 22, \quad 23, \quad 31, \quad 32, \quad 33,$$

permits $3! = 6$ exchanges of colors and an additional $2! = 2$ exchanges of position. Altogether, there are $3! \times 2! = 12$ symmetries. An encoder's symmetric strategy can be represented by the codes 11 and 12. Thus the set of codes

$$11, \quad 22, \quad 33$$

and the set of codes

$$12, \quad 13, \quad 21, \quad 23, \quad 31, \quad 32$$

---

[1] A formal description of all this can be found in K. R. Pearson, Reducing two person, zero sum games with underlying symmetry, *Journal of the Australian Mathematical Society, Ser. A*, **33** 1982, pp. 152-161.

are selected with equal probability. If one restricts the encoder to symmetric strategies, then the decoder can restrict his attention on his first move to the guesses 11 and 12.

For calculating the minimax strategies for 3^2 Mastermind we can use the techniques that we used in our investigations of le Her and baccarat, where the amount of effort will be of about the same order of magnitude.[2] For the reduced game one obtains the following minimax strategy:

- the encoder decides equiprobably between the two representatives 11 and 12. Thus the codes 11, 22, and 33 are chosen each with probability 1/6, while the other codes, 12, 21, 13, 31, 23, and 32, are chosen with probability 1/12 each.

- the decoder first guesses the code 12.

    - If the reply is two black sticks, the decoder has arrived at his goal.

    - With no stick or two white sticks, the decoder can immediately recognize the chosen code, namely, 33 or 21, respectively. The decoder thus achieves his goal on the second move.

    - With one white stick, 31 and 23 remain as possible codes. If one of these, say code 23, is guessed on the next turn, then in the worst case the goal is reached on the third turn.

    - If the encoder answers with one black stick, then there remain the four possibilities 11, 13, 32, and 22. On the second turn, the decoder guesses code 11 with probability 3/4, and code 13 with probability 1/4. Except for one case, namely, no stick as answer to the guess 11, the game is over on the second or third turn. In this one exceptional case there remain two codes, 22 and 32, after the second turn, and the decoder guesses code 22 on the third turn.

The game value, that is, the expected number of turns from the collision of this pair of minimax strategies, can be easily calculated directly. One simply goes through the nine existing codes and determines for each the expected number of turns to guess it:

---

[2]In the work cited in the previous footnote, 3^2 Mastermind is reduced to a 2×5 matrix.

Code	Expectation
11	$\frac{3}{4} \times 2 + \frac{1}{4} \times 3 = \frac{9}{4}$
22	$\frac{3}{4} \times 3 + \frac{1}{4} \times 3 = 3$
33	2
12	1
21	2
13	$\frac{3}{4} \times 4 + \frac{1}{4} \times 2 = \frac{7}{2}$
31	$\frac{1}{2} \times 2 + \frac{1}{2} \times 3 = 3$
23	$\frac{1}{2} \times 2 + \frac{1}{2} \times 3 = 2$
32	$\frac{3}{4} \times 3 + \frac{1}{4} \times 3 = 3$

Thus the expected number of moves required for decoding is

$$\frac{1}{6} \times \frac{29}{4} + \frac{1}{12} \times \frac{29}{2} = \frac{29}{12} \approx 2.417.$$

And what does the minimax strategy for 6^4 Mastermind look like? It is clear that the encoder's minimax strategy can be described by five probabilities, reflecting the mixing relationships for the classes of equivalent codes that can be represented by the codes 1111, 1112, 1122, 1123, 1234. These five codes also represent the choices that the decoder has for his first move.

How information about the minimax value of 6^4 Mastermind could be obtained at a significantly reduced cost over that of a complete minimax analysis was shown in 1986 by Merrill Flood.[3] Flood began with search strategies found in the course of other investigations whose authors had attempted to minimize the expected number of moves for an equiprobably distributed code.[4] Such a strategy can be slightly modified in a number of

---

[3]Merrill M. Flood, Mastermind strategy, *Journal of Recreational Mathematics* **18**, 1985–1986, pp. 194–202.

Flood was one of the pioneers of game theory. In 1950, he invented, together with Melvin Dresher, a very instructive 2 × 2 two-person game without zero-sum character that was later given the name "prisoner's dilemma" by Albert Tucker. In 1980, Robert Axelrod organized a computer tournament for this game, to which he invited game theorists to offer strategies for a series of games, where decisions could be based on the outcomes of the previous games. However, one should note that such a series of games, often called a supergame, has a significantly different character from that of an individual game.

[4]These are the works of Knuth, Irving, and Neuwirth presented in Chapter 30. The publication of Koyama and Lai mentioned in that chapter, whose results are significantly better, appeared later.

Code Class		Expected Number of Moves in Strategies of					
Example	Number	Flood (Nr. 1)	Flood (Nr. 2)	Flood (Nr. 3)	Flood (Nr. 4)	Knuth	Irving
1111	6	3.8333	3.8333	3.8333	3.8333	4.1667	3.6667
1112	120	4.3667	4.3667	4.3667	4.3750	4.4500	4.3667
1122	90	4.3667	4.3778	4.3667	4.3667	4.4444	4.3667
1123	720	4.3681	4.3667	4.3667	4.3667	4.4833	4.3764
1234	360	4.3667	4.3667	4.3694	4.3667	4.4833	4.3667
	1296	4.3650	4.3650	4.3650	4.3650	4.4761	4.3688

Table 45.1. The effects of Flood's modified search strategy.

ways without changing the expected number of moves. For example, this can most easily be done by guessing one or another code first in situations in which only two codes are possible. If such a strategy modification relates to codes that are not symmetric to one another, then the two search strategies will differ in their expected numbers of moves, which depend on the five code classes. That is, depending on which code class the decoder has chosen, either one or the other search strategy is slightly better suited for decoding. The decoder thereby obtains a means of taking action against the strategic influence of the encoder within this limited framework.

In detail, Flood determined for four variants of a search strategy the conditional expectations for the number of moves in the five classes of equivalent codes. These expectations are collected in Table 45.1, where for comparison, the corresponding values for the strategies given by two other authors are given.[5]

It can be seen clearly from Table 45.1 that four of the five code classes can be recognized equally efficiently by their four Flood strategies. Flood achieves a minimax-based improvement by randomly mixing his four decoder strategies in proportion 24 : 3 : 12 : 4, which corresponds to the minimax strategy in the case that the decoder is allowed to use only these four strategies; the encoder's associated minimax strategy provides for a random mix of the code classes in proportion 0 : 4 : 3 : 24 : 12.

The search strategy thus found for 6^4 Mastermind is not optimal. Nonetheless, Flood's approach is of great interest from the viewpoint of practical play, since it provides the decoder with a relatively easily realizable strategy that reduces the minimax value of 6^4 Mastermind to at most 4.3674. To bound the minimax value more strongly from above in an analogous way, the analysis must be repeated on a broader basis with additional good

---

[5]Since the strategies of Knuth and Irving were not more precisely specified due to the nature of their original purpose, this was completed by Flood.

search strategies, such as the strategy found by Koyama and Lai, which is optimal for the expected number of moves, namely, $5625/1296 = 4.340$.[6]

According to a 1995 newsgroup memo from Michael Wiener, the minimax value of 6^4 Mastermind is in fact $5600/1290 = 4.341$. He obtained the following minimax strategies using a computer calculation that took several months:

- the encoder chooses his code equiprobably from among the 1290 codes that contain at least two colors.

- the decoder plays exactly as specified by Koyama and Lai, by which the encoder selects his code equiprobably from among all codes. In particular, the decoder uses a pure strategy!

For the strategy found by Koyama and Lai, it remains to show that one-color codes can be recognized in 25/6 moves on average. Since the minimax value v corresponds to the conditional expectation that arises for codes with at least two colors, it satisfies the equation

$$\frac{6}{1296} \times \frac{25}{6} + \frac{1290}{1296}v = \frac{5625}{1296},$$

which yields the value $v = 5600/1290$.

---

[6]See Chapter 32.

Index